Marine Fishes

Other Related Books

Comparative Anatomy of the Vertebrates	*Eaton T.H.*
Culture, Feeding and Diseases of Commercial Flatfish Species	*Liewes E.W.*
Farming Marine Fishes and Shrimps	*Korringa P.*
Fishermans Workbook	*FAO*
Freshwater Aquaculture, 3rd Ed.	*Rath R.K.*
Freshwater Fish Pond Culture and Management	*Sharma B.M.*
Medical and Veterinary Virology	*Abraham A.M.*
Medical Entomology	*Tyagi, B.K.*
Medical Microbiology and Immunology	*Reddy S.M.*
Ocean and Coastal Ecology (Textbook Series: 21st Century Biology and Agriculture)	*Kathiresan K.*
On farm Feeding and Feed Management in Aquaculture	*FAO*
Principles of Systematic Zoology	*Mayr E.*
Utilization of Wastewater in Agriculture and Aquaculture	*Kaul, S.N.*

Marine Fishes

Chemical Composition and Processing Properties

Editor

V.P. BYKOV

United Book Prints

First reprint in India, 2017

Published by:
United Book Prints
4806/24, Ansari Road,
Daryaganj,
New Delhi - 110 002
Tel.: 011-41511055
E-mail: unitedbookprints@gmail.com

Print: 2017

ISBN: 978-93-83692-72-9

Printed in India

UDC 664.951.022.14

This monograph describes both the well known and relatively less known marine fishes reported from the Soviet catches up to 1970. It includes information on their length, total weight, weight proportions of individual body parts and chemical composition. Quality of fish meat, viz. color, taste and smell, and consistency of the processed products are briefly described, followed by the recommendations for processing.

LIST OF ABBREVIATIONS USED IN THE BOOK

Abbreviation	*Expanded Form*	*English Trans.*
AzcherNIRO	Azovško-Chernomorskii Nauchno-issledovatel'skii Institut Morskogo Rybnogo Khozyaistva i Okeanografii	Azov-Black Sea Research Institute of Marine Fisheries and Oceanography
AtlantNIRO	Atlanticheskii Nauchno-issledovatel'skii Institut Morskogo Rybnogo Khozyaistva i Okeanografii	Atlantic Research Institute of Marine Fisheries and Oceanography
VNIRO	Vsesoyuzno Nauchno-issledovatel'skii Institut Morskogo Rybnogo Khozyaistva i Okeanographii	All-Union Research Institute of Marine Fisheries and Oceanography
PINRO	Polyarnyi Nauchno-issledovatel'skii Institut Morskogo Rybnogo Khozyaistva i Okeanografii in N.M. Kni'povicha	N.M. Kni'povicha Polar Research and Planning Institute of Marine Fisheries and Oceanography
TINRO	Tikho-Okeanskii Nauchno-issledovatel'skii Institut Rybnogo Khozyaistva i Okeanografii	Pacific Research Institute of Fisheries and Oceanography

CONTENTS

INTRODUCTION

During the past decade, the fishing industry of the Soviet Union has made considerable advances, resulting in increases in the catch of marine fish and other organisms from 35 million qtl in 1960 to 77 million qtl in 1970. In the process, marine fish from different or previously unknown regions of the world's oceans have acquired great significance. The fishing industry has had to become familiar with many new species of fish in the catches; this has required studying chemical and processing properties and peculiarities to evaluate these fish as raw material for processing.

In recent years, the Research Institutes of the Ministry of Fisheries of the USSR (VNIRO, TINRO, AtlantNIRO, AzcherNIRO, PINRO) have accumulated a large volume of experimental data. Under instructions from the Ministry of Fisheries of the USSR, the VNIRO and the other coastal institutions have worked on the systematization and generalization which have been presented in this handbook.

The description of fish is presented in a systematic manner (their families, genera and species). For convenience in using the handbook, an alphabetic index of the scientific and common names of fish is also given. The handbook consists of two sections: the first is a description of cartilaginous fish and the second of bony fish.

The handbook describes 452 species of fish belonging to 91 families and includes diagrams, common and scientific names, commercial significance, area of distribution, size parameters, weight proportion of individual body parts, chemical composition of meat and other parts of the fish body along with a brief assessment of the organoleptic properties of fish meat and recommendations for processing and consumption. However, in the descriptions of less known fish, in the absence of required data, some of this information could not be given.

Both scientific names of the fish and common ones have many synonyms. In preparing the handbook only those names were used which could be found in local literature on fish technology.

The handbook indicates whether a particular species is a commercial fish or is only incidental in the catch. However, in some cases this information has not been given as these species have not been studied in detail and their commercial value is undetermined. Moreover, the industrial value of one constituent in a fishery may change as a result of its study and use.

The size characteristics of fish, length and weight, are given in the descriptions. The maximum depth and maximum body thickness are also indicated. The ranges and the predominant values of length and weight, where available, are indicated. In some cases the maximum body sizes of the fish are given. For those species which have not been studied in detail, size parameters are given only for the fish specimens which were examined.

The weight proportions of different body parts are indicated as percentages of the total weight and where there is considerable variation the range of the proportions is also given. Where considerable differences occur in the proportions of individual body parts of fish from different regions or fishing seasons the data are presented separately.

In indicating the weight proportions of different body parts, we note that the total of the constituent components of the body, in many cases, is less than 100%. This shortage in weight is due to unavoidable losses during fish dressing.

Average chemical composition, moisture, fat, protein and mineral matter in the edible parts (meat, liver, roe) and in the offal is also given. In case of significant variations, the limits are indicated. When deviations in the composition of fish of different sizes, caught from different regions and in different periods are considerable, the data are presented separately.

In the descriptions of inadequately studied fish the results of analyses of individual fish specimens are presented.

In determining the chemical composition of fish we note that the sum total of the substances (moisture, fat, protein, mineral matter) constituting meat and other body parts is usually a little less than 100% and this is especially true for liver. Thus, in the meat and especially in the liver, there is a small quantity of matter, for example, carbohydrates, which is not estimated in the chemical composition.

In the description of some fish data are given on the vitamins in the meat, liver and viscera of the fish and about the mineral composition of the meat.

This handbook has been compiled by T.I. Makarova, V.P. Bykov, Senior Scientist T.V. Sergeeva (VNIRO), Dr. I.P. Levanidov, V.S. Gordievskaya, O.M. Melnikova, Senior Scientist M.N. Vakhrusheva (TINRO), V.N. Podsevalov (AtlantNIRO), L.P. Minder (PINRO) and G.K. Kovalchuk (AzcherNIRO).

The compilers of this handbook received great help from K.V. Shust of VNIRO in determining the correct names of different families and individual fish species. In preparing the handbook for printing, senior scientist M.N. Eremeeva and engineer E.A. Burmenkov of VNIRO played an active part.

CHARACTERISTICS OF FISH

SUPERORDER SELACHOIDEI—SHARKS

Sharks are distributed throughout the world's oceans. They vary in size considerably. Giant size sharks may reach a length of 20 m. About 290 shark species are known of which only a few are of commercial value.

Sharks are an appreciable constituent in fisheries in tropical, temperate and Arctic waters.

Sharks are used as food (in Europe, America, Africa and Australia) and in preparing fish meal as livestock feed. Liver oil is also extracted from sharks. The following sharks have the highest gustatory properties: porbeagle, mako, blacktip shark, gray shark and picked dogfish (catran).

The meat of many shark species is white or slightly pinkish, juicy and agreeable in appearance. However, the taste differs considerably from the taste of other marine fish meat.

Shark meat contains up to 2.5% urea and has a high nitrogen content in the form of volatile bases, ammonia and trimethylamine. It therefore has an unpleasant specific odor and pungent-acid taste; it cannot be used as a food without special processing.

A large quantity of edible sharks caught by the fishing industry of the Soviet Union is exported.

In accordance with the "Inter-Republican Technical Conditions" which have been approved by the Ministry of Fisheries of the USSR, the following species of sharks are exported as food:

Dusky shark—*Carcharhinus obscurus* (*C. albimarginatus*), mackerel sharks—*Isurus glaucus*, whitetip shark—*Carcharhinus longimanus*, spiny dogfish—*Squalus acanthias*, whale shark—*Rhinocodon typus*, lesser spotted dogfish—*Scyliorhinus caniculus*, smooth dogfish—*Mustelus canis*, smooth hammerhead—*Sphyrna zygaena*, thresher—*Alopias vulpinus*, small-toothed sawfish—*Pristis microdon*, Greenland shark—*Somniosus microcephalus*, porbeagle—*Lamna cornubica*, gray shark—*Carcharhinus johnsoni*, reef shark—*Triaenodon obesus*, sharp nosed shark—*Scoliodon walbeehmi*.

Family CARCHARHINIDAE—REQUIEM SHARKS

Sharks of this family are found in tropical, subtropical and temperate latitudes in coastal and distant waters. Specimens as long as 550 cm have been recorded. This family consists of the largest number of genera (15), species and varieties.

Sharks belonging to three genera of this family are of great interest: *Carcharhinus, Scoliodon* and *Mustelus*. From the genus *Carcharhinus* blacktip shark, whitetip shark and gray shark have been studied; from the genus *Scoliodon* sharp nosed shark and scoliodon and from *Mustelus* smooth dogfish and other dogfishes.

Genus *Carcharhinus*

Sharks of this genus live in the Atlantic, Indian and Pacific Oceans. They grow in weight to 400 kg and attain sexual maturity at a length of 180 cm.

Blacktip shark—*Carcharhinus limbatus*. Length of a fish caught in March in the eastern Central Atlantic 190 cm, maximum depth 36 cm, maximum thickness 36 cm, weight 95 kg.

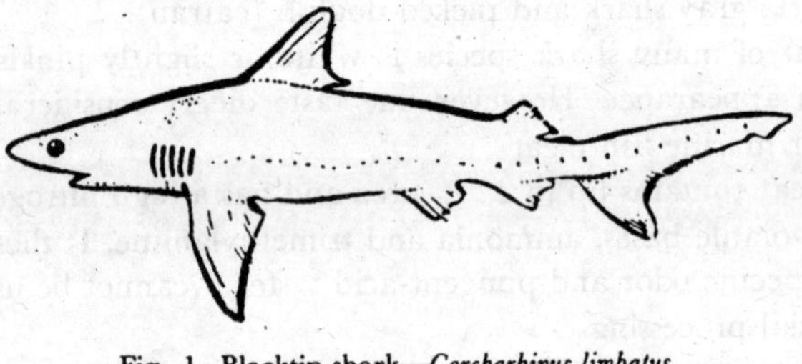

Fig. 1. Blacktip shark—*Carcharhinus limbatus*.

Weight proportions of individual body parts (%): head 20, trunk 52, fins 10, viscera 18, including liver 5. Output of fillets from gutted and dressed fish 69%.

Chemical composition of meat (%): moisture 74, protein 22, fat 0.3, ash 1.7.

Chemical composition of liver (%): moisture 37, oil 52, ash 1.

The meat of this shark is tough. After soaking, it is suitable for the preparation of culinary items and canned products in tomato sauce.

Dusky shark—*Carcharhinus obscurus* (*Carcharhinus albimarginatus*). Lives in the Atlantic Ocean and some parts of the Indian Ocean; a closely related form of this fish is found in the Pacific Ocean. Length up to 3.6 m. Length of a shark caught off the coast of North America 117 cm, weight 56 kg.

Weight proportions of individual body parts (%): head 20, trunk 62, viscera 18. Output of fillets (with skin) 70% of the trunk.

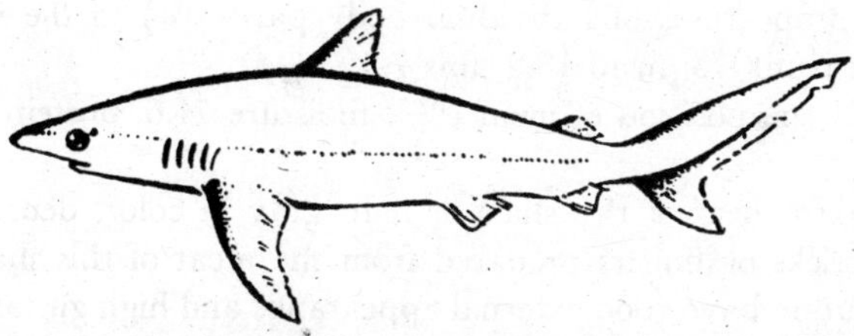

Fig. 2. Dusky shark—*Carcharhinus obscurus.*

Chemical composition of meat (%): moisture 74–76, protein 20–22, fat 0.3–0.6, ash 1.5.

Conserves and culinary items from this shark have not become popular because its meat has a bitter taste. It should be used for fish meal.

Whitetip shark—*Carcharhinus longimanus*. Lives in tropical and subtropical waters of the Atlantic, Indian and Pacific Oceans and is also found in temperate waters. Length up to 4 m. This shark has high commercial value presently.

Weight of a shark (immature) caught in January in the Indian Ocean 20 kg, standard length 106 cm. Maximum depth of body 24 cm and thickness 23 cm.

Weight proportions of individual body parts (%): trunk 48, head 29.3, viscera 15.8, including liver 4, fins 6.7, meat 42, skin 4.2, cartilage 1.9.

TABLE 1. CHEMICAL COMPOSITION, %

Body part	Moisture	Protein	Fat	Ash
Meat	76.9	21.4	0.4	1.3
Head	77.3	19.9	0.3	2.5

After boiling, the meat is light colored, soft, sour, unpleasant to taste. It should be used for fish meal.

Black shark—*Carcharhinus melanopterus*. Distributed in Indian and Pacific Oceans. Length up to 3 m, usually less.

Weight of a shark caught in spring in the tropical part of the Indian Ocean, 71.4 kg.

Fig. 3. Black shark—*Carcharhinus melanopterus.*

Weight proportions of individual body parts (%) to the weight of the dressed fish: trunk 73, head 18.2, fins 7.3.

Chemical composition of meat (%): moisture 74.6, protein 23.6, fat 0.4, ash 1.3.

The cooked meat of this shark is light gray in color, dense, pleasant in taste. Fish sticks or fingers prepared from the meat of this shark soaked in mustard solution have good external appearance and high gustatory qualities.

Eulamia or white checked shark—*Carcharhinus dussumieri* (*Eulamia dussumieri*). Distributed in the coastal waters of Indonesia, Australia, Burma, India, China, the Philippines and in the Mariana Islands. Body length not exceeding 120 cm.

Fig. 4. Eulamia shark—*Carcharhinus dussumieri.*

Weight proportions of individual body parts caught in the Andaman Sea (%): trunk 63.2, head 17.6, viscera 9.8, including liver 4.3, fins 9.4.

Chemical composition of meat (%): moisture 79.5, protein 17.7, fat 0.8.

The meat of a fresh Eulamia shark has a fairly unpleasant specific odor which is characteristic of other shark species also. Cooked, the meat of this shark is white, soft and juicy in consistency, sour in taste (generally a characteristic of all sharks). The specific odor of cooked meat is milder than fresh meat. Fried Eulamia meat is pleasant in appearance and has a soft consistency. After soaking, it is suitable for canned and culinary products.

Genus *Mustelus*

Smooth hound or gummy shark—*Mustelus manazo*. Found off the Pacific coast of Japan, in the Yellow Sea, Indian Ocean and Arabian Sea. Body length up to 220 cm.

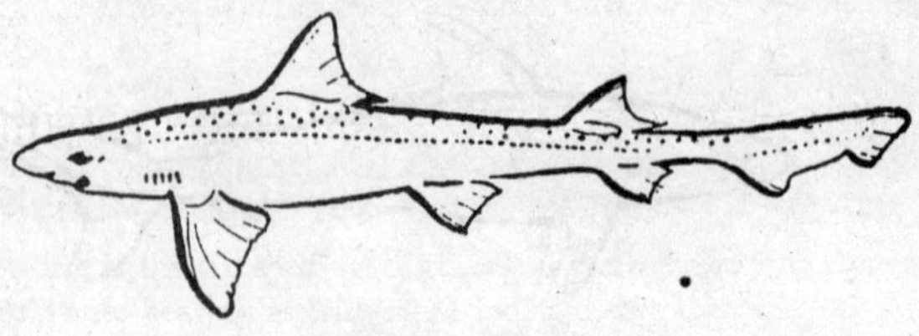

Fig. 5. Smooth hound—*Mustelus manazo.*

Weight proportions of individual body parts (%) of a fish caught in February in the Arabian Sea weighing about 1 kg: head 17.7, trunk 60.9, fins 5.5, viscera 15.9, including liver 4.2.

The cooked meat of the smooth hound is grayish with an acid pungent taste. After soaking, the meat can be used for culinary products.

Genus *Scoliodon*

Sharp nosed shark—*Scoliodon walbeehmi.* Distributed in the eastern Atlantic Ocean (along the coast of Africa up to Madeira), in the Indian Ocean, along the coast of South Africa, in the Red Sea, near the coast of Indonesia and in the western Pacific Ocean from Australia to Japan. Length up to 1.2 m.

Fig. 6. Sharp nosed shark—*Scoliodon walbeehmi.*

Length of a fish caught in the northwestern Atlantic 50 cm, the maximum depth of the body 8 cm and maximum thickness 8 cm; weight of the fish 1.1 kg.

Weight proportions of individual body parts (%): head 17, trunk 52, viscera and fins 31.

Chemical composition of meat (%): moisture 78, protein 18, fat 0.8, ash 1.5.

After soaking and frying, the meat has satisfactory taste.

Because of the poor taste qualities its use is recommended for fish meal.

Palasorrah shark—*Scoliodon palasorrah.* Distributed along the coasts of India, Indonesia, China and Japan. Body length about 70 cm. The body length of fish caught in June in the Indian Ocean near Cape Fartak ranged from 56–63 cm in length and 1.9–2.4 kg in weight.

Weight proportions of individual body parts (%): head 17–22, trunk 54–65, including meat 39.6, skin 6, fins 4, viscera 15–22, including liver 10–12 and testes 0.8.

Chemical composition of meat (%): moisture 74.8, protein 23.5, fat 0.3, ash 1.4.

The meat is gray with the smell of rotten eggs. It is recommended for fish meal.

Yellow dogfish—*Scoliodon sorrakowah.* Distributed in the eastern Atlantic Ocean, in the Indian Ocean, along the coast of South Africa, in the Red Sea,

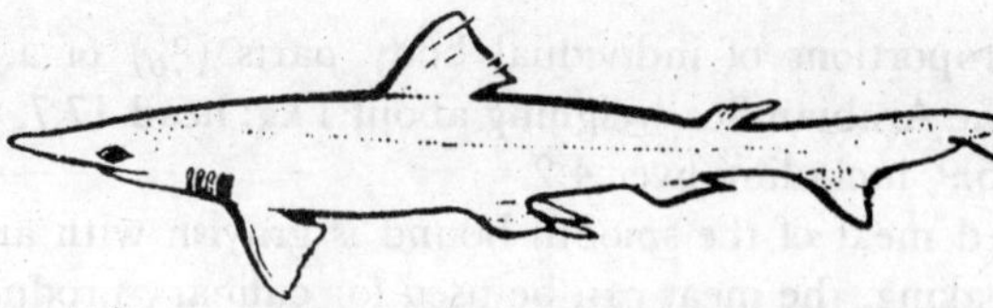

Fig. 7. Yellow dogfish—*Scoliodon sorrakowah.*

along the coast of Indonesia and in the western Pacific Ocean from Australia to Japan. Body length about 90 cm.

Body length of the fish in catches in February in the Gulf of Aden, 45–68 cm and weight 0.8–2.6 kg.

Weight proportions of individual body parts (%): head 16.5, trunk 62.6, fins 4.5, viscera 16.4, including liver 11.1.

Chemical composition of meat (%): moisture 75.8, protein 21.5, fat 0.5.

The meat when cooked is white, pleasant in appearance, with soft consistency and a specific smell and sour-pungent taste. Fried meat of this shark has a softer consistency and a little better taste than boiled meat. After soaking, the meat becomes suitable for canned products and culinary purposes.

Family LAMNIDAE—MACKEREL SHARKS

These are distributed in the tropical and temperate waters of all oceans. They grow to 5 m in length. This family consists of five species; thresher shark, blue shark, mako and porbeagle have been investigated.

Genus *Alopias*

Thresher shark—*Alopias vulpinus.* Distributed in oceans and seas almost everywhere, except in the high latitudes. Length of fish usually encountered 3 m, maximum length 7 m. Length at birth 0.7–1.2 m.

Length of a specimen caught in March in the eastern Central Atlantic 190 cm, maximum body depth 40 cm, maximum thickness 30 cm. Weight of the fish 140 kg. Weight proportions of individual body parts (%): head 14, trunk 57, fins 14, viscera 15, including liver 5. Chemical composition of meat (%): moisture 84, protein 14, fat 0.7, ash 1.

Weight of thresher shark caught in spring in the tropical Indian Ocean

Fig. 8. Thresher shark—*Alopias vulpinus.*

71.4 kg. Weight proportions of individual body parts (%) to weight of the dressed fish: head 15.4, trunk 70.4, fins 13. Chemical composition of meat (%): moisture 77.4, protein 21.0, fat 0.4, ash 1.1.

The high content of volatile bases (215 mg%) in the meat of this shark makes it unsuitable as food. The canned products and culinary preparations are pungent in taste and are considered inedible; therefore, the thresher shark should be used for fish meal.

Genus *Isurus*

Blue shark or bonito shark—*Isurus glaucus*. Lives in the Indian and Pacific Oceans and in tropical and temperate waters. Attains a length of 6 m.

The weight of a fish caught in the spring in the tropical part of the Indian Ocean was about 54 kg. Weight proportions of individual body parts (%) to

Fig. 9. Blue shark—*Isurus glaucus*.

the weight of the dressed fish: head 17.2, trunk 76.2, fins 5.5. Chemical composition of meat (%): moisture 75, protein 23.3, fat 0.4, ash 1.2.

The meat of this shark caught in the Indian Ocean along the southeastern coast of Sumatra in March had the following chemical composition (%): moisture 77, fat 6.1, protein 15.7. The cooked meat does not have unpleasant taste or smell.

The meat of the blue shark differs from the meat of other shark species by its milky white color and soft consistency. It can be used for culinary products.

Mako—*Isurus oxyrinchus*. Lives in tropical and warm waters around the world. Maximum length 3.7 m.

Investigations were carried out on frozen, beheaded, dressed fish weighing

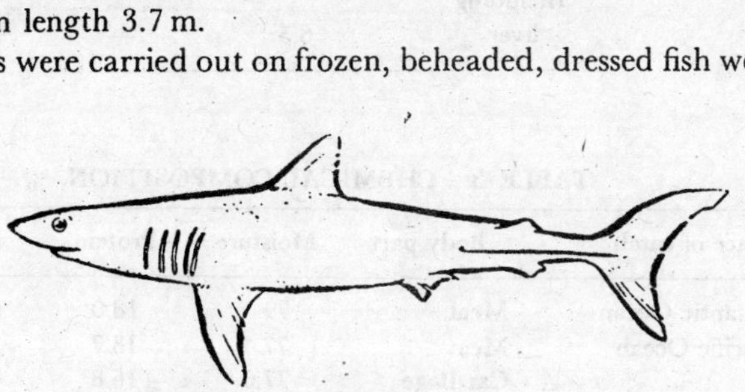

Fig. 10. Mako—*Isurus oxyrinchus*.

29 kg caught in the Atlantic Ocean along the coast of North America. Chemical composition of meat (%): moisture 75, protein 21, fat 0.4, ash 1.4.

The products canned in tomato sauce and culinary preparations made at the Kaliningrad Fish Processing Plant from pieces of this fish soaked for an additional two hours had good external appearance and satisfactory taste.

Genus *Lamna*

Porbeagle—*Lamna nasus*. Found in the Atlantic and Pacific Oceans and absent in the Arctic and Antarctic Oceans. In the northern Pacific Ocean and along the coasts of Australia and Argentina, subspecies or closely related species of the porbeagle are found. It reaches a length of 3–4 m, but the fish usually caught are 2.0–2.5 m in length. Weight 100–150 kg.

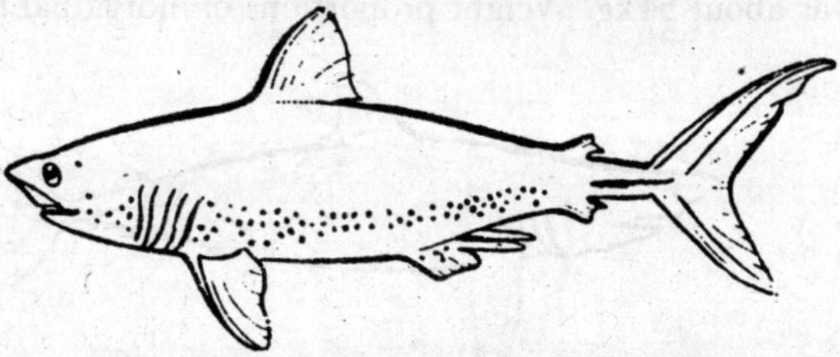

Fig. 11. Porbeagle—*Lamna nasus*.

TABLE 2. WEIGHT PROPORTIONS OF INDIVIDUAL BODY PARTS, %

Body part	Northeastern Atlantic Ocean	Northeastern Pacific Ocean
Trunk	55	48.8
Head	25	16.9
Fins	6	5.2
Viscera	14	28.2
Including liver	6.5	—

TABLE 3. CHEMICAL COMPOSITION, %

Place of catch	Body part	Moisture	Protein	Fat	Ash
Atlantic Ocean	Meat	79	18.0	0.8	—
Pacific Ocean	Meat	77.4	18.7	2.1	1.7
	Cartilage	77.6	16.8	1.8	3.8

Culinary preparations made from the meat of this fish after the necessary soaking and heat treatment have good external appearance and satisfactory taste. Canned products also can be made from it.

Family SPHYRNIDAE—HAMMERHEAD SHARKS

These sharks generally live in tropical and subtropical waters, but are sometimes found in temperate latitudes. There is only one genus *Sphyrna* in this family. Two species of this genus have been examined. Some of these sharks reach a length of 9 m.

Smooth hammerhead shark—*Sphyrna zygaena*. Distributed in the tropical and subtropical waters of the Pacific, Indian and Atlantic Oceans. Attains a length of 4 m, in individual cases up to 6 m, weight 500 kg. Total length of a hammerhead shark caught in March in the eastern Central Atlantic was 300 cm, with standard length 230 cm, maximum depth 50 cm and thickness 35 cm. Standard length of a hammerhead shark caught in the Indian Ocean was 350 cm.

Fig. 12. Smooth hammerhead shark—*Sphyrna zygaena*.

Weight proportions of individual body parts of a shark weighing 130 kg caught in the Atlantic Ocean (%): head 25, trunk 46, fins 12, viscera 17, including liver 6.

TABLE 4. CHEMICAL COMPOSITION OF MEAT, %

Place of catch	Moisture	Protein	Fat	Ash
Atlantic Ocean	73	23	0.4	1.5
Indian Ocean	73.2	24.5	1.0	1.4

Chemical composition of the liver of common hammerhead shark caught in the Atlantic Ocean (%): moisture 38, protein 10, fat 50.

After cooking, the meat becomes gray, rubbery and is not tasty. It is recommended that this fish be used for fish meal.

Arrow-headed hammerhead shark—*Sphyrna blochii*. Distributed in the Red Sea and Indian Ocean, from the Persian Gulf to Australia, in the Pacific

Fig. 13. Arrow-headed hammerhead shark—*Sphyrna blochii.*

Ocean from the Australian coast to the Philippines. Attains standard length of 120 cm.

The length of three specimens of this shark caught in March in the Indian Ocean on the western shelf of India was 88–106 cm and weight 9.0–14.6 kg.

Weight proportions of individual body parts (%): head 14.9, trunk 63.6, fins 5.6, viscera 15.9, including liver 4.5.

Chemical composition of meat (%): moisture 72.8, protein 24.7, fat 1.0.

Cooked meat of this shark is gray, unpleasant in color, tough, with strong sour-bitter taste. It is recommended that this fish be used for fish meal.

Family SQUALIDAE—DOGFISHES

This family is represented by 11 genera and many species. The sharks of this family are relatively small fish with a maximum length of 260 cm. They are distributed in all oceans and seas except the polar regions. Three species of dogfishes have been studied: spiny dogfish, spinax shark and hypoprion.

Genus *Squalus*

Spiny dogfish—*Squalus acanthias* (*Acanthias acanthias*). Distributed in the temperate and subarctic latitudes of all oceans. Large stocks of this fish found in the northwestern Atlantic. Maximum length 1.2 m.

The meat of this fish does not have any specific odor so typical of other species of sharks but it has more fat. In England and Sweden, the meat of this shark is considered to be a higher quality than the meat of saithe and some other fish. This shark is marketed in chilled, frozen, salted and smoked

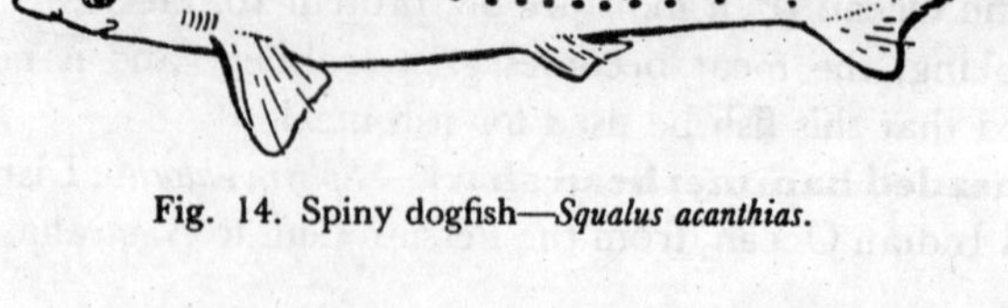

Fig. 14. Spiny dogfish—*Squalus acanthias.*

TABLE 5. SIZE PARAMETERS

Place of catch	Month of catch	Total length, cm	Standard length, cm	Maximum depth, cm
Northwestern Atlantic	September	58–72	53–67	7–10
North Sea	October	60–74	59–69	8–11
	March	48–66	35–50	8–9

TABLE 6. WEIGHT PROPORTIONS OF INDIVIDUAL BODY PARTS, %

Place of catch	Month of catch	Weight of fish, kg	Trunk	Head	Viscera	Including liver
Northwestern Atlantic	September	0.9–2.4	48	20	30	8
	October	1.1–2.7	60	18	22	—
North Sea	March	0.4–1.0	55–57	19–25	22–24	—

TABLE 7. CHEMICAL COMPOSITION, %

Place of catch	Month of catch	Body part	Moisture	Protein	Fat	Ash
Northwestern Atlantic	September	meat	72	18.0	9	1
	October	„	70	17.6	11.1	1.3
	October	„	66.8	19.3	12.8	1.1
North Sea	March	„	71.0	—	9.4	—
Northwestern Atlantic	October	liver	30	8	60	1
	October	roe	47	27	23	0.6

forms. Canadians use the meat of this fish for fish fingers; the Japanese prepare national dishes from it; in Turkey cured balyks are made.

At AtlantNIRO smoked balyks were prepared from a specimen of this shark with the following chemical composition (%): moisture 43, protein 26, fat 20, ash and salt 11. In a wide range of degustations it was recorded that the preparations had good external appearance and high taste qualities. This fish has been recommended as a food.

Genus *Etmopterus*

Spinax shark—*Etmopterus spinax*. Distributed in the eastern Atlantic Ocean from Norway to the Azores, Madeira Islands, Cape Verde, in the

Fig. 15. Spinax shark—*Etmopterus spinax.*

TABLE 8. CHEMICAL COMPOSITION, %

Body part	Moisture	Protein	Fat	Ash
Meat	71.9	17.3	9.4	1.3
Liver	32.3	7.2	58.0	0.6
Head	81.0	12.5	0.8	5.7
Cartilage	73.7	18.1	2.5	5.7

Mediterranean Sea and in the Gulf of Mexico at depths of 70 to 1110 m [*sic*].

This shark is small in size, up to 1 m in length. In June, in the northwest Atlantic, near the shelf of Ireland, a spinax shark was caught, 77 cm in length, weighing 2.1 kg.

Weight proportions of individual body parts (%): head 12.5, trunk 56.9 including meat 43.3, viscera 28.9, fins 1.2, skin 6.7, cartilage 6.1, liver 12.7.

Meat of the spinax shark contains a large quantity of fat, more than 9%.

Genus *Hypoprion*

Hypoprion shark—*Hypoprion palasorrah*[1]. Chemical composition of a hypoprion shark which was caught in the Indian Ocean on the western shelf of India (%): moisture 76.6, fat 0.5, protein 21.4.

Cooked meat is white, pleasant in appearance, with soft consistency, specific shark-like smell and slightly acid-pungent taste. The fried meat is softer in consistency with a slightly better taste than boiled meat. After soaking, canned products and culinary preparations can be made from the meat.

Family RHINOBATIDAE—GUITAR FISHES

Spotted guitar fish—*Rhinol atus djiddensis*. In taxonomy, spotted guitar fish occupies an intermediate position between sharks and rays. Guitar fish are distributed in the coastal waters of the tropics and subtropics. In catches on the shelf of western India, there were guitar fish weighing less than 1 kg and some weighing tens of kilograms. Chemical composition of meat of a guitar fish caught in this region (%): moisture 80.5, fat 0.6, total nitrogen 3.6.

[1] *Hypoprion palasorrah* is now considered a synonym of *Scoliodon palasorrah*—Technical Editor.

The meat of the guitar fish resembles the meat of sharks in various characteristics and has an acid-bitter taste.

The guitar fish should be frozen and subsequently used for culinary preparations.

SUBCLASS TELEOSTEI—BONY FISHES

Family ALBULIDAE—BONEFISHES

Bonefish—*Albula vulpes* (*Albula conorhynchus*). Large size pelagic fish found in warm waters. Distributed in the eastern Indian Ocean. Also found in the Atlantic Ocean along the coast of Africa.

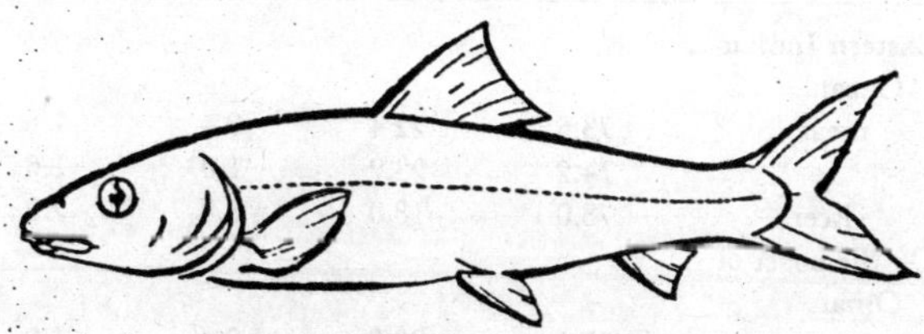

Fig. 16. Bonefish—*Albula vulpes.*

In August in catches off the coast of Oman, individuals 60 cm (57.7–66.8) in length and 3,600 g (2800–4088) in weight were predominant; the maximum depth was 13.6 cm (13.5–14.0) and the maximum thickness as 9.2 cm (8.5–9.9).

TABLE 9. WEIGHT PROPORTIONS OF INDIVIDUAL BODY PARTS, %

Body parts	Eastern Indian Ocean	Off the coast of Oman	Atlantic Ocean
Head	18.7–19.0	20.1–21.9*/20.8	17
Trunk	69.0–70.3	64.0–67.4/65.0	73
Meat with skin	60.7–61.3	—	57**
Scales	4.3–4.9	44–54/5.0	—
Bones	8.3–8.8	—	—
Fins	1.2–2.2	1.0–1.8/1.4	3

1	2	3	4
Viscera	4.3–4.9	5.7–8.8 / 7.8	7
Including liver	0.8	—	—

* In this Table and in subsequent Tables the range of weight proportions of individual body parts of fish are given in the numerators and their average values are given in the denominators.

** Output of fillets while dressing the trunk.

TABLE 10. CHEMICAL COMPOSITION, %

Body parts and place of catch	Moisture	Protein	Fat	Ash
Eastern Indian Ocean				
meat	73.6	22.4	0.4	1.5
	74.2	23.2	1.0	1.6
viscera	73.0	18.0	5.8	2.9
Off the coast of Oman				
meat	73.1	22.8	2.4	1.2
Atlantic Ocean along the coast of Africa				
meat	73–76	22–24	1.1–1.2	1.3–1.6
head, bones, fins	60–62	20–21	7–9	10.6–10.8
viscera	69–79	17–18	3–12	1.8–2.5

In the eastern Indian Ocean, individuals 32–64 cm in length and 350–3160 g in weight were predominant. In the Atlantic Ocean along the coast of Africa, specimens 30–35 cm in length and 0.4–0.5 kg in weight were caught.

The meat of bonefish caught in different regions rated different grades. The meat of bonefish caught in the Indian Ocean off the coast of Oman and in the Atlantic Ocean along the African coast was more compact. In boiled and fried form it had better taste. However, the meat of bonefish caught in the eastern Indian Ocean had somewhat different taste.

Bonefish should be used in canned products, semi-finished preparations and culinary items as well as for salt-cured products.

Family ANOPLOPOMIDAE—SABLEFISHES

These are distributed in the northern Pacific Ocean. This family has two genera: *Anoplopoma* and *Erilepis*.

Black cod or sablefish—*Anoplopoma fimbria*. An important commercial fish distributed in the northeastern Pacific Ocean from Bristol Bay to San Diego and in the eastern Bering Sea; also found in the western Bering Sea and in the Olutor and Kamchatka Gulfs. In the Gulf of Alaska the usual length of black cod is 70 cm and weight is 3.9 kg, but larger specimens are also found. The length of black cod varies from 42.5–65.0 cm, weight from 780–5,950 g.

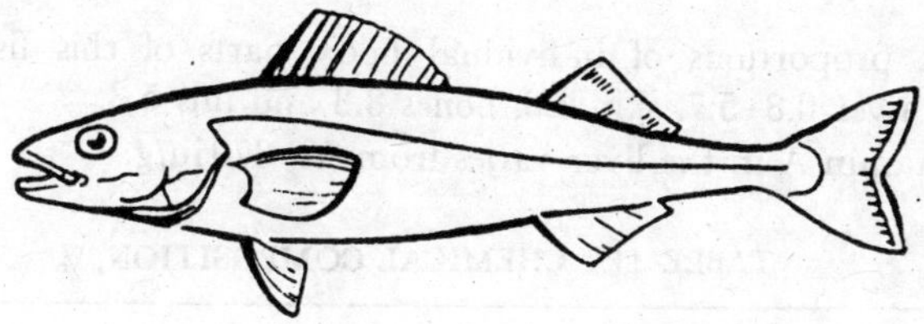

Fig. 17. Black cod or sablefish—*Anoplopoma fimbria*.

Weight proportions of individual body parts (%): head 10.1–24.6, meat with skin 54.1–65.0, fins and bones 6.7–10.0, viscera 5.7–17.1, including liver 1.0–2.3.

Chemical composition of meat (%): moisture 62.8–81.9, protein 10.8–13.5, fat 4.9–24.2, ash 0.8–1.5. As a rule the smaller specimens contain less fat.

This fish has a high output of meat. It is distinguished by relatively low protein and high fat content. Cooked meat is white, soft and juicy in consistency with good taste even after frying. After hot or cold smoke curing, black cod has high taste qualities. When caught this fish may be frozen or salted on board ship for subsequent marketing.

Family ANARHICHADIDAE—WOLFFISHES

These are fairly large marine fish which live in the northern Pacific and Atlantic Oceans. The family is represented by three genera. The genus *Anarhichas* includes only three species of commercial value, common wolffish, spotted wolffish and blue wolffish.

Common or Atlantic wolffish—*Anarhichas lupus*. Distributed in the northern Atlantic Ocean and adjoining seas, in the Barents Sea up to Goose Bay and around Bear Island and the southwestern coasts of Spitsbergen. The southern boundary of its distribution is along the northwestern coast of France. Found in the White and Baltic Seas, in the west around Iceland and in Davis Strait. The southern boundary passes near Cape Cod.

Attains a length of 125 cm, off the coast of Canada rarely less than 150 cm. In catches, usual length of this wolffish is 70 cm, with a weight of about 4 kg and usually netted in small numbers with other demersal fish.

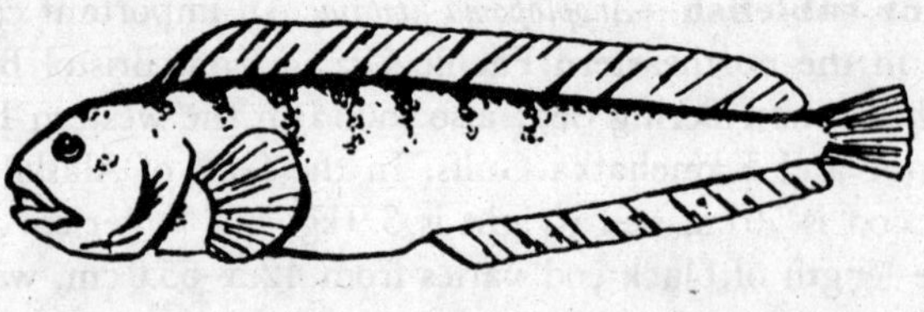

Fig. 18. Common or Atlantic wolffish—*Anarhichas lupus.*

Weight proportions of individual body parts of this fish (%): meat 42.6–50.0, liver 0.8–5.7, skin 4.6, bones 8.3 and fins 5.3.

The vitamin A in the liver varies from 58–370 iu/g.

TABLE 11. CHEMICAL COMPOSITION, %

Body part	Moisture	Protein	Fat	Ash
Meat	69.8–82.3 / 77.2	13.3–18.3 / 16.0	0.7–13.1 / 5.0	— / 1.1
Liver	51.1–77.9 / 64.1	9.3–14.5 / 11.8	4.6–38.7 / 21.0	1.1–1.9 / 1.0
Head	69.7–75.0 / 72.4	14.3–17.8 / 15.8	2.7–6.4 / 5.3	4.4–5.2 / 4.7
Bones	58.0–73.5 / 67.0	14.0–16.2 / 15.0	5.7–17.4 / 10.5	5.2–8.1 / 6.8
Viscera	76.9–92.0 / 84.3	49–14.0 / 10.2	1.0–8.0 / 3.5	0.6–1.7 / 1.1

Because of the high protein content, boiled and fried meat of this wolffish has good consistency. It is a valuable fish, suitable for culinary and smoked products.

Spotted wolffish—*Anarhichas minor*. Distributed in the northern Atlantic Ocean. Found in particularly large numbers in the Barents Sea up to Goose Bay in the east and up to Bear Island and the southwestern coasts of Spitsbergen, Scotland in the north. The southern boundary of its distribution stretches from Bergen, Scotland in the east to Massachusetts Bay in the west.

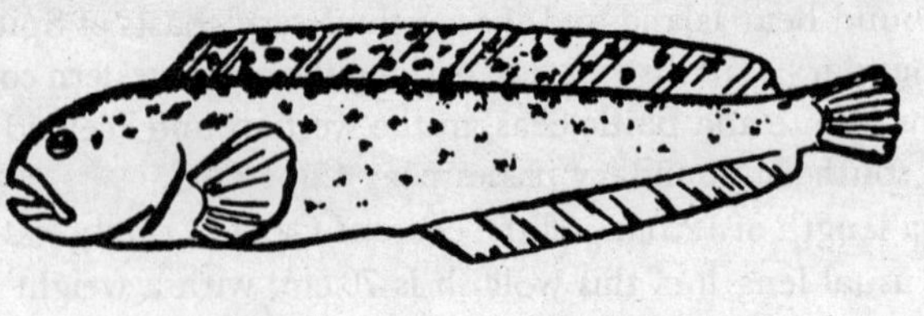

Fig. 19. Spotted wolffish—*Anarhichas minor.*

The spotted wolffish attains a length of 135 cm and weight of 20 kg.

Weight proportions of individual body parts of the spotted wolffish are characterized by relatively low percentage of meat, 41–49. The head severed by a straight cut together with the pectoral girdle constitutes 24% and the head separated by the commercial method along with the skull bones, 15 to 19% (average 17%). The spotted wolffish has thick and strong skin which on an average constitutes 5.5%. The size of the liver varies depending upon the fatness of the fish: from 1.3–1.9% in lean fish in the winter to 5.6% at the end of the summer. The total quantity of edible parts on an average is 54%.

The potency of vitamin A in the liver of spotted wolffish varies from 0–1,600 iu/g.

The spotted wolffish is a valuable fish. It is good for smoking and many other fine culinary preparations and canned products can be made.

Blue wolffish—*Anarhichas latifrons*. Distributed in the northern Atlantic from the Kanin bank in the east to the American coasts in the west. Southern boundary of distribution from the Lofoten Islands to Cape Cod.

Attains a length of 134 cm and weight of more than 24 kg.

The potency of vitamin A of the liver oil varies from 150–3,600 iu/g.

The meat of this fish is characterized by low food and feed value because of the low protein and fat contents. Because of the high water content of its

TABLE 12. CHEMICAL COMPOSITION, %

Body part	Moisture	Protein	Fat	Ash
Meat	74.4–82.8 / 78.3	13.3–17.7 / 14.7	2.3–9.0 / 5.3	0.7–1.2 / 1.0
Liver	46.6–75.0 / 62.1	8.2–14.2 / 11.9	9.1–42.0 / 23.1	0.8–1.6 / 1.1
Roe	67.9–80.8 / —	10.3–19.1 / —	5.8–18.2 / —	1.0–1.8 / —
Testes	— / 79.0	— / 10.8	— / 8.3	— / 1.2
Head	64.4–80.5 / 75.0	11.7–14.0 / 12.7	4.4–17.7 / 7.8	1.7–3.7 / 3.0
Bones	61.5–76.1 / 66.9	9.2–14.2 / 11.5	6.7–26.1 / 15.5	3.7–6.1 / 5.0
Fins	78.6–83.3 / 80.9	12.2–15.6 / 14.0	0.2–0.9 / 0.4	2.5–5.1 / 3.6
Viscera	80.5–87.4 / 84.0	7.2–13.0 / 10.6	0.6–5.8 / 2.2	1.0–2.6 / 1.4

TABLE 13. WEIGHT PROPORTIONS OF INDIVIDUAL BODY PARTS OF BLUE WOLFFISH (INVESTIGATIONS ON SIX MALES IN STAGE II OF MATURATION AND ONE FEMALE IN STAGE III OF MATURATION), %

Length, cm	Weight, g	Head	Meat	Skin	Bones	Fins	Viscera		
							Total	Including	
								Liver	Others
83–129 / 104	6,540	11.5	51.7	4.1	4.6	2.6	9.9	1.0	8.3
—	24,000	18.6	58.7	10.3	8.5	4.2	18.0	3.1	10.6
—	11,704	15.2	54.6	7.1	6.5	3.3	12.8	2.2	8.4

TABLE 14. CHEMICAL COMPOSITION OF BLUE WOLFFISH, %

Body parts	Moisture	Protein	Fat	Ash
Meat	89.6–93.3 / 90.9	3.6–7.2 / 5.9	0.2–2.1 / 1.2	0.6–1.2 / 0.9
Skin	82.9–90.6 / 87.1	11.4–15.6 / 11.4	0–1.3 / 0.4	0.7–1.3 / 1.0
Liver	52.4–74.2 / 69.2	8.9–11.5 / 8.5	10.7–37.0 / 21.2	1.0–1.4 / 1.1
Head	81.0–88.1 / 85.1	6.5–9.6 / 8.1	1.6–6.8 / 3.9	1.9–2.4 / 2.3
Bones	75.0–86.9 / 80.8	5.7–8.9 / 6.8	4.6–12.5 / 8.2	2.6–4.0 / 3.3
Viscera	89.1–90.8 / 89.9	6.4–7.3 / 6.8	0.1–0.9 / 0.6	0.6–1.2 / 0.9

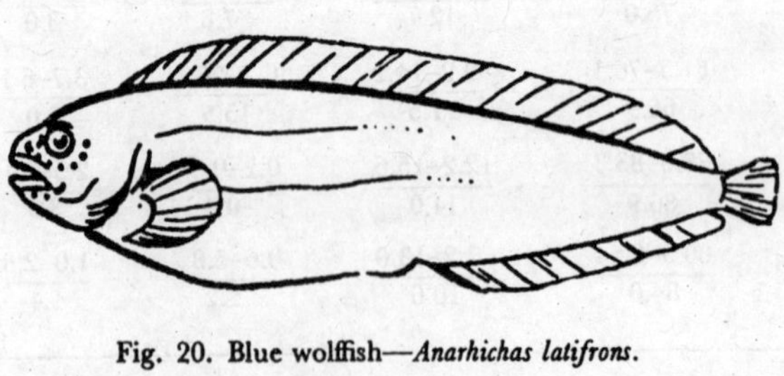

Fig. 20. Blue wolffish—*Anarhichas latifrons*.

meat, it cannot be used for culinary purposes. It is recommended that it be used for fish meal.

Family ANGUILLIDAE—EELS

This family is represented by one genus, *Anguilla,* and several species which are distributed in the Atlantic, Indian and Pacific Oceans. Only one species has commercial value and this has been investigated, European eel (*Anguilla anguilla*).

European eel—*Anguilla anguilla.* Distributed along the coast of Europe from the Black Sea to the White Sea. Attains a length of 1.5 m and weight

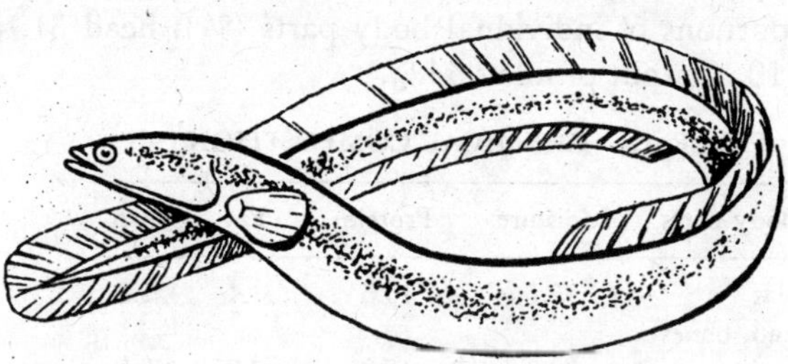

Fig. 21. European eel—*Anguilla anguilla.*

TABLE 15. SIZE PARAMETERS AND WEIGHT PROPORTIONS OF INDIVIDUAL BODY OF DRESSED EEL FROM KURSK BAY

Size of eel	Length, cm	Depth, cm	Thickness, cm	Weight of fish, kg	Trunk, %	Head, %	Viscera, %
Large	61–69	4.7–5.7	3–4	0.5–0.6	83	9	8
Medium	51–60	3.3–4.7	2.4–2.8	0.3–0.4	82	9	9
Small	41–48	2.4–3.3	1.8–2.1	0.1–0.4	—	—	—

TABLE 16. CHEMICAL COMPOSITION OF EEL CAUGHT IN MAY AND JUNE, %

Size of eel	Body parts	Moisture	Protein	Fat	Ash
Large, 61–69 cm	Head	52–54	14	31–33	0.8–0.9
	Middle region	48–49	13–14	36–40	0.7–0.9
	Tail	47–49	12–13	37–40	0.7–0.9
Medium, 51–60 cm	Head	56–62	15–17	21–29	0.9
	Middle region	50–57	14–16	27–36	0.8–0.9
	Tail	50–56	13–14	28–35	0.9
Small, 41–48 cm	Head	59–70	16–17	12–35	0.9–1
	Middle region	57–68	15–17	14–28	0.8–1
	Tail	54–71	14–18	11–31	0.9

of 6 kg. Assumes commercial significance in the areas of the Baltic Sea and its gulfs.

The meat of this eel is characterized by high fat content. The highest fat content has been recorded in a large eel in the middle and tail regions.

Delicatessen specialties are made from this eel after hot smoking; they are in great demand, including the high quality canned product, jellied eel.

Family APOGONIDAE—CARDINALFISHES

Cardinalfish—*Apogonus telescopus*. Found in commercial catches in the Atlantic Ocean, particularly along the coast of northwest Africa. A small fish, those caught range in length from 13–18 cm and in weight from 50–100 g.

Weight proportions of individual body parts (%): head 31, trunk with fins 59, viscera 10. Output of fillets 44%.

TABLE 17. CHEMICAL COMPOSITION, %

Body parts	Moisture	Protein	Fat	Ash
Meat	77.7	19.6	1.4	1.3
Head, bones and viscera	70.7	15.9	7.9	5.5

The meat is white, juicy, soft and tasty. The fish can be used for culinary products.

Cheilodipterus saltator. Caught in February in the Atlantic Ocean near the Patagonian shelf.

Length of fish ranges from 64–65 cm, maximum body depth 12.6–14.1 cm, maximum thickness 6.4–7.8 cm. Weight of fish ranges from 2.5–3.4 kg. Weight proportions of individual body parts (%): head 24.7, meat 50.9, viscera 13.5, including liver 1.0, roe 0.8, fins 2.5, skin 1.7, bones 5.5, scales 0.5.

This is a good table fish which can be used for culinary products and conserves.

TABLE 18. CHEMICAL COMPOSITION, %

Body parts	Moisture	Protein	Fat	Ash
Meat	70.1	20.9	7.7	1.2
Roe	74.5	17.8	5.9	1.4
Liver	66.3	16.0	12.1	1.4
Head	66.9	—	9.5	6.3
Bones	55.7	16.9	15.2	12.0
Skin	61.8	26.0	6.5	1.8
Fins	53.1	19.4	17.1	8.7

Family ARGENTINIDAE—ARGENTINES OR HERRING SMELTS

This family belongs to the suborder Salmonoidei, characterized by the presence of adipose fin. It has one genus, *Argentina*. The fish belonging to this genus live in the northern Atlantic Ocean, in the northern and southern Pacific Ocean and along the southern coast of Africa in the Indian Ocean.

Atlantic argentine—*Argentina silus*. Distributed in the northern Atlantic north of 50°N. Large catches recorded near Iceland. Also constituting commercial catches in the northwest Atlantic near Georges and Sable Banks. Length attained in this region up to 55 cm (off the coast of Europe it is more than 40–50 cm and less along the coast of America) and weight a little more than 1 kg. Standard length 24–38 cm, maximum body depth 5.0–7.6 cm, maximum thickness 3.0–5.2 cm and weight 150–630 g.

Weight proportions of individual body parts of this fish vary slightly throughout the year mainly because of the changes in meat and gonads: the higher the proportion of gonads, the lower the proportion of meat. For example, when the weight proportions of gonads were 5, 8 and 21%, the proportions of meat were 62.4, 59.6 and 50.5%, respectively.

TABLE 19. CHEMICAL COMPOSITION, %

Body parts	Moisture	Protein	Fat	Ash
Whole fish	75.5*	15.2*	3.0*	2.3*
Meat	$\frac{75.4–81.1}{78.6}$	$\frac{17.2–18.5}{17.6}$	$\frac{0.4–4.0}{2.0}$	$\frac{1.3–1.5}{1.3}$
Roe	$\frac{63.1–82.1}{69.2}$	$\frac{11.7–22.8}{18.0}$	$\frac{4.8–14.0}{10.1}$	$\frac{0.8–1.4}{1.2}$
Testes	$\frac{82.4–86.2}{84.0}$	$\frac{11.8–14.5}{13.1}$	$\frac{0.8–1.2}{1.0}$	$\frac{1.6–1.8}{1.7}$
Liver	$\frac{63.9–79.3}{71.7}$	$\frac{14.6–15.0}{14.8}$	$\frac{7.0–14.6}{11.9}$	$\frac{1.6–1.7}{1.6}$
Head	$\frac{77.0–80.0}{78.2}$	$\frac{9.3–13.2}{10.7}$	$\frac{2.8–6.4}{5.2}$	$\frac{4.2–4.9}{4.4}$
Bones	$\frac{66.1–73.2}{68.8}$	$\frac{15.0–16.1}{15.4}$	$\frac{3.1–9.0}{7.1}$	$\frac{7.4–8.0}{7.7}$
Viscera	$\frac{76.7–81.8}{80.2}$	$\frac{10.1–11.5}{10.7}$	$\frac{2.4–8.0}{7.6}$	$\frac{1.0–1.9}{1.5}$
Fins	$\frac{56.0–58.0}{57.0}$	$\frac{14.5–16.0}{15.2}$	$\frac{1.7–3.7}{2.7}$	$\frac{2.2–2.3}{2.3}$

* Average values.

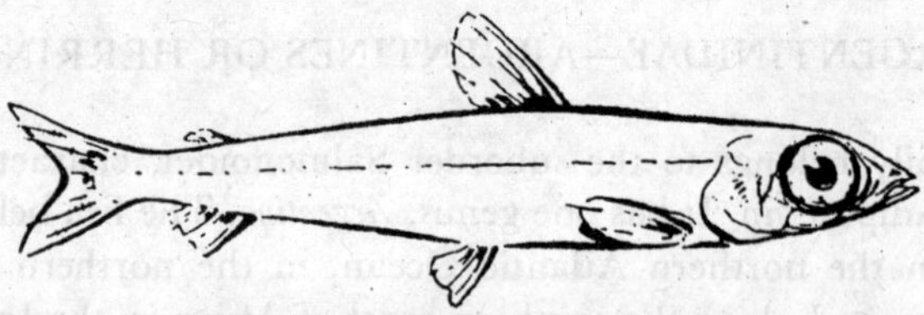

Fig. 22. Atlantic argentine—*Argentina silus*.

The weight of the head when severed by a straight cut is 18–19% and when severed along the skull bones, skull portion only, on an average 14.8%. The proportions of other parts are: roe 21%, testes 8.2%, liver 0.8%, swimbladder 0.6–1.6% with an average of 0.9%, fins 0.8–0.9% and scales 0.4–0.8%.

The meat of this fish is white, soft and tasty. It is, therefore, considered a fairly valuable fish and can be used for first and second courses; smoked and canned products are good.

Family ARRIPIDAE—AUSTRALIAN SALMON

Australian salmon—*Arripis trutta esper*. Distributed in the Pacific Ocean along the coast of southeast Australia in the region adjoining Victoria and Tasmania and along the coast of New Zealand. Found in small numbers in the commercial catch. Length of the specimens caught in December-January 45–52 cm, weight 1.2–1.6 kg.

Weight proportions of individual body parts (%): head 19.7, trunk 72.2, including bones 5.3, fins 2.6, viscera 5.5, including liver 1.6.

Chemical composition of meat with skin (%): moisture 70.0, protein 21.8, fat 7.1, ash 1.1. The liver of this fish contains from 3.2 to 13.2% oil with a fairly high vitamin A potency.

The Australian salmon belongs to the category of table fish. When fried or boiled, the meat has dense consistency and typically pleasant taste.

Australian salmon should be frozen for subsequent production of culinary preparations and hot smoked items. It is also possible to prepare light salted products and natural conserves from the fresh raw fish.

Family ARIIDAE*—SEA CATFISHES

Sea catfishes are commercially important. They are found in coastal, tropical and subtropical waters. Their greatest distribution is in the seas of

*Here and elsewhere in the text spellings of generic names were verified through authoritative sources. All generic names bearing asterisk mark indicate that their spelling was wrong in the Russian original—General Editor.

Southeast Asia, in the Bay of Bengal, northwestern Indian Ocean, along the coast of South America and in the Atlantic Ocean. They grow in length up to 1 m.

Catfish is a good table fish and the swimbladders are used for high quality aviation glue. Six species of catfish belonging to this family are the most widely distributed.

Giant catfish—*Netuma* thalassinus* or *Arius thalassinus.* A large commercial fish. In catches, fish 35 cm in length generally predominate (range 23–80 cm) weighing 1,000 g (range 225–7,640 g), with maximum body depth 7.2 cm (range 4.6–17.0 cm) and maximum body thickness 7.0 cm (range 4.2–15.0 cm).

TABLE 20. CHEMICAL COMPOSITION, %

Body parts	Moisture	Protein	Fat	Ash
Meat	68.8–78.0 74.5	18.4–21.0 19.6	0.4–12.0 4.5	1.3–1.5 1.4
Inedible parts of the body (entire offal)	64.5–69.2 66.8	16.1–18.2 17.2	5.8–10.6 8.1	6.6–7.5 7.0

Weight proportions of individual body parts (%): head 28.5 (21.4–32.4), trunk 56.5 (50.3–62.3), including meat with skin 47.9 (39.7–53.0), fins 2.8 (1.1–4.0), whole viscera 12.2 (9.1–17.3).

After cooking, the meat of this catfish is soft, juicy and very tasty. Soup prepared from this fish is rich and yellow with droplets of fat, with a pleasant mildly acid taste. The meat in fried form is soft and tasty.

This fish should be used as a table fish for first and second courses and for frozen fillets and canned products in tomato sauce (boneless).

Sometimes this catfish has an iodine taste in the meat (for example, in catches from the Persian Gulf); this depends upon the peculiarities of the feeding habit. The occurrence of iodine taste should not be considered a defect since iodine is essential for normal functioning of the human thyroid and its presence in the meat of marine fishes increases the food value.

Sometimes catfish have a strong smell of rotten eggs emanating from the viscera (for example, in the catches near Madagascar Island). Such fish should be gutted and carefully cleaned and washed before freezing; if the smell of rotten eggs comes from the viscera and from the meat, such fish should be used for fish meal.

Small eyed catfish—*Pseudarius jella*. A commercial fish. In catches in October along the coast of Pakistan individuals about 30 cm in length (26–38 cm) and about 500 g in weight (340–1,075 g) were predominant.

Weight proportions of individual body parts (%): head 32.4 (30.9–34.1), trunk 55.1 (53.0–57.8), including meat 40.0 (38–43.5), skin 6.7, bones 8.4, fins 2.6, entire viscera 9.9 (6.4–10.7).

TABLE 21. CHEMICAL COMPOSITION, %

Body parts	Moisture	Protein	Fat	Ash
Meat	78.5	19.1	1.0	1.3
Inedible parts of body (entire offal)	67.6	18.8	3.8	9.5

After cooking, the meat becomes gray, soft and tasty. The soup is tasty; it is a good table fish. It is recommended for first and second courses, processing into frozen fillets and canned products in oil (without bones).

"Gagoriodes" sea catfish—*Arius gagoriodes*. A commercial fish. In the catches taken on the shelf off Bombay in March individuals 60 cm in length weighing 3,200 g were predominant.

Weight proportions of individual body parts (%): head 34.1, trunk 52.8, including meat with skin 40.0, bones 12.8, fins 5.3, entire viscera 5.8. Chemical composition of meat (%): moisture 79.8, protein 18.1, fat 0.8.

After cooking, the meat acquires a somewhat dry consistency and a mild fishy odor. It is recommended as a table fish for second courses, for processing into frozen fillets (from large fish) and for culinary preparations.

"Goniaspis" sea catfish—*Arius goniaspis*. A commercial fish. In catches taken in January in the Persian Gulf and in May in the Gulf of Aden, individuals with a length of 45 cm (33–53 cm) and weight of 2,000 g (750–3,320 g) predominate.

Weight proportions of individual body parts (%): head separated by an oblique cut 24.4 (23.3–27.5), trunk 60.5 (56.6–61.7), including meat with skin 48.0 (43.7–48.7), bones 10.5, fins 2.2 (1.7–2.9), entire viscera 12.8 (11.4–13.2), including liver about 1.8.

TABLE 22. CHEMICAL COMPOSITION, %

Body parts	Moisture	Protein	Fat	Ash
Meat	$\frac{73.8–80.0}{76.8}$	$\frac{19.1–20.3}{20.1}$	$\frac{1.1–2.5}{1.7}$	$\frac{1.5–1.6}{1.5}$
Inedible parts (entire offal)	66.4	16.8	10.3	6.5

After cooking, the meat becomes soft with the typical taste of table fish. The soup is lemon colored, rich and tasty. The meat in natural canned products and those canned in oil is pink; its consistency is soft, taste pleasant, but the bones are not thoroughly softened.

This fish should be used as a table fish for first and second courses, for processing into fillets (from large fish) and for canning (without bones).

"Dussumier's" sea catfish—*Ariodes dussumieri*. A commercial fish. In catches taken in November–December in the Red Sea, there was a predominance of individuals 35 cm in length (27–75 cm), weighing about 1,400 g (376–6,495 g).

Maximum body depth of this catfish on an average 6.5 cm (5.7–7.0 cm), maximum thickness on an average 4.8 cm (5.0–6.4 cm).

Weight proportions of individual body parts showing significant fluctuations (%): head 29.7 (23.6–37.6), trunk 58.1 (43.9–60.2), fins 1.4 (1.3–2.0), entire viscera 10.8 (8.8–12.9).

TABLE 23. CHEMICAL COMPOSITION, %

Moisture	Protein	Fat	Ash
74.8–76.0 / 75.0	17.4–20.8 / 19.5	1.7–4.6 / 4.2	1.1–1.4 / 1.3

Cooked meat is soft and easily fried. The taste is pleasant. The soup has a peculiar (iodine) taste.

It is recommended for use as a table fish for second courses and for processing into frozen fillets.

Serrate sea catfish—*Arius serratus*. A commercial fish. In catches taken in June near Sri Lanka, a predominance of individuals were 35 cm in length (33–41 cm), weighing 850 g (705–1,100 g), maximum depth 7.5 cm (7.0–8.5 cm) and maximum thickness 6.7 cm (6.5–7.0 cm).

Weight proportions of individual body parts (%): head 28.7 (27.3–30.4), trunk 57.0 (55.3–59.0), including meat and skin 48.6 (45.1–52.2), bones 8.4 (6.3–10.1), fins 3.3, entire viscera 10.6 (9.2–11.8).

Chemical composition of meat (%): moisture 77.4, protein 18.6, fat 1.5, ash 1.5.

After cooking, the meat is grayish, soft and very tasty. The soup is lemon color and pleasant tasting. This is a good table fish.

It is recommended that this fish be used for first and second courses and for processing into fillets and canned products in tomato sauce (without bones).

"Filiceps" sea catfish—*Tachysurus filiceps*. A commercial fish. In catches in October along the coast of Pakistan, fish 30 cm in length and 600 g in weight predominate.

Weight proportions of individual body parts showing significant fluctuations (%): head on an average 35.1 (31.1 in males, 38.9 in females), trunk 47.2 (45.5–47.4), including meat 34.5 (28.6–41.4), skin 6.4, bones 6.3 (4.7–8.8), fins 6.3, entire viscera 11.4 (8.6–15.5).

TABLE 24. CHEMICAL COMPOSITION, %

Body parts	Moisture	Protein	Fat	Ash
Meat	79.2	17.7	1.2	1.3
Inedible parts of body (entire offal)	64.3	16.2	7.0	10.9

The mince meat of raw catfish is outwardly similar to beef; however, it has the unpleasant odor of rotten eggs. This is not noticeable in the finished products.

After cooking, the meat becomes soft; it does not differ in taste from the meat of table fish. The soup has a fleshy or meatlike (non-fish) taste. This species of catfish is recommended for use as a table fish, for first and second courses and production of frozen fillets.

Tachysurus dussumiera. A commercial fish. In catches in August in the Persian Gulf, there is a predominance of individuals 31 cm in length (30–32), weighing 507 g (460–525 g), maximum depth 5.9 cm (5.5–6.0 cm) and maximum thickness 4.7 cm (4.5–5.5 cm).

TABLE 25. CHEMICAL COMPOSITION, %

Body parts	Moisture	Protein	Fat	Ash
Meat	78.6	17.8	0.9	1.4
Inedible parts of body (waste during gutting)	55.7	20.1	9.6	12.1

Weight proportions of individual body parts (%): head 30.1 (27.7–32.6), trunk 55.1 (51.0–59.0), including meat with skin 43.9 (41.4–45.7), bones 11.2, fins 3.1, entire viscera 9.7.

After cooking, the meat is white, soft and tasty. The soup is yellow and tasty. After frying, the meat becomes white, dense and tasty.

The fish is recommended for use as a table fish for first and second courses and for production of frozen fillets.

Family BERYCIDAE—ALFONCINOS

Genus *Beryx*

Red bream—*Beryx decadactylus*. Caught in the Atlantic off the coast of northwest Africa.

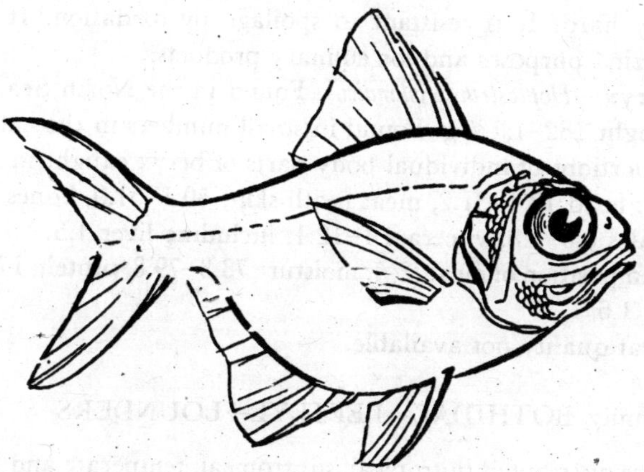

Fig. 23. Red bream—*Beryx decadactylus.*

Weight proportions of individual body parts of fish with a length of 25 cm and weight of 0.5 kg (%): head 30, trunk with fins 62, viscera 8. Output of fillets 40%.

Chemical composition of meat (%): moisture 77, protein 21.5, fat 0.2, ash 1.1.

It is marketed by the name scup and is suitable for processing in frozen form as well as for preparing culinary products.

Red snapper—*Beryx gerardi.* A commercial fish. Distributed in the Great Australian Bight. Length 35–49 cm, weight 1,150–2,470 g (average 1,580 g)

Weight proportions of individual parts (%): head 32–44.6, trunk 43.1–56.6, including meat with skin 31.1–42.4, bones 8.9–11.8, fins 2.6–4.0, viscera 3.4–5.0, including liver 0.6, scales 2.7–4.3.

Chemical composition of meat with skin (%): moisture 76.9–78.2, protein 19.0–20.4, fat 1.2–1.4, ash 1.1–1.2.

This is an excellent table fish. After cooking, the meat of red snapper is white, soft and pleasantly sweet in taste. It is recommended for processing into frozen form and for preparation of culinary products.

Common beryx—*Beryx decapterus.* Lives in the northwestern Atlantic Ocean. Does not have commercial significance. Length of common beryx taken in August–September was 32.5–36.0 cm, weight 905–1,945 g.

Weight proportions of individual body parts (%): head 20.5–21.6, trunk 60.0–63.2, including meat with skin 50.9–54.3, bones 7.8–8.1, fins 1.8–2.1, scales 3.2–3.8, viscera 9.4–13.1, including liver 1.5–2.0.

Chemical composition of meat (%): moisture 72.9–76.1, protein 18.4–19.2, fat 3.1–6.7, ash 1.4–1.5.

This is a good table fish. Cooked beryx meat is white but when fried it becomes slightly hard. It is resistant to spoilage by oxidation. It is more suitable for freezing purposes and for culinary products.

Iceland beryx—*Hoplostethus islandicus.* Found in the North Sea. Length 18.8–36 cm, weight 182–1,130 g. Found in small numbers in the catches.

Weight proportions of individual body parts of beryx caught in August–September (%): head 18.1–31.2, meat (with skin) 50.1–51.9, bones 8.8–9.7, fins 1.7–1.8, scales 0.3–3.8, viscera 4.1–12.1, including liver 1.5.

Chemical composition of meat (%): moisture 73.3–79.8, protein 17.4–19.6, fat 0.4–6.0, ash 1.6.

Data on meat quality not available.

Family BOTHIDAE—LEFTEYE FLOUNDERS

These fish are distributed in tropical, subtropical, temperate and northern seas. In many regions they constitute an important object of fisheries. This family has 38 genera.

Genus *Rhombus*

Turbot—*Rhombus maximus.* Inhabits the waters off the European coast from the Black Sea to the western Barents Sea. Attains 1 m in length but is usually about 40 cm. Length of a turbot investigated 57 cm, maximum body depth 51 cm, thickness 10 cm, weight 8,710 g.

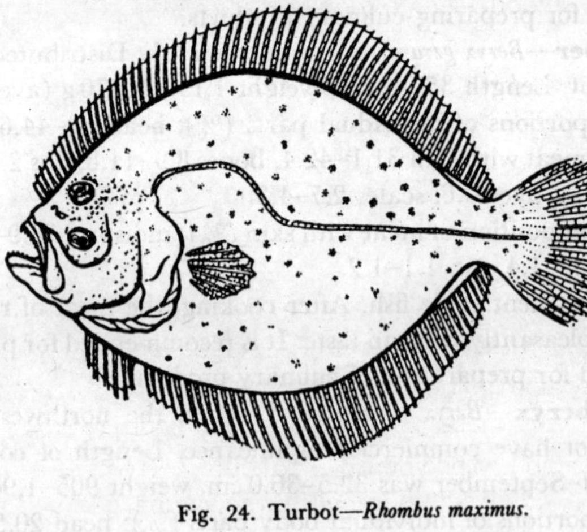

Fig. 24. Turbot—*Rhombus maximus.*

Weight proportions of individual body parts (%): head 18.8, trunk 62.0, meat 40.5, fins 2.8, viscera 16.4.

Chemical composition of meat (%): moisture 80.9, protein 17.5, fat 0.2, ash 1.0.

Vitamin content in this fish: B_1 in juvenile fish 43 γ/g, riboflavin in the roe 9.1 γ/g, vitamin E in the fat obtained from the trunk 450 γ/g.

Genus *Paralichthys*

Bastard halibut—*Paralichthys olivaceus*. Lives in the eastern Indian Ocean and is found in small numbers in catches.

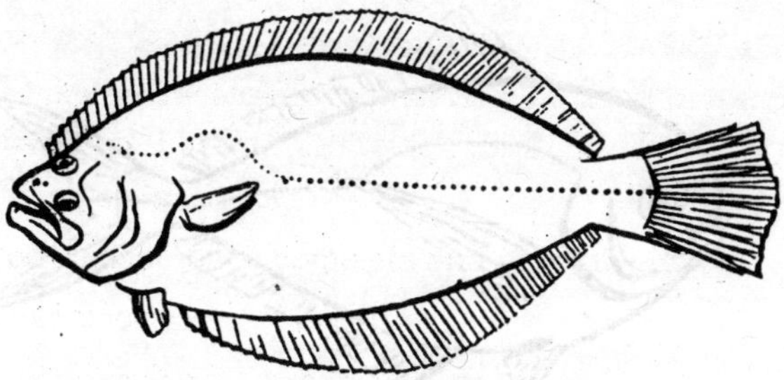

Fig. 25. Bastard halibut—*Paralichthys olivaceus*.

Length of the specimen caught in August 47.0 cm and weight 1,740 g. Weight proportions of individual body parts (%): head 15.2, trunk 70.4, viscera 13.1, including liver 2.2, fins 1.1. Output of fillets 63.5%.

This species has tasty and juicy meat with high fat content. The meat is tastiest in fried form. After heating, the meat is separated along the myosepta. The meat is slightly coarser than in northern halibut, but it is softer than in true flat fish. It is recommended for use after freezing with subsequent preparations of culinary dishes.

Brazilian flounder—*Paralichthys brasiliensis*. This is a constituent of commercial fishing in the northwest Atlantic, in the region of the Patagonian shelf. The local name of this fish is lenguado. Fish caught in March were 53–55 cm long and weighed 2.1–3.0 kg.

Average weight proportions of individual body parts (%): head 19, trunk 69, viscera 10, fins 2. Output of fillets 52%.

Chemical composition of meat (%): moisture 77.7, protein 20.8, fat 0.2, ash 1.3.

Fried meat is white, soft and tasty. It is a good table fish.

Family BRAMIDAE—POMFRET

Genus *Brama*

Ray's bream—*Brama raii**. Widely distributed in the Atlantic Ocean in the open sea as well as on the continental slope. Usual length of this fish 20–50 cm, sometimes up to 70 cm. Those caught in the southwestern Atlantic near Wolffish Bay, in the eastern Central Atlantic and in the northwest Atlantic off the coast of Greenland had the following measurements:

Length, cm	20	30	40	48
Weight, kg	0.7	1.4	3.0	3.5

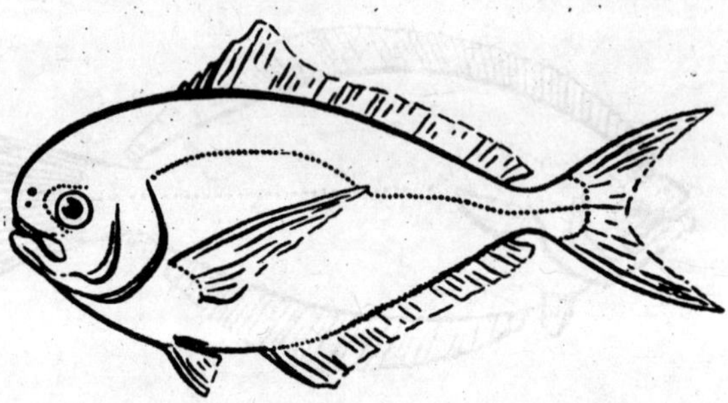

Fig. 26. Ray's bream—*Brama raii**.

TABLE 26. CHEMICAL COMPOSITION, %

Body parts	Moisture	Protein	Fat	Ash
Meat	65	15.8	13	5.5
Head, bones, fins	64	18	13	5
Viscera	78–81	13.6–16.8	1–4	1.3–1.8

Weight proportions of individual body parts (%): head 20–28, viscera 7–13, skin 4, bones 7, fins 1.3–3.2. Output of fillets was 50–65%.

Chemical composition of meat (%): moisture 63–74, protein 13.0–21.5, fat 2–21, ash 1.3–1.4. The mineral composition of the meat (mg%): potassium 355, calcium 146, magnesium 37, phosphorus 206 and iron 4.1.

The meat of fried and boiled fish is white. The taste qualities of the lean fish are quite satisfactory, but those of the fatty fish are good, resembling those of sturgeon. The head can be used for soup. Smoked products from fatty fish are high in quality.

This fish should be processed for freezing in ungutted form with subsequent use in culinary products and smoked preparations.

Genus *Taractes*

Big scale pomfret—*Taractes longipinnis*. Found in small numbers in mixed catches from the southeastern Atlantic. Average size of individuals caught in

Fig. 27. Big scale pomfret—*Taractes longipinnis*.

TABLE 27. CHEMICAL COMPOSITION, %

Body parts	Moisture	Protein	Fat	Ash
Meat	79.2	18.6	0.8	1.4
Head, bones, fins	78.1	15.5	1.2	5.2
Viscera	83.6	12	2.7	1.7

July (cm): length 37, depth 14, thickness 3. Weight 1.3 kg.

Weight proportions of individual body parts (%): head 26, trunk with fins 68, viscera 6. Output of fillets 50%.

Fried meat of this fish has satisfactory taste. Otherwise, meat is suitable for culinary preparations.

Family CARANGIDAE—JACKS, SCADS AND POMPANOS

This family has more than 200 species belonging to more than 30 genera. The fish are widely distributed in temperate and tropical waters of the Pacific, Indian and Atlantic Oceans.

Most of the fish have commercial value.

Genus *Alectis*

These fish are found in small numbers in commercial catches. Two species are known: the Indian threadfish and the cobblerfish.

Indian threadfish—*Alectis indica*. Caught in small numbers along with other fish. Among the fish caught in March in the Persian Gulf and in September in the Red Sea, there was a predominance of individuals 30 cm long (21–45), 900 g in weight (195–1,600), maximum body depth 16 cm (10–21) and maximum thickness 4 cm (2.4–6.0).

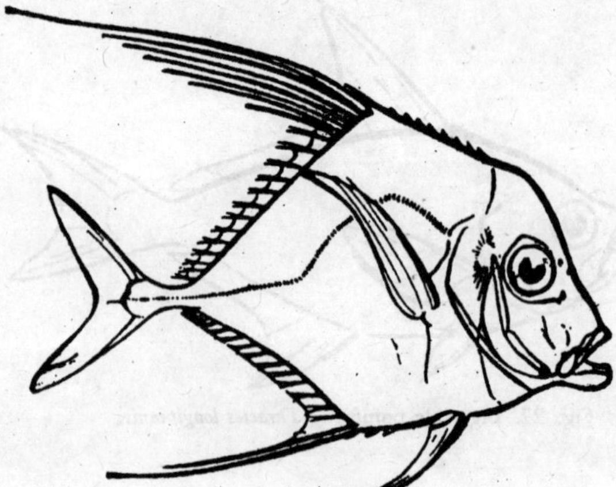

Fig. 28. Indian threadfish—*Alectis indica.*

TABLE 28. CHEMICAL COMPOSITION, %

Body parts	Moisture	Protein	Fat	Ash
Meat	76.9–78.6 77.2	19.2–21.1 20.2	0.8–1.5 1.2	1.4
Inedible parts of the body (entire offal)	66.8	17.1	6.4	9.0

Weight proportions of individual body parts (%): head 27.3 (25.0–29.5), trunk 61.5 (60.1–62.8), including meat with skin 51.1 (50.0–52.3), bones 10.4 (10.0–10.5), scales (scutes) 0.4, fins 2.1 (1.3–3.1), entire viscera 8.0 (6.9–9.2).

Chemical composition of the meat of sexually immature and mature Indian threadfish caught from different regions is very similar.

After cooking, the meat is gray and soft in consistency and has a pleasant taste typical of table fish. The soup is light yellow with a pleasant light, sweet taste. Fried meat is also tasty. This is a good table fish recommended for first and second courses.

Longfin Alectis—*Alectis ciliaris*. A large fish, reaching about 1 m in length and caught along with other fish.

In the Gulf of Aden (July) there were specimens in catches 70 cm long (53–87), weighing 5,000 g (2,315–9,920), with a maximum depth 24.2 cm (21.5–28) and maximum thickness 7.2 cm (4–10).

Average weight proportions of individual body parts (%): head 26.5

TABLE 29. CHEMICAL COMPOSITION, %

Body parts	Moisture	Protein	Fat	Ash
Meat	77.8	18.4	0.9	1.3
Viscera	79.7	16.1	2.4	1.2
Testes	81.5	14.1	1.6	1.7

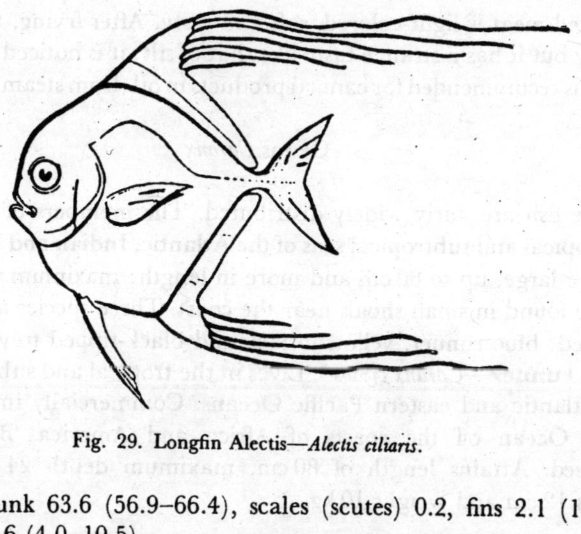

Fig. 29. Longfin Alectis—*Alectis ciliaris.*

(25.6–27.6), trunk 63.6 (56.9–66.4), scales (scutes) 0.2, fins 2.1 (1.8–2.4), entire viscera 5.6 (4.0–10.5).

After cooking, the meat is tough, slightly sour with a specific taste typical of common carangids. It is recommended for smoked and canned products in oil.

Genus *Atule*

Yellowtail jack—*Atule djeddaba.* Found in the northwestern and southwestern Indian Ocean. A small fish.

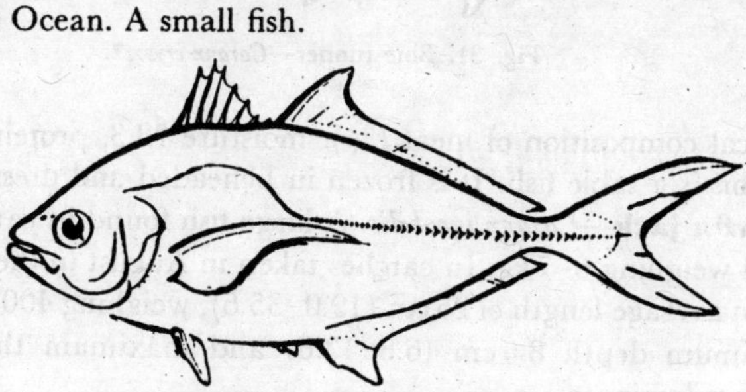

Fig. 30. Yellowtail jack—*Atule djeddaba.*

In catches taken in August from the southwestern Indian Ocean most of the individuals were 18 cm long (16–21), weighing 124 g (100–195), with maximum depth 6.2 cm (5.7–7.3) and maximum thickness 2.5 cm (2–3).

Average weight proportions of individual body parts (%): head 28.3, trunk 56.6, including meat with skin 44.4, bones 12.2, scales 2.4, fins 3.2, entire viscera 7.3.

Chemical composition of meat (%): moisture 75.3, protein 21.6, fat 1.4, ash 1.6.

Cooked meat is light colored, soft and tasty. After frying, the meat is soft and tasty but it has a strange taste like that of silt; it is noticed with difficulty. This fish is recommended for canned products in oil (from steam blanched fish).

Genus *Caranx*

These fish are fairly widely distributed. The members of this genus live in the tropical and subtropical seas of the Atlantic, Indian and Pacific Oceans. Many are large, up to 60 cm and more in length; maximum weight 16.5 kg. They are found in small shoals near the coast. Three species are most widely distributed: blue runner, yellowfin jack and black-tipped trevally.

Blue runner—*Caranx crysos**. Lives in the tropical and subtropical waters of the Atlantic and eastern Pacific Oceans. Commercially important in the Atlantic Ocean off the coasts of Africa and America. Body long and compressed. Attains length of 80 cm, maximum depth 24 cm, maximum thickness 12 cm and weight 10 kg.

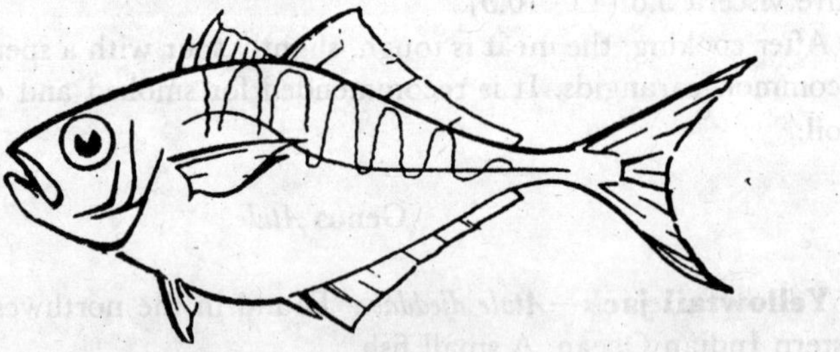

Fig. 31. Blue runner—*Caranx crysos**.

Chemical composition of meat (%): moisture 78.3, protein 19.8, fat 0.2, ash 1.7. This is a table fish. It is frozen in beheaded and dressed form.

Yellowfin jack—*Caranx ignobilis*. A large fish found in catches; 75 cm in length and weighing 6–7 kg. In catches taken in August in the Gulf of Aden, fish with an average length of 25 cm (19.0–35.6), weighing 400 g (179–1,214), with maximum depth 8.6 cm (6.8–13.6) and maximum thickness 3.6 cm (2.8–5.0) predominate.

Weight proportions of individual body parts showing significant fluctu-

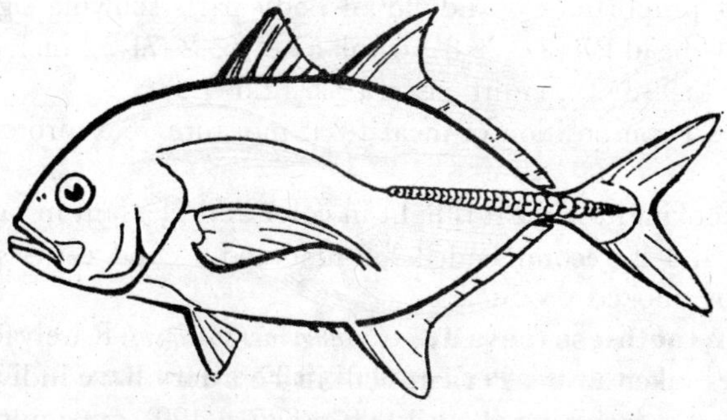

Fig. 32. Yellowfin jack—*Caranx ignobilis*.

ations (%): head 30.0 (26.2–35.1), trunk 60.0 (55.7–64.0), including meat 50.5, scales (scutes) 1.3 (1.0–1.8), fins 1.8 (1.5–2.3), entire viscera 6.3 (5.1–8.1).

Chemical composition of meat (%): moisture 75.2, protein 21.3, fat 1.4, ash 1.5. The meat is tasty in boiled and fried forms. It is a table fish.

Black-tipped trevally—*Caranx melampygus* (*Caranx stellatus*). Caught mixed with other fish. Catches taken in July near Sri Lanka and in September in the Red Sea had a predominance of fish 40 cm long (28–56), weighing 1,600 g (500–3,200), with maximum depth 14.1 cm (9–20.4) and maximum thickness 5.7 cm (3.5–8.7).

Weight proportions of individual body parts with significant fluctuations (%): head 30.1 (26.2–33.5), trunk 59.5 (55.8–62.8) including meat with bones 49.2 (47.3–55.0), scales (scutes) 0.9 (0.5–1.1), fins 2.2 (1.6–2.9), entire viscera 6.6 (4.8–7.9).

Chemical composition of meat (%): moisture 77.2 (77.0–77.4), protein 20.7 (20.4–20.9), fat 0.9 (0.8–1.0), ash 1.3 (1.2–1.4).

After cooking, the meat is white, tasty and soft in consistency; the soup is tasty and slightly sweet.

This is a good table fish for first and second courses as well as for products made by hot smoking.

Cases of poisoning have been reported after eating this fish ("Ciguatera" type of poisoning). Therefore, this fish should be used as a food only after toxicological tests.

Genus *Carangoides*

Members of this genus are widely distributed in coastal waters. Six species are commonly known.

Two-thread trevally—*Carangoides dinema*. Caught with other fish. Catches in September in the Gulf of Aden had individuals 27 cm in length (24.5–30.5) and 500 g in weight (360–705).

Weight proportions of individual body parts showing significant fluctuations (%): head 27 (24–29.8), trunk 67.0 (65.2–71.2), including meat with skin 48.6 (45.7–54.0), entire viscera 3.6 (2.6–4.5).

Chemical composition of meat (%): moisture 76.3, protein 20.9, fat 1.2, ash 1.4.

After cooking, the meat is light in color and pleasant in taste. It is a good table fish and is recommended for first and second courses as well as for making hot smoked products.

Golden toothless trevally—*Carangoides speciosus*. Rarely found in catches.

Catches taken in the Persian Gulf in February have individuals 32 cm in length (30–34), weighing about 1,000 g (885–1,190), maximum depth 13.3 cm (12.4–14.0), maximum thickness 5.9 cm (5.5–6.5). Weight proportions of individual body parts (%): head 35.8 (33.8–39.2), including meat of the occiput and on an average 4.5, trunk 53.4 (50.8–54.9), including meat with skin 45.1 (42.5–48.4), bones 8.3 (6.6–8.9), scales (scutes) 0.6, fins 2.0, entire viscera 7.5 (5.9–9.9), including roe on an average 2.1, liver 1.0.

TABLE 30. CHEMICAL COMPOSITION, %

Body parts	Moisture	Protein	Fat	Ash
Meat	75.8	20.3	2.1	1.3
Inedible parts of the body (entire offal)	66.8	17.3	6.8	7.7

Cooked meat is light gray, dense, with a pleasant flavor with a mild acid taste. The soup is greenish yellow and very sour. The fried meat is light in color, somewhat dry with a light sour taste. Its taste and appearance in canned products with oil are good; in tomato sauce the meat is hard and not tasty.

This fish should be used for smoke curing and preparation of canned products in oil.

Armed trevally or longfin jack—*Carangoides armatus*. A large fish caught along with other fish. In catches taken in April in the Gulf of Kutch (Arabian Sea) most fish were 53 cm long (50–57), weighing about 3,500 g (3,080–3,950), with maximum depth 19–20 cm and maximum thickness 7–8.5 cm.

Weight proportions of individual body parts showing only slight fluctuations (%): head 33.2 (32.4–34.0), including meat of occiput and nape 5.5, trunk 56.3 (54.5–57.5), including meat with skin 45.8 (44.5–46.8), bones 10.5 (10.1–10.7), scales 0.4, fins 2.6, entire viscera 7.1 (6.8–7.4) including liver 2.

Chemical composition of meat (%): moisture 73.1, fat 3.6, protein 22.2, ash 1.2.

After boiling, the meat is white, dense and very tasty. The soup is rich

Fig. 33. Armed trevally or longfin jack—*Carangoides armatus.*

with the taste of meat (nonfish) and is mildly sour. In cooked form the meat is very tasty; it is a good table fish. It is recommended for first and second courses, frozen fillets (from very large fishes) and hot smoked products.

Naked-breast trevally or bludger—*Carangoides gymnostethoides.* Caught along with other fish. In catches taken in August from the Red Sea, fish 40 cm long (36.7–45.7), weighing about 900 g (635–1,230), with maximum depth 9.8 cm (8.7–10.3) and maximum thickness 5.6 cm (5.0–6.3) predominate.

Weight proportions of individual body parts showing significant fluctuations (%): head 22.9 (19.5–24.8), trunk 62.0 (58.5–63.8), scales 0.4 (0.3–0.5), fins 2.5 (1.0–3.4), entire viscera 9.1 (7.8–11.4), losses during gutting 3.1.

Chemical composition of meat (%): moisture 75.9, fat 0.6, protein 21.8, ash 1.3.

After cooking, the meat is dark with a mildly sour but pleasant taste. This fish should be used for making products by hot smoking and canning in oil.

Yellow-back trevally—*Carangoides fulvoguttatus.* A large fish living in the Indian Ocean. Appearing as an incidental catch.

According to the findings of AzcherNIRO, in catches in September in the Gulf of Aden and in the Arabian Sea, there was a predominance of individuals 45 cm in length (32–60) and weighing 2,400 g (768–3,950); according to the reports of TINRO (area of catch not precisely known), the individuals were 23–30 cm in length. According to the data of AzcherNIRO, the weight proportions of the individual body parts fluctuated considerably (%): head 28.9 (24.9–32.8), trunk 60.9 (56.6–64.9), including meat with skin

TABLE 31. CHEMICAL COMPOSITION, %

Source	Moisture	Protein	Fat	Ash
AzcherNIRO	75.4–76.6 / 76.0	19.8–22.2 / 21.0	0.5–2.2 / 1.4	1.3–1.5 / 1.4
TINRO	74.1	20.1	4.9	1.0

50.4 (47.3–53.5), scales 0.8, fins 2.2 (1.5–2.8), entire viscera 6.8 (5.5–8.1), bones 14.9. The chemical composition of the meat fluctuated slightly.

After cooking, the meat is dark in color. Its taste is mediocre. The soup is sour. This fish should be used for hot smoking and for canned products in oil after steam blanching of the fish.

Big scaled trevally—*Carangoides adsensionis*. Appears as an incidental catch. In catches in October in the Gulf of Aden, individuals 40 cm in length and weighing 680 g predominate.

Weight proportions of individual body parts on an average (%): head 31.2, trunk 59.2, including meat with skin 44.7, scales 0.2, fins 2.7, entire viscera 6.2.

Chemical composition of meat (%): moisture 76.7, protein 20.1, fat 1.4, ash 1.3.

After cooking, the meat is white, pleasant in taste with consistency varying from dense to hard. It is a good table fish. Its soup also has a pleasant taste and smell.

This fish should be used for first and second courses, products of hot smoking and canning as "smoked in oil".

Ferdau's trevally—*Carangoides ferdau*. Forming an incidental catch. Attains 75 cm in length.

Fig. 34. Ferdau's trevally—*Carangoides ferdau*.

In catches in January on the shelf of western India fish about 24 cm in length (23–37), weighing about 400 g (314–526) predominate.

Weight proportions of individual body parts (%): head 27.6, trunk 62.0, including meat with skin 51.0, fins 2.0, entire viscera 6.4.

Chemical composition of meat (%): moisture 76.8, protein 18.8, fat 1.1, ash 1.4.

After cooking, the meat has a slightly hard consistency and pleasant taste. It should be used for hot smoked products and canned preparations in oil.

Dusky trevally—*Carangoides chrysophrysoides*. Distributed in the Indian Ocean, constituting an incidental catch.

According to the data of AzcherNIRO, in catches in September in the Gulf of Aden there was a predominance of individuals 25 cm in length (24–28), weighing around 400 g (305–470). According to reports of TINRO, some fish were 34 cm long, weighing 885 g.

Weight proportions of individual body parts (%): head on an average 34.5, trunk separated by a circular cut 59.0, including meat with skin 43.7, fins 1.6, whole viscera 3.3 (AzcherNIRO); head 28.2, bones 10, viscera 11.8, fins 2.2. Output of fillets 47.4% (TINRO).

Chemical composition of meat (%): moisture 76.9, protein 20.1, fat 1.6, ash 1.4 (AzcherNIRO); moisture 70.3, protein 20.0, fat 7.8, ash 2.0 (TINRO).

This is a good table fish. After cooking, the meat is light colored, soft and tasty. The soup is very tasty as is the boiled meat. After mild salting and partial drying it has a good taste. Because of the thin body, sun curing of this fish is rapid. Its fat is resistant to oxidation. After hot smoking, this fish acquires very good external appearance.

This fish should be used as a table fish for first and second courses, production of sun cured products after hot smoking and canned products such as "smoked fish in oil".

Genus *Chorinemus*

Members of this genus are widely distributed in the coastal tropical zones of the Indian and in parts of the Pacific Oceans. These fish are relatively large. Their body length ranges from 60–120 cm.

Queenfish—*Chorinemus lysan*. A large fish widely distributed in the Indian Ocean, constituting only an incidental catch.

In catches in January and March in the Persian Gulf, in August in the Gulf of Aden and in September in the Red Sea, fish 60 cm long (35–86.5), weighing around 3,000 g (565–6,210) with maximum depth 16 cm (11–22) and maximum thickness 5.6 cm (3.2–7.6) were frequently found.

Weight proportions of individual body parts with significant fluctuations (%): head 17.9 (14.6–21.0), trunk 70.0 (63.4–75), including meat and skin 54.8 (51.0–59.3), fins 1.8 (1.5–2.3), entire viscera 9.2 (4.6–16.0).

Taste qualities of meat of small fish (length 35–45 cm) were significantly better than those of the meat of large fish.

Cooked meat of sexually immature fish is light in color with soft consistency and the typical taste of table fish. Soup prepared from these fish is greenish and tasty. Fried meat is dense and tasty. After salting, it becomes seasoned slowly.

Cooked meat of large sexually mature fish is torn, with dense consistency and the slight sour taste typical of mediocre table fish. The soup is colorless and tasty. The meat has a pleasant taste after salting.

TABLE 32. CHEMICAL COMPOSITION, %

Place of catch	Body parts	Moisture	Fat	Protein	Ash
Western Indian Ocean	Meat	74.6–77.7 / 77.2	0.2–0.7 / 0.5	20.3–21.3 / 20.9	1.2–1.5 / 1.4
	Inedible parts of the body (entire offal)	68.2	3.8	19.4	8.2
Eastern Indian Ocean	Meat	70.4–74.0	1.5–4.4	22.4–23.0	1.3–1.5

This fish is recommended for cured fillets, smoked products and canned preparations in oil. The small fish (up to 45 cm in length) can also be used as a table fish for first and second courses.

Genus *Decapterus*

The members of this genus live in the warm waters of the Atlantic, Pacific and Indian Oceans.

These are medium size commercial fish; some attain a length of 40 cm.

Japanese or round scad—*Decapterus maruodsi**. Found in the Gulf of Tonkin. Length of fish caught in May–August 15.0–23.5 cm and weight 40–165 g.

Weight proportions of individual body parts (%): head 21.2–26.3, trunk 60.0–66.1, gut 3.4–6.0, bones and fins 11.6–14.1, gonads 1.1–4.4, liver 0.7–1.8. Output of fillets 50.6–57.0%.

Chemical composition of meat (%): moisture 71.8–75.1, protein 20.7–23.3, fat 0.4–3.5, ash 1.6–2.4.

This fish should be frozen on board the fishing vessel and then supplied to processing plants for preparation of culinary products or canning in its own juice.

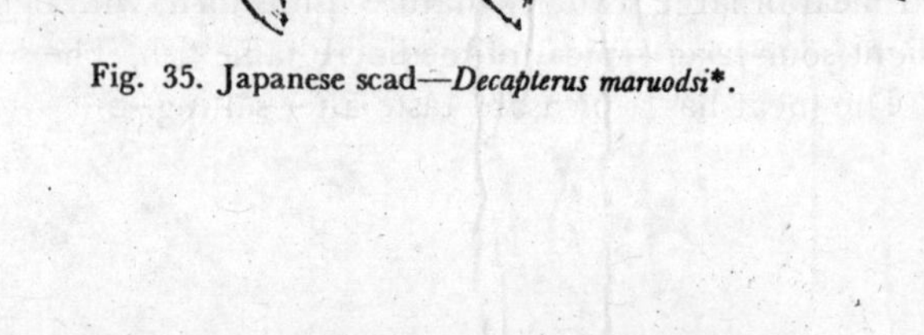

Fig. 35. Japanese scad—*Decapterus maruodsi**.

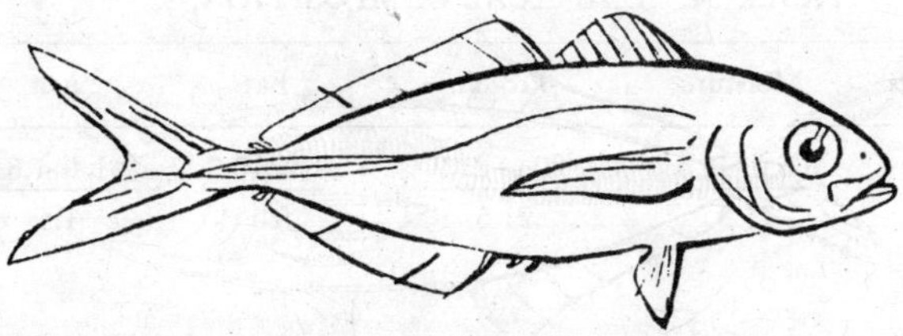

Fig. 36. Ten-finned horsemackerel—*Decapterus ronchus.*

Ten-finned horsemackerel—*Decapterus ronchus*. Found in the Atlantic Ocean all along the coast of Africa, a fish of considerable commercial significance.

Size characteristics

Length, cm	19	25	31
Maximum depth, cm	—	6	8
Maximum thickness, cm	—	3	4.5
Weight, g	140	230	490

Weight proportions of individual body parts (%): head 21, trunk with fins 71, viscera 8. Output of fillets 53%.

TABLE 33. CHEMICAL COMPOSITION, %

Body parts	Size of fish	Moisture	Protein	Fat	Ash
Meat	Large	71–74.8	20.7–21.5	2.9–5.1	1.6
	Small	76.8	19.3	2.1	1.8
Head, bones, viscera	—	67.1	20.7	2.9	9.3

The taste qualities of this fish, unlike the common carangids, are good. The slightly acid taste is absent. This fish is recommended for use in canning, preparation of culinary and smoked products.

Indian scad—*Decapterus kiliche*. A large fish widely distributed in the Indian Ocean.

In catches in February and March in the Persian Gulf and in August in the Gulf of Aden, fish 18 cm long (15.2–23.4), weighing about 110 g (65–185) with maximum depth 4.0 cm (3.6–5.0) and maximum thickness 2.5 cm (2.3–3.0) were found.

Weight proportions of individual body parts with significant variations

TABLE 34. CHEMICAL COMPOSITION, %

Body parts	Moisture	Protein	Fat	Ash
Meat	66.0–75.5 / 71.9	20.6–21.9 / 21.5	1.2–10.7 / 5.0	1.4–1.6 / 1.5
Inedible parts of the body (entire offal)	70.1	15.1	5.4	7.4

(%): head 17.9 (14.6–21.0), trunk 70.0 (63.4–75), including meat with skin 54.8 (51.0–59.3), fins 1.8 (1.5–2.3), gut 1.9 (1.0–2.8), whole viscera 10.8 (7.6–13.2).

In the chemical composition of the meat of this fish we find great differences in moisture and fat contents. Fish caught in August in the Gulf of Aden were much fattier (10.7% fat) but the meat had a low moisture content (66.0%).

After cooking, the meat is dark with a mildly sour taste. The soup is greenish with a pleasant but mildly acid taste. Products canned in oil had good appearance and taste. Products canned in tomato sauce did not have satisfactory taste. This fish should be used for smoke curing and products canned in oil.

California scad—*Decapterus polyaspis*. Inhabits the Gulf of California. Length of the individuals caught in February–July 49.0–55.0 cm, weighing 1.5–1.8 kg.

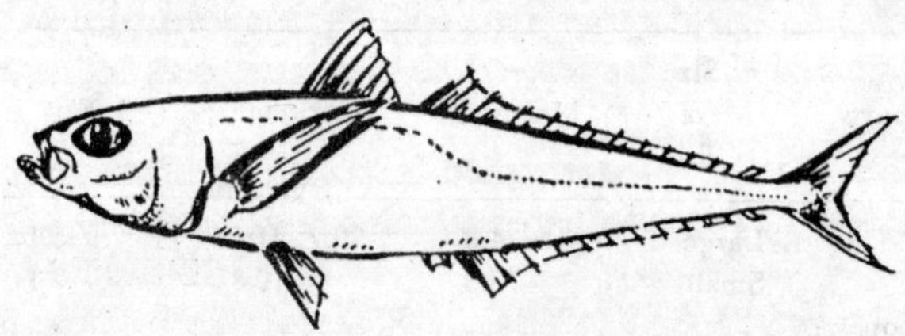

Fig. 37. California scad—*Decapterus polyaspis*.

Weight proportions of individual body parts (%): head 18.2, trunk 67.4, guts 6.7, bones 6.5, fins 5.7, liver 2.4. Output of fillets 56.4%.

Chemical composition of meat (%): moisture 71.3, protein 22.5, fat 4.6, ash 1.3.

The meat is white, soft and juicy. The sterilized canned preparations from fresh scad, both "natural" and "blanched in oil" are high grade.

Round scad—*Decapterus russelli*. Lives in the Indian Ocean. Usual length 28–33 cm and weight 340–415 g.

Weight proportions of individual body parts of fish caught in February–

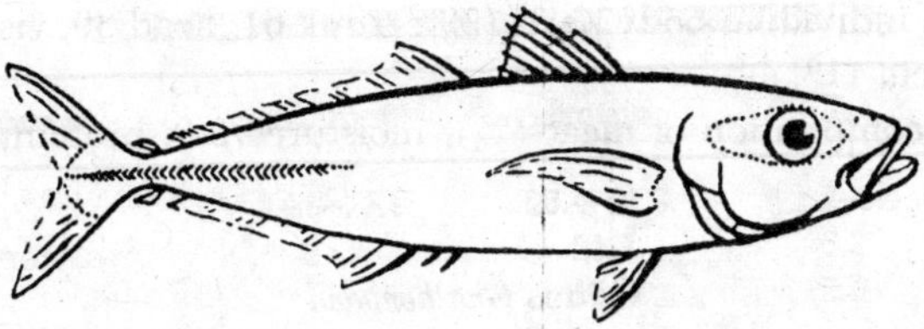

Fig. 38. Round scad—*Decapterus russelli.*

July (%): head 22.7–27.9, trunk 63.1–66.1, viscera 4.7–8.3, fins 1.5–2.2.

Chemical composition of meat (%): moisture 73.8–77.5, protein 20.0–23.8, fat 0.2–0.8, ash 1.7–2.3.

The meat is dense, slightly dry and mildly acid. The best use of frozen scad is in hot smoked products.

Genus *Elagatis*

Rainbow runner—*Elagatis bipinnulatus.* Distributed in the Atlantic and Indian Oceans and in the tropical waters of the Pacific Ocean. Attains a length of 110 cm. In appearance very similar to the members of the genus *Seriola.*

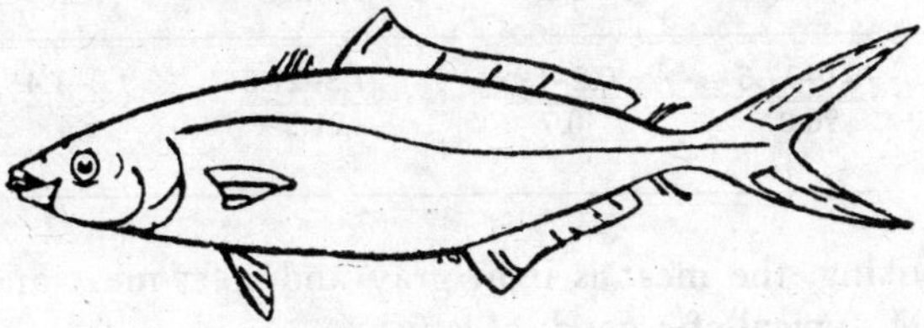

Fig. 39. Rainbow runner—*Elagatis bipinnulatus.*

This fish constitutes a rare catch in commercial fishing in the eastern Central Atlantic.

Weight proportions of individual body parts of this fish with 71 cm length, 17 cm depth, 9 cm thickness and weighing 4.2 kg (%): trunk with fins 72, head 16, viscera 12. Output of fillets 54%.

Chemical composition of meat (%): moisture 74.1, protein 20.6, fat 3.8, ash 1.5. The meat is not tasty; it is dark and hard. This fish should be used for fish meal.

Genus *Hynnis**

*Hynnis** *goreensis.* Fish caught off the coast of Africa from Senegal to the Gulf of Guinea. A fish of small commercial value.

Length of fish caught in February 29–36 cm and weight 0.6–1.0 kg. Weight proportions of individual body parts (%): trunk 61, head 30, viscera 7, fins 2. Output of fillets 41%.

Chemical composition of meat (%): moisture 79.0, protein 19.3, fat 0.2, ash 1.5.

Genus *Gnathonodon*

The members of this genus are large, with considerable fullness. They are found in the Red Sea and in the Gulf of Aden.

Golden trevally—*Gnathonodon speciosus*. Caught in small numbers along with other fish. In catches in September in the Gulf of Aden and the Red Sea most fish were 65 cm long (63.0–66.5), weighing 6,000 g (4,690–5,290) [*sic*], with maximum depth 21 cm and maximum thickness 9.7 cm.

Weight proportions of individual body parts showing significant fluctuations (%): head 31.3 (27.9–34.6), trunk 60.4 (57.6–63.3), including meat with skin on an average 47, scales 1.3, fins 2.2 (2.1–2.3), entire viscera 5.2 (4.2–6.3).

TABLE 35. CHEMICAL COMPOSITION, %

Moisture	Fat	Protein	Ash
$\frac{76.0\text{–}77.6}{76.8}$	$\frac{0.2\text{–}1.1}{0.7}$	$\frac{21.3\text{–}21.6}{21.5}$	$\frac{1.3\text{–}1.4}{1.4}$

After cooking, the meat is light gray and soft; meat and soup have a pleasant taste, typical of a good table fish.

This fish should be used for first and second courses and for frozen fillets (from large fish).

Genus *Lichia*

Striped garrick—*Lichia vadigo*. Found in the eastern Atlantic Ocean, all along the African coast. The fish caught were relatively large in body length, 30–60 cm, but larger fish up to 80 cm were also caught.

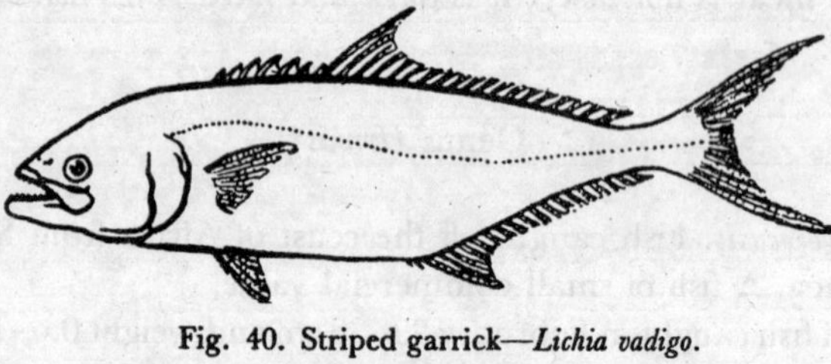

Fig. 40. Striped garrick—*Lichia vadigo*.

TABLE 36. CHEMICAL COMPOSITION, %

Body parts	Moisture	Protein	Fat	Ash
Meat	74.3–77.1	20.4–21.9	1.2–3.4	1.2–1.3
Head, bones, fins	64.9	21.3	6.9	7.9
Viscera	79.1	16.7	2.5	1.7

In this species, fish 30–46 cm long, with maximum depth 10–15 cm and maximum thickness 4.5–8.0 cm, weighing 0.7–2.3 kg had weight proportions of individual body parts as under (%): head 18–20, trunk 70–72, viscera 6–10, bones 7.6, skin and fins 5, scales 1.7. Output of fillets 54–56%.

The meat of boiled and fried fish is white with a pink tinge, mildly sweet and tasty. This is a good table fish. It can be used for culinary preparations and semi-finished products.

Genus *Megalaspis*

These are large pelagic fish. The species *Megalaspis cordyla* is widely known.

Finny scad or Torpedo trevally—*Megalaspis cordyla*. A good commercial fish. Found in the Indian and Pacific Oceans. Caught off the eastern coast of Africa from Madagascar to Australia, the Philippines and Hawaiian Islands. Body length up to 80 cm.

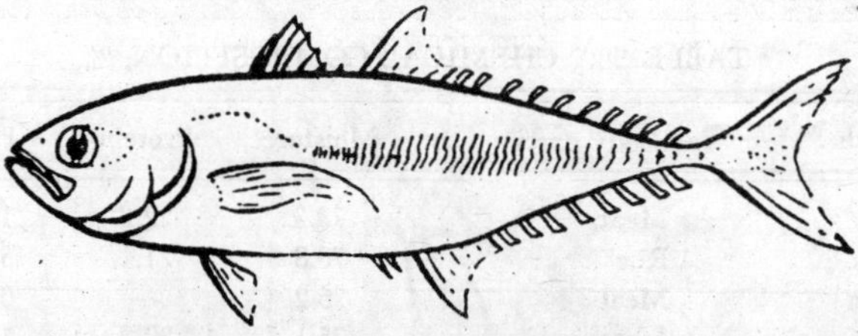

Fig. 41. Finny scad—*Megalaspis cordyla*.

In catches in July from the Gulf of Aden, large fish predominated, about 60 cm long (58–65), weighing 4,000 g (3,250–4,065) with maximum depth 16.2 cm (15.1–17.4), and maximum thickness 9.3 cm (8.3–10.3). The lengths of fish caught in the Gulf of Tonkin ranged from 23.5–31.0 cm and they weighed 165–375 g; for the fish caught in the Persian Gulf the values were 17.6–19.2 cm and 110–140 g. The boiled meat is light gray, its consistency varying from soft to hard, with a mildly acidic taste, typical of scads. The fish is recommended for preparation of smoked and canned in oil products.

TABLE 37. WEIGHT PROPORTIONS OF INDIVIDUAL BODY PARTS OF *MEGALASPIS CORDYLA* (%)

Body parts	Gulf of Aden	Gulf of Tonkin	Persian Gulf
Head	26.6–28.8 27.8	26.4–28.8	25.0
Trunk	62.1–64.9 63.0	58.3–60.2	60.0
Fillet	—	44.5–48.5	52.5
Scales (scutes)	0.4–0.8 0.6	—	0.5
Bones	—	14.4–16.4 (bones and fins)	7.5
Fins	2.0–2.1 2.0		1.5
Viscera	5.3–6.4* 5.9	3.6–3.8**	10.5
Gonads	—	2.2–4.2	—
Liver	—	1.4	—

* Viscera as whole

**Viscera without gonads and liver

TABLE 38. CHEMICAL COMPOSITION, %

Place of catch	Body parts	Moisture	Protein	Fat	Ash
Gulf of Aden	Meat	73.3	22.3	1.6	1.4
Gulf of Aden	Roe	70.8	21.1	5.6	1.8
Gulf of Tonkin	Meat	76.2	—	0.3	—
Persian Gulf	Meat	75.1	20.3	1.0	1.6
Persian Gulf	Waste after dressing	67.7	18.6	5.2	7.6

Genus *Olepis*

Scad—*Olepis mata*. Found in the waters off northern Australia. Length of fish caught in February–July 17–20 cm and weight 92–165 g.

Weight proportions of individual body parts (%): head 21.4, trunk 68.5, fins 4.2, viscera 5.7.

The meat of this fish is similar to the meat of other scads: light in color and pleasant in taste, mildly acid. It should be frozen before culinary products are prepared.

Genus *Selar*

In the Soviet waters the members of this genus are found in all the warm seas. Only two species are of commercial value; they are relatively small, 30–35 cm in length. Some species, however, attain larger sizes: in the Indian Ocean 40 cm and in the western Atlantic Ocean 60 cm.

Scad—*Selar boops*. Inhabits the Great Australian Bight. Length of fish caught in February–July 6–19 cm and weight 50–125 g.

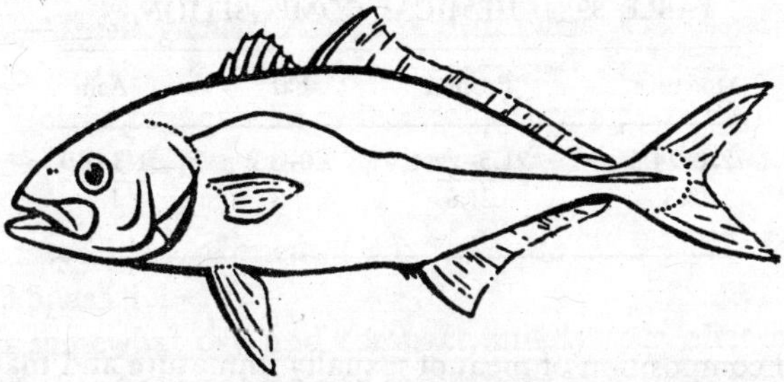

Fig. 42. Scad—*Selar boops*.

Weight proportions of individual body parts (%): head 25.9–30.1, trunk 61.6, viscera 3.6–6.4, scales 2.8–3.8 and scutes 1.8. Output of fillets 50.0–54.4%.

Chemical composition of meat (%): moisture 76.1, protein 20.9, fat 0.9, ash 1.5.

This scad has light meat, tasty in boiled as well as in fried forms. It should be frozen before preparation of culinary or smoked products.

Bigeye scad—*Selar crumenophthalmus*. A commercial fish. Distributed along the coasts of India, Africa, China, Japan and islands of the Indo-Australian archipelago and off the Philippines and Solomon Islands. Attains body length of 30 cm.

In autumn catches in the Gulf of Aden and in spring near Masira Island, there is a predominance of fish 10–25 cm long (average 17), weighing 14–430 g (average 150), with maximum depth 5.5 cm (3.4–8.0) and maximum thickness 3.2 cm (2.0–4.5).

Weight proportions of individual body parts with some fluctuations: head 29.5 (28.6–30.6), trunk 58.2 (55.7–60.4), including meat with skin 48.7 (45.5–50.3), bones 10.4 (8.3–12.6), scales (scutes) and fins 3.0 (2.0–3.9), entire viscera 8.1 (6.1–10.3), including roe in the sexually mature fish 1.6 and liver 1.3.

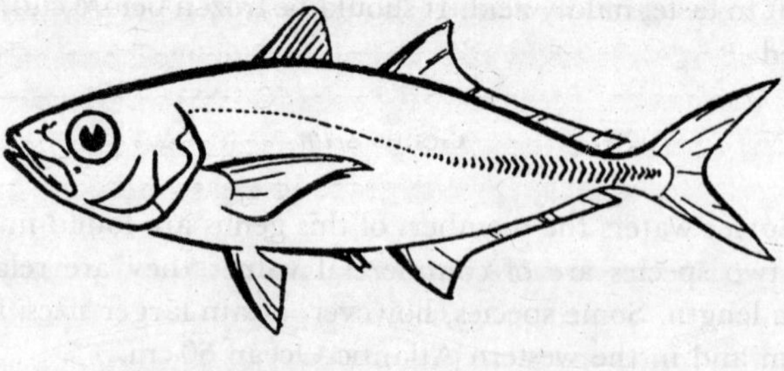

Fig. 43. Bigeye scad—*Selar crumenophthalmus.*

TABLE 39. CHEMICAL COMPOSITION, %

Moisture	Protein	Fat	Ash
72.4–74.5 73.3	21.5–22.4 21.9	2.0–3.2 2.8	1.3–1.6 1.4

Chemical composition of meat of sexually immature and mature fish from different regions is fairly consistent.

After cooking, the meat is gray, dense or hard, with acid taste (this is more pronounced in small sexually immature fish). The soup is yellow, with a fish taste. Fried meat is dense and sour.

This fish is suitable for products prepared by hot smoking and by canning in oil (with steam blanching).

Genus *Selaroides*

Slender scaled scad—*Selaroides leptolepis*. Area of distribution, the Great Australian Bight. Length of fish caught between July–February 11–15 cm, weight 40–85 g. Weight proportions of individual body parts (%): head 23.5–29.1, trunk 62.5–64.7, viscera 2.1–12.0, fins 4.2.

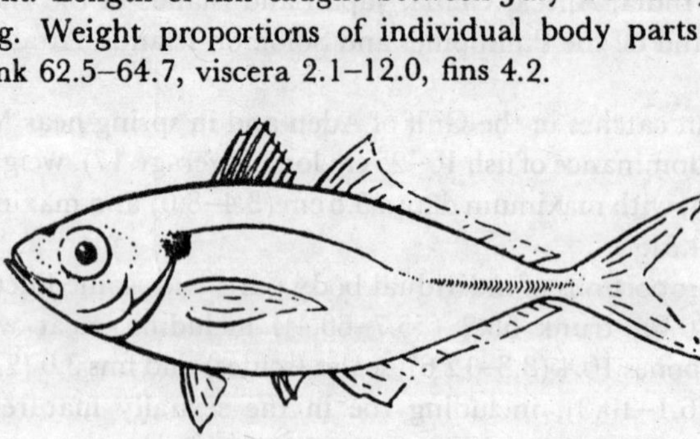

Fig. 44. Slender scaled scad—*Selaroides leptolepis.*

Chemical composition of meat (%): moisture 75.9, protein 21.5, fat 1.2, ash 1.8.

This species of scad has light colored meat which is quite tasty, but mildly sour. Culinary products and canned preparations can be made from raw meat. The meat should be frozen onboard the fishing vessels.

Genus *Seriola*

The members of this genus live in the tropical and subtropical waters of the Atlantic, Indian and Pacific Oceans. The fish of this genus have commercial value and are distinguished by large size. The members of certain species reach 100–180 cm in length and 50 kg in weight.

Yellow tail—*Seriola grandis*. Area of distribution, the Great Australian Bight. Length of specimens caught in December–May 49–63 cm, weight 2,200–2,860 g. Weight proportions of individual body parts (%): 24.1–24.9, trunk 65.3, bones 7.3–8.5, fins 2.1–2.7, viscera 7.6–9.4, scales 0.3. Output of fillets 55.2%.

Chemical composition of meat (%): moisture 73.3–76.2, fat 0.7–1.9, protein 21.5–23.5, ash 1.1–1.6.

The meat is somewhat dry and compact, mildly sour after boiling and frying; it is tasty after hot smoking and satisfactory in canned products. This fish should be used for sterilized canned products "blanched in oil" or for freezing onboard.

The meat becomes rancid quickly because of oxidation of the fat; it should therefore be glazed during freezing.

Amber jack—*Seriola lalandi*. Attains a length of 2 m and weight of 50 kg. Rarely found in the Atlantic Ocean in commercial catches along the coasts of Africa and America.

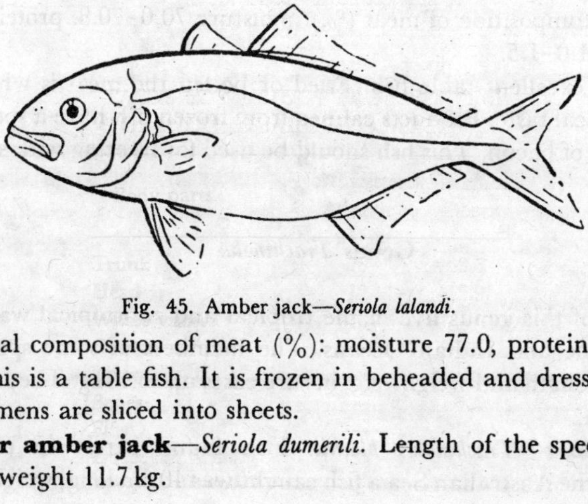

Fig. 45. Amber jack—*Seriola lalandi*.

Chemical composition of meat (%): moisture 77.0, protein 20.1, fat 1.4, ash 1.5. This is a table fish. It is frozen in beheaded and dressed form. The large specimens are sliced into sheets.

Greater amber jack—*Seriola dumerili*. Length of the specimen caught 92 cm and weight 11.7 kg.

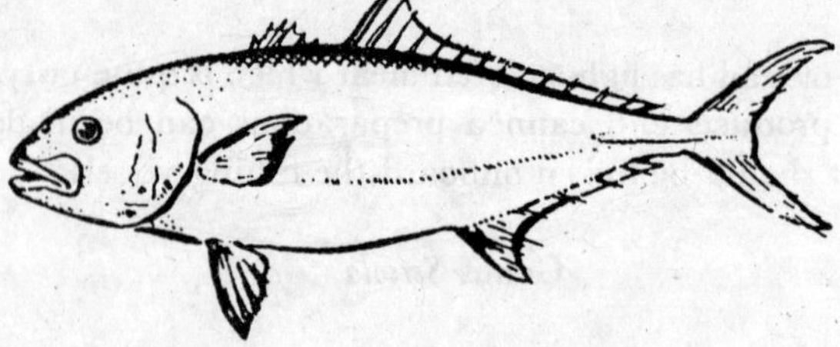

Fig. 46. Greater amber jack—*Seriola dumerili.*

TABLE 40. CHEMICAL COMPOSITION, %

Body parts	Moisture	Fat	Protein	Ash
Meat	74.6	1.6	22.5	1.3
Viscera	79.9	2.3	16.4	1.4
Bones, skin and fins	61.6	10.7	20.6	7.1

Genus *Seriolella*

Seriolella or sea bream—*Seriolella brama.* Distributed near southern Australia, Tasmania and New Zealand. Length of specimens caught in June–July, 47–58 cm and weight 2.5–4.1 kg.

Weight proportions of individual body parts (%): head 16.6–17.6, meat with skin 48.6–56.0, bones 6.0–6.8, fins 1.5–2.0, viscera without gonads and liver 6.8–8.0, gonads 10.4–12.3, liver 1.3–16.6.

Chemical composition of meat (%): moisture 70.0–70.8, protein 19.4, fat 7.9–10.5, ash 1.0–1.5.

This is an excellent table fish. Fried or boiled the meat is white, dense, with good typical taste. Products canned from frozen fish have a specific smell similar to that of bacon. This fish should be used for freezing with subsequent canning.

Genus *Trachinotus*

Members of this genus live in the tropical and subtropical waters of the Atlantic, Pacific and Indian Oceans. The members of some species reach 90–100 cm in length and weigh 7 kg, in rare cases up to 18 kg. They are caught with other fish.

Bloch's dart—*Trachinotus blochii.* Fish found rarely in catches. In September in the Australian Sea a fish caught was 40 cm long, weighing 560 g.

TABLE 41. WEIGHT PROPORTIONS OF INDIVIDUAL BODY PARTS, %

Body parts	Arabian Sea	Gulf of Tonkin
Head	19.8	27.3–30.7
Trunk	70.9	56.6–62.3
Meat	54.2	46.6–52.3
Skin	3.8	—
Bones	12.9	13.3–13.4
Fins	2.3	—
Viscera, entire	6.4	—
Including		
gonads	0.9	1.4–1.6
liver	2.7	0.8–1.4

TABLE 42. CHEMICAL COMPOSITION OF MEAT, %

Components	Arabian Sea	Gulf of Tonkin
Moisture	73.4	71.1–76.4
Protein	23.8	22.1–22.7
Fat	1.3	0.6–1.5
Ash	1.6	—

Maximum depth was 12 cm and maximum thickness 2.8 cm. Length of a fish caught in the Gulf of Tonkin 9.0–18.5 cm and weight 35–172 g.

Fish caught in different regions differed slightly in weight proportions of individual body parts.

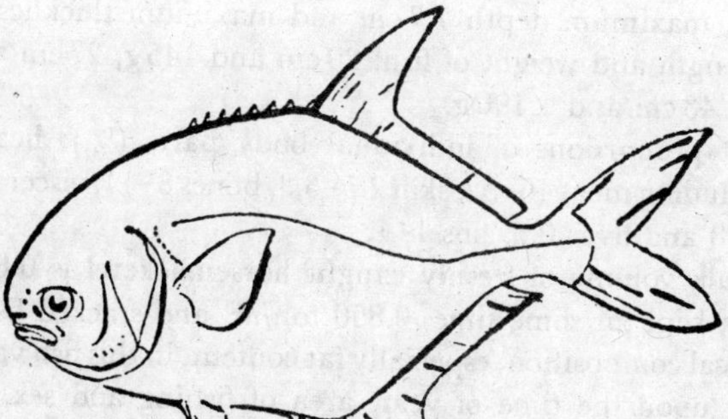

Fig. 47. Bloch's dart—*Trachinotus blochii.*

Chemical composition of meat of fish caught in different zones hardly differed.

Boiled meat is slightly tough with an acid taste, unpleasant in appearance. Should be used mainly for fish meal.

Russell's dart—*Trachinotus russelli**. Distributed in the Indian Ocean from the coast of Africa to Australia and New Guinea. Length 22.0–27.5 cm and weight 250–472 g. Weight proportions of individual body parts (%): head 32.0, guts 3.2, bones and fins 12.7, gonads 2.4, liver 1.1. Output of fillets 42.0%. Chemical composition of meat (%): moisture 77.3, protein and mineral substances 21.7, fat 1.3.

Round pompano—*Trachinotus falcatus*. Found in small numbers with other fish along the American and African coasts in the tropical zones of the Atlantic Ocean. Usual length 20–40 cm. Length of fish caught in the Amazon estuary 22 cm, depth 7.3 cm, thickness 1.8 cm and weight 100 g. Weight proportions of individual body parts (%): trunk with fins 70, head 25, viscera 5. Chemical composition of meat (%): moisture 74.5, fat 3.2, protein 20.8, ash 1.5.

Data on taste qualities not available.

Genus *Trachurus*

Among the members of this genus, horsemackerel *Trachurus trachurus* with its subspecies *Trachurus capensis* have maximum commercial value.

Horsemackerel or maasbanker—*Trachurus trachurus*. Constitutes an important fish in the Atlantic Ocean especially along the coast of northwest Africa near Dakar, Takoradi, Cape Blanc. Also caught in the North Sea, in the Strait of La Manche and along the coast of South America near the Uruguay shelf. Also found in catches in the Indian Ocean near St. Francis and in the shelf of Agulhas. Length of fish varies from 18–39 cm, weight from 120–730 g, maximum depth 7.7 cm and maximum thickness 4.5 cm. Ratios between length and weight of fish: 20 cm and 145 g, 27 cm and 350 g, 34 cm and 770 g, 43 cm and 1,190 g.

Weight proportions of individual body parts (%): head 22–32, trunk 57–73, including meat 46–52, skin 2.5–3.8, bones 8–11, viscera 7–14, including gonads 1–8 and liver 0.6, fins 1–4.

The bulk volume of freshly caught horsemackerel is 0.813 ton/m^3 while that of fish kept for some time, 0.850 ton/m^3 and stacked fish, 0.891 ton/m^3.

Chemical composition, especially fat content, in this fish varies significantly depending upon the time of year, area of fishing and sex. The highest fat content was found in fish caught from the Strait of La Manche and the lowest values were in fish caught off the African coast.

TABLE 43. CHEMICAL COMPOSITION, %

Parts of the body and region of fishing	Moisture	Fat	Protein	Ash
Meat				
Off the coast of Africa	72–76	1.3–3.4	18.0–21.7	1.2–2.2
Off La Manche Strait	66–74	6.0–13.6	18–21	0.6–1.6
Off the coast of South America	74	5.4	18.4	1.4
North Sea	75	4.3	17.6	2.3
Indian Ocean (region of St. Francis and Agulhas shelf)	$\frac{73.2\text{–}75.2}{73.8}$	$\frac{3.7\text{–}6.4}{5.6}$	$\frac{19.3\text{–}19.6}{19.4}$	$\frac{1.0\text{–}1.3}{1.1}$
Waste				
Atlantic Ocean				
heads	74.2	6.1	13.1	6.6
heads, bones and viscera	71.6	8.0	13.3	6.7
Viscera, entire				
Indian Ocean	$\frac{63.2\text{–}65.1}{64.1}$	$\frac{9.6\text{–}13.5}{11.8}$	$\frac{15.9\text{–}17.5}{16.8}$	$\frac{5.8\text{–}8.0}{7.1}$

Fat content in meat of horsemackerel caught in the Strait of La Manche depending upon the month of catch, (%)

September	8.2–9.8
November	10.0–11.8
December	13.6–12.9
January	6.0–8.4

The most fatty fish from the La Manche Strait were caught in December.

The chemical composition of meat of this fish caught from the St. Francis region and from the shelf of Agulhas and that of the inedible parts of the body do not fluctuate much. However, it has been noted that in July the fat in the meat of this fish is lower (3.7%) than in March (5.8–6.4%).

The fat content in the meat of this fish from the La Manche Strait is higher in females than in males (Table 44).

The meat of horsemackerel from the La Manche Strait has better taste qualities. After boiling, it is light gray, dense and pleasant in taste. Whether boiled or fried, it is slightly sour in taste. The soup is lemon colored and rich with a pleasant acid taste. The products canned in oil from fish which had been steam blanched have good appearance, soft consistency and pleasant taste. This fish is recommended for products canned in oil and for smoke curing. The freshness of the raw meat plays an exceptionally great role in determining the quality of products canned from horsemackerel. Therefore, canned products of this fish should be prepared at sea onboard the fishing vessels.

TABLE 44. FAT CONTENT, %

Size of fish, cm	Females	Males
24	8.1	5.6
25	6.0	5.4
26	8.6	4.2

Horsemackerel (ssp)—*Trachurus capensis*. An important commercial fish. In catches in December in the Indian Ocean, at the southern tip of Africa, fish 17 cm long (16–20), weighing around 110 g (65–150) predominate, maximum depth 4.5 cm (3.5–5.0) and maximum thickness 2.4 cm (2.0–2.8).

Weight proportions of individual body parts showing significant fluctuations (%): head on an average 29.7, trunk 58.8 (53.8–69.0) including meat with skin 49.9 (42.1–62.5), bones 8.9 (7.9–9.9), scales (scutes) 0.9 (0.6–1.2), entire viscera 6.5 (5.5–7.5).

This subspecies shows some difference in chemical composition of meat and inedible parts from that of the species. In the taste qualities, there are no differences. As the species, this subspecies is recommended for smoke curing and canning in oil.

TABLE 45. CHEMICAL COMPOSITION, %

Body parts	Moisture	Fat	Protein	Ash
Meat	74.9	3.9	19.7	1.4
Inedible parts of the body (entire offal)	65.9	10.1	17.1	6.3

Japanese horsemackerel—*Trachurus japonicus*. Lives in the Pacific Ocean, off the coast of southeast Asia. A warmth-loving tropical fish. Length 12–35.5 cm, weight 25–547 g [*sic*]. Weight proportions of individual body parts (%): head 13.5–23.6, trunk 56.3–69.5, bones and fins 8.0–13.9, whole viscera 4.5–11.0, including liver 1.0 and gonads 1.9–5.5. Output of fillets 52.0–57.5%.

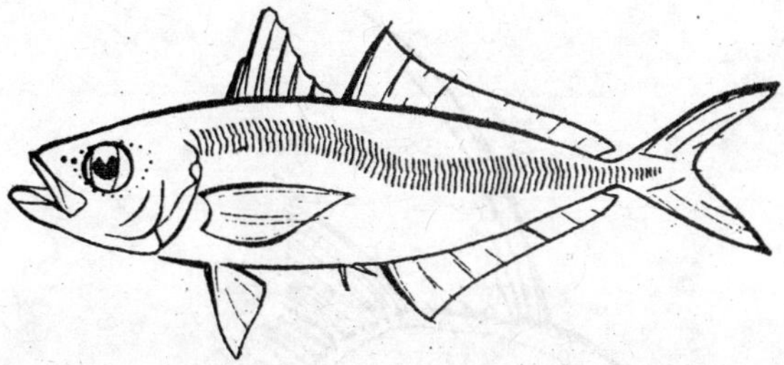

Fig. 48. Japanese horsemackerel—*Trachurus japonicus.*

Chemical composition of meat (%): moisture 68.0–78.1, protein 18.6–22.0, fat 1.0–10.0, ash 1.3–1.7.

The meat of Japanese horsemackerel is slightly sour, dense, grayish. The fat in the tissues is quickly oxidized and becomes rancid; therefore, frozen fish should be carefully glazed. Frozen fish can be used for preparing products after hot smoking and canned preparations smoked in oil. Canned products blanched in oil are better prepared from fresh fish.

Genus *Vomer*

Moonfish—*Vomer setapinnis.* Lives in the tropical zones of the Atlantic and Pacific Oceans, off the coasts of Africa and America. Length 15–50 cm, weight up to 4 kg. Chemical composition of meat (%): moisture 75.8, fat 2.9, protein 20.0, ash 1.3. The meat is tasty; it is prepared in frozen form without gutting, with subsequent utilization in culinary preparations.

Family CENTROPOMIDAE—SNOOKS

Glass perch—*Ambasis kopsii**. Found in the East China Sea. Length of fish caught in February 15–19 cm and weight 90–120 g. Weight proportions of individual body parts (%): head 19.5–23.2, trunk with fins 69.8–74.3, viscera 3.5–6.3. Chemical composition of meat (%): moisture 79.8, protein 17.6, fat 0.6, ash 1.8.

The meat is not tasty; however, it can be used for culinary preparations.

Family CHAENICHTHYIDAE—GLASSFISHES

This family includes nine genera and about 17 species which live in the Antarctic region. The blood of these fish is unique among vertebrates in being colorless, since it does not contain erythrocytes or hemoglobin.

Four species of this family have been studied.

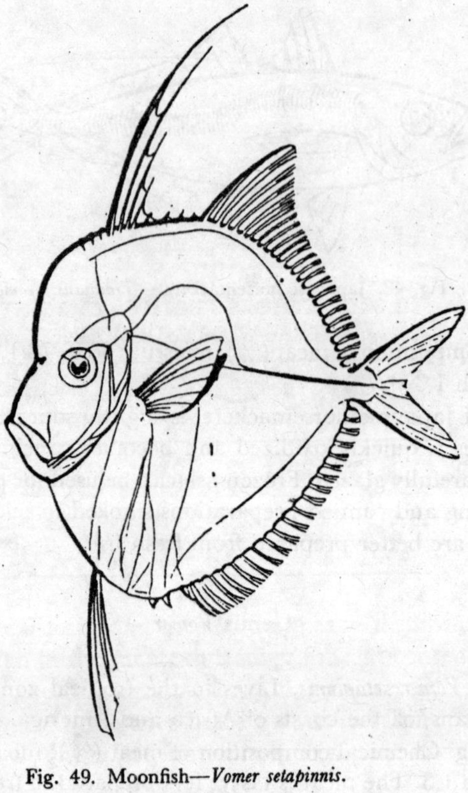

Fig. 49. Moonfish—*Vomer setapinnis.*

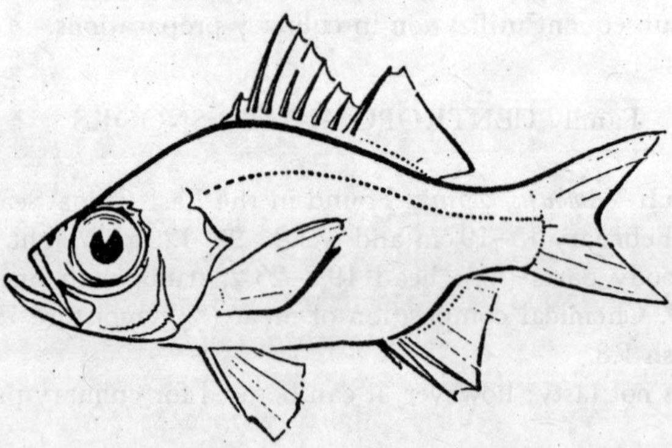

Fig. 50. Glass perch—*Ambasis kopsii**.

Genus *Pseudochaenichthys**

Dusky glassfish—*Pseudochaenichthys georgianus**. Caught in small quantities along with other fish; distributed in the South Atlantic near South Georgia

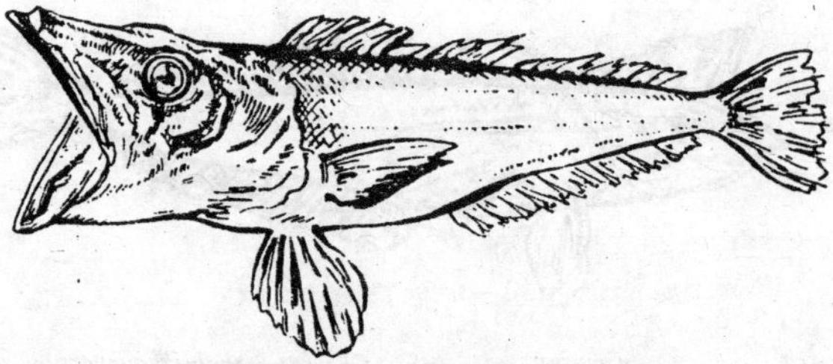

Fig. 51. Dusky glassfish—*Pseudochaenichthys georgianus**.

Island. In catches, a few individuals of 57 cm length but most of them 43–56 cm in length, weighing 750–2,150 g.

Weight proportions of individual body parts of fish caught in February–March (%): head 31.6, meat 38.8, skin and fins 8.8, bones 5.8, viscera 14.8, including liver 5.1.

Chemical composition of meat (%): moisture 75.5, protein 17.5, fat 7.7, ash 1.3.

This species has no significant commercial value. Its meat is edible. It can be used for culinary preparations (first and second courses).

Genus *Chaenocephalus**

Crocodile glassfish—*Chaenocephalus aceratus*. A commercial fish. Distributed in the Scotia Sea. In catches, some are 72 cm long, weighing 3.7 kg. Weight proportions of individual body parts (%): head 34.2, meat 37.4, skin and fins 6.8, bones 6.9, viscera 12.7, including liver 2.1.

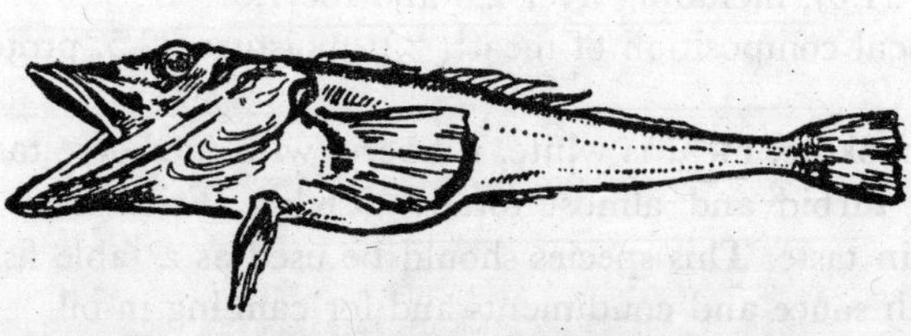

Fig. 52. Crocodile glassfish—*Chaenocephalus aceratus*.

Chemical composition of meat (%): moisture 80.6, protein 17.1, fat 1.0, ash 0.7.

The meat of the crocodile glassfish is white, dense, juicy, slightly sweet and fairly tasty. It is a good table fish. It can be used for first and second courses.

Genus *Champsocephalus**

Pike glassfish—*Champsocephalus gunnari*. Distributed in the South

Fig. 53. Pike glassfish—*Champsocephalus* gunnari.*

Atlantic, near the South Georgia and Kerguelen Islands. Also found near the South Orkneys and South Shetland Islands. Length 64 cm, weight little more than 2 kg. Chemical composition of meat of fish caught in February in the Scotia Sea (%): moisture 80.3, protein 17.3, fat 0.9, ash 0.9.

This fish can be used for culinary products (first and second courses).

Genus *Chaenichthys*

Pike glassfish—*Chaenichthys rhinoceratus*. Caught in small quantities along with other fish. It is found in the Antarctic regions.

The meat, viscera and gills are white-marble in color. The roe is orange, the skin spotted.

In catches in February near Kerguelen Islands, fish 40 cm in length (39–42), weighing around 500 g (475–550), with maximum depth 5.8 cm (5.4–7.0) and maximum thickness 4.9 cm (4.5–5.0) were predominant.

Weight proportions of individual body parts (%): head on an average 34.5 (30.0–38.4), including meat of the occiput, averagely, 5.9, trunk 48.0 (45.7–52.1), including meat with skin 41.3, bones 6.7, fins 1.6, entire viscera 15.3 (8.7–24.0), including liver 2.6 and roe 1.9.

Chemical composition of meat (%): moisture 79.5, protein 19.0, fat 1.2, ash 1.3.

After cooking, meat is white, rubbery, with mediocre taste. The broth is somewhat turbid and almost totally tasteless. Fried meat is rubbery and mediocre in taste. This species should be used as a table fish for the second course with sauce and condiments and for canning in oil.

This species is not suitable for natural canned products and canning in tomato sauce (the natural canned products have an unpleasant taste and the canned preparation in tomato sauce undergoes color change).

The taste quality of roe in mild salted form is low.

Family CHEILODACTYLIDAE*—MORWONGS

Genus *Cheilodactylus*

Besugo—*Cheilodactylus bergi.* A significant fish in the southwest Atlantic. Length 25–30 cm, weight 0.4–0.5 kg.

Size characteristics

Length, cm	23–32
Depth, cm	9–11
Thickness, cm	3.0–4.4
Length of head, cm	6.8
Weight of fish, kg	0.25–0.65

Weight proportions of individual body parts, %

Trunk	50–57
Head	24–28
Viscera	12–16
Fins	2.2–6.0
Meat	38–45
Skin	2.1–2.8
Bones	6.4–6.8
Scales	1.6–2.5
Gonads	1.2–4.8
Liver	1.9–3.4

Chemical composition of meat, %

Moisture	70.0–72.9
Fat	5.6–8.9
Protein	18.7–21.3
Ash	1.0–1.6

TABLE 46. MOISTURE AND FAT CONTENTS IN MEAT OF FISH IN DIFFERENT MONTHS, %

Parameters	January	February	March	April	September	October
Moisture	70.5	70.6	72.2	72.5	72.9	70.0
Fat	8.5	8.3	5.9	5.6	6.1	8.9

TABLE 47. CHEMICAL COMPOSITION OF INDIVIDUAL BODY PARTS OF THE FISH, %

Body parts	Moisture	Fat	Protein	Ash
Head	58–62	12–16	12–15	8–10
Bones	47–50	16–20	15–17	14–15
Skin	59–61	11–14	27–30	1.5–2.2
Viscera	71–79	3–12	10–14	2–5
Roe	79–82	1.5–3.5	14–15	1–1.5
Testes	79.5	1.5	16	2
Liver	56–74	5–28	12–16	1.4–1.7

The fish are processed in frozen form. The meat has relatively high fat content, 6–9%. This fish should be used for smoked products and culinary preparations.

Genus *Nemadactylus*

Jackass fish—*Nemadactylus macropterus*. **Blue morwong**—*Nemadactylus valenciennesi*. These are commercial fish. They are distributed near eastern and southern Australia, Victoria, Tasmania and New Zealand.

TABLE 48. RATIO OF INDIVIDUAL BODY PARTS BY WEIGHT, %

Body parts	*Nemadactylus macropterus*	*Nemadactylus valenciennesi*
Head	25.5–27.0	31.0
Meat	50.5–59.8	49.4
Bones	6.2–7.9	11.9
Fins	2.0–4.7	2.1
Intestines	3.1–4.9	1.3
Including liver	1.0	—

TABLE 49. CHEMICAL COMPOSITION OF MEAT WITH SKIN, %

Parameters	*Nemadactylus macropterus*	*Nemadactylus valenciennesi*
Moisture	74.5	75.8
Fat	3.8	4.9
Protein	21.6	19.0
Ash	1.8	1.2

The lengths of jackass fish which were caught in May and September were 24–50 cm and weights 0.5–1.6 kg and those of blue morwongs were 80–82 cm and 8–8.5 kg, respectively, caught in December–May.

These are valuable table fish. The meat is white, tasty, sweet and dense in consistency. The jackass fish has slightly soft meat. The meat of these fish should be frozen and subsequently used for culinary preparations.

Family CHIROCENTRIDAE—WOLF HERRINGS

Dorab—*Chirocentrus dorab*. Caught in small numbers along with other fish in the eastern Indian Ocean, in the Gulfs of Tonkin and Siam.

Length of dorabs caught in September in the Indian Ocean 44.5–55.0 cm, body depth 8.4 cm, width 3.1 cm, average weight 670 kg (512–950 g). Length of dorabs caught in the Gulfs of Tonkin and Siam 80 cm and weight 2,800 g.

Weight proportions of individual body parts of dorab (%): head 11.9–13.1, trunk 76.4–80.5, scales [illegible].3, bones 10.0–20.8, viscera 6.7–8.5, including liver 1.0–1.5 and gonads 0.5–0.7. Output of fillets 59.7–64.7%.

Fig. 54. Dorab—*Chirocentrus dorab*.

The meat of dorab has low fat content but it is tasty. It has many thin and sharp bones.

TABLE 50. CHEMICAL COMPOSITION, %

Body parts	Moisture	Fat	Protein	Ash
Meat	76.2–76.3	1.4–1.6	20.1–20.9	1.6–1.8
Viscera	74.2	8.0	—	—

Cooked meat of dorab has satisfactory taste. Fried cutlets are better in taste. It can also be used for products canned in oil and for culinary products.

Wolf herring—*Chirocentrus nudus*. Weight of this wolf herring caught in February–March in the Arabian Sea 410 g. Weight proportions of individual body parts (%): meat 64.9, head 13.7, skin 0.7, fins 1.0, bones 10.0, viscera 8.5, including liver 1.0 and gonads 0.5.

Chemical composition of meat (%): moisture 75.9, fat 1.7, protein 20.6, ash 1.7. The meat of this wolf herring can be used for culinary preparations (fried fish and cutlets).

Family CENTRACANTHIDAE*—PICARELS

Bigeye picarel—*Smaris* macrophthalmus*. A small fish rarely exceeding 30 cm in length. Distributed in the eastern Central Atlantic, an important constituent of trawl catches especially near Cape Blan and San Luis.

Weight proportions of individual body parts of this fish with body length

TABLE 51. CHEMICAL COMPOSITION, %

Body parts	Moisture	Fat	Protein	Ash
Meat	74.0	4.1	20.6	1.3
Head, bones and fins	65.3	10.5	17.3	6.9

15–19 cm, depth 5–7 cm, thickness 2–3 cm and weight 80–170 g (%): head 20, trunk 71, viscera 9. Output of fillets 46%.

The taste qualities of the meat of this fish are quite satisfactory. This fish is marketed in America as "Atlantic scup".

Family CLUPEIDAE—HERRING

This family includes about 190 species. Herring are distributed in tropical, subtropical and temperate waters. Most herring are marine fish, about 30 species are fresh water fish and about 30 species are migratory. In terms of gross weight herring occupy second place in the world fish catch.

In this large family of herring, the most important constituents of the world fisheries are the oceanic herring of genus *Clupea,* the total catch of which accounts for 49.9% of the total catch of herring, followed by sardines, which are members of the genera *Sardina, Sardinella* and *Sardinops,* the catches of which constitute 24.8% and menhaden, members of genus *Brevoortia,* 11.6%. The catches of all other members of this family account for 13.7%.

The commercial types of herring are: Atlantic herring, alewife, shad, menhaden, thread herring, African shad, gilt sardine, flat sardine, sardinops, pilchard, Mexican sardine, round herring, Baltic herring and sprat.

Genus *Alosa*

These are marine and migratory fish. They are represented by five to six species with many subspecies in the Caspian, Black and Azov Seas and by two to four species with a few subspecies along the European and American coasts of the Atlantic Ocean, in the Baltic, Mediterranean and partly in the Black Seas as well as in the rivers draining into these waters.

African shad—*Alosa africana.* Caught along with other fish in the eastern Central Atlantic near the Moroccan shelf. A migratory fish similar to the Caspian black, back shad, but larger.

A large fish. Length of individuals observed 35–50 cm, weight 1.3–3.5 kg. Weight proportions of individual body parts (%): trunk 74, head 13, viscera 11, fins 2. Output of fillets 66%.

Chemical composition of meat (%): moisture 60.8, fat 21.2, protein 16.3, ash 1.7.

In some countries of Europe, African shad is considered a delicious fish. Because of the high fat content in the meat, it is used to make smoke cured balyks. Such balyks do not have the typical herring smell so they can be included in high quality products.

American shad—*Alosa sapidissima.* Distributed in the Atlantic Ocean along the coast of North America from the Gulf of St. Lawrence to Florida and ascending into the rivers. Through acclimatization it inhabits the Pacific Ocean where it is found along the western coast of North America. Recently some specimens of this shad have been caught in the Strait of Corfu and in the Hvaler Bay and on the eastern coast of Kamchatka. It is a migratory fish. It probably remains in the sea near the estuaries of rivers.

In catches individual fish 16–70 cm in length have been caught.

Fish were examined which were caught in the northwestern Atlantic and in the Pacific Ocean near the coasts of Canada and the USA.

Fish caught in December in the northwestern Atlantic ranged in length between 30–35 cm and in weight between 450–600 g while those from the Pacific Ocean were 41–51 cm in length and 890–1,195 g in weight.

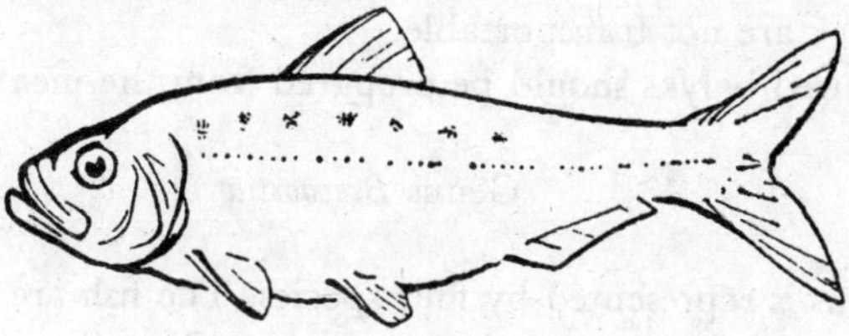

Fig. 55. Shad—*Alosa sapidissima.*

TABLE 52. WEIGHT PROPORTIONS OF INDIVIDUAL BODY PARTS, %

Body parts	Northwestern Atlantic	Pacific Ocean
Trunk	78	43.6
Head	12	15.9
Whole viscera	10	—
Gonads	—	0.5
Liver	—	2.3
Bones	—	4.2
Fins	—	2.4
Output of fillets	60	—

TABLE 53. CHEMICAL COMPOSITION, %

Object of analysis	Months of catch	Moisture	Fat	Protein	Ash
Shad from the northwestern Atlantic					
meat	April	65.7	15.2	16.8	2.0
meat	June	72.7	6.2	19.7	1.4
head, bones and viscera	June	62.6	13.5	18.2	5.8
Shad from the Pacific Ocean					
meat	—	66.1	14.3	18.0	1.3

The meat of fried fish is soft, pink and tasty but it contains many small bones. The fish contain the most fat in September. After salting and sun curing the meat is tasty but there are so many intramuscular bones that it is difficult to eat it. Canned products of this fish, prepared by blanching in oil, are good in taste but the consistency of the meat in conserves is poor and, therefore, they are not transportable.

Smoke cured balyks should be prepared from the meat of this fish.

Genus *Brevoortia*

This genus is represented by four species. The fish are distributed in the Atlantic Ocean along the American coast from New Scotland to the coast of Brazil and northern Argentina. The usual length of the fish is 30–38 cm, sometimes up to 50 cm, weight 300–450 g, sometimes 840 g.

Atlantic menhaden—*Brevoortia tyrannus*. Lives along the Atlantic coast of America from New Scotland to Brazil. A distinguishing feature of this species is that the head constitutes 30% of the total body length. Usual length of the fish, 17–30 cm.

Fig. 56. Atlantic menhaden—*Brevoortia tyrannus*.

TABLE 54. CHEMICAL COMPOSITION, %

Body parts	Moisture	Fat	Protein	Ash
Meat	59.7–75.0	5.0–21.0	16.8–19.0	1.3–1.4
Viscera	54.1	31.3	9.9	3.4

Weight proportions of individual body parts of a menhaden 17–36 cm in length and weighing 120–740 g (%): trunk 63–78, head 17–28, guts 13–16, bones 5.7–6.4, fins 0.8–0.9, gonads 0.5–0.8, liver 2.0–3.3. Output of fillets 55–56%.

Smoked products should be prepared from this menhaden.

Genus *Clupea*

Baltic herring—*Clupea harengus membras*. The main fishery of this fish is the Baltic Sea. A subspecies of the Atlantic herring, but smaller in size.

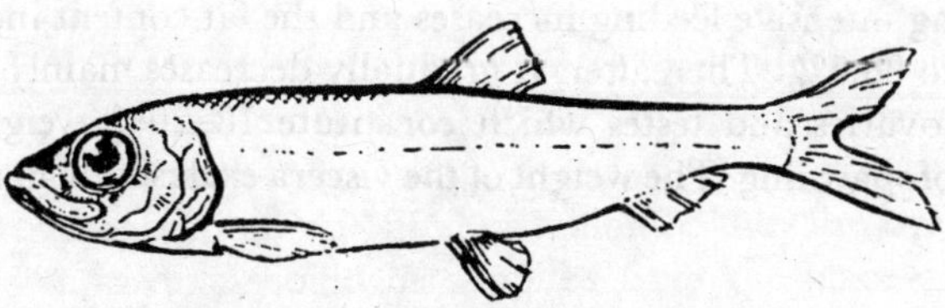

Fig. 57. Baltic herring—*Clupea harengus membras*.

TABLE 55. WEIGHT PROPORTIONS OF INDIVIDUAL BODY PARTS, %

Quarter of the year	Head	Trunk with fins	Viscera
I	21	63	16
II	21	57–62	16–20
III	21	68	11
IV	19	68	13

Distributed in the central, northern and eastern Baltic Sea, as well as in the Gulfs of Riga, Finland and Bothnia. Average length 15.4 cm, weight 24.1 g. Lengths of fish caught in the Gulfs of Finland and Riga, 11–19 cm, weights 15–50 g. The Baltic herring from the Kaliningrad Bay is larger, 15–22 cm in length and weighing 15–80 g.

The Baltic herring can be marketed in chilled, frozen, marinated, salted and smoked forms and after canning as "Baltic sardine".

Atlanto-Scandinavian herring—*Clupea harengus harengus.* This herring in the Norwegian and Barents Seas lives in the waters of Atlantic origin which penetrate through the Faeroes-Shetland Bay between Iceland and Greenland, Iceland and the Faeroe Archipelago and near their confluence with the polar (Arctic) waters and the waters of continental original.

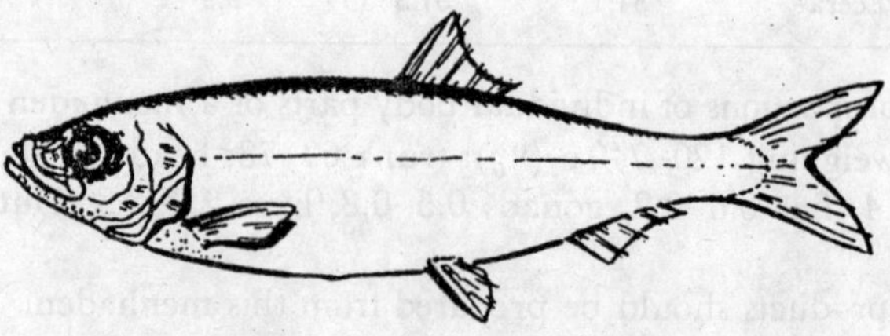

Fig. 58. Atlanto-Scandinavian herring—*Clupea harengus harengus.*

The seasonal changes in the fattiness of this herring significantly influence the weight proportions of individual body parts (Fig. 59). The weight of the trunk during intensive feeding increases and the fat content increases rapidly from 81.0 to 88.5%. Thereafter, it gradually decreases mainly because of the growth of ovaries and testes which constitute 18–20% weight toward the beginning of spawning. The weight of the viscera exclusive of gonads fluctuates

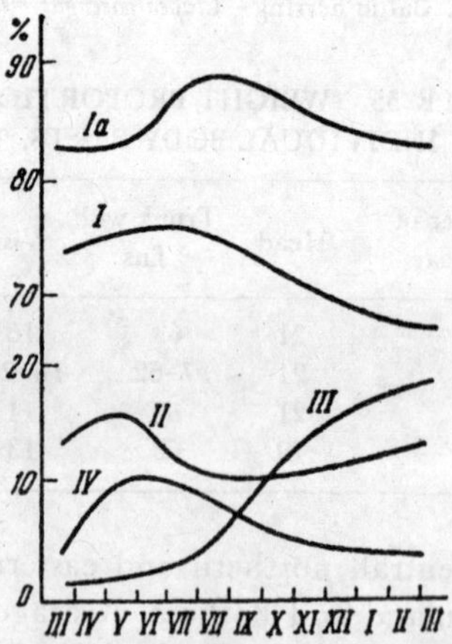

Fig. 59. Changes in the weight of individual body parts of Atlanto-Scandinavian herring in different periods of the year:

I—trunk as % of the whole fish; *Ia*—trunk as % of dressed fish; *II*—head; *III*—gonads; *IV*—viscera.

TABLE 56. CHEMICAL COMPOSITION OF MEAT, %

Place of catch	Month of catch	Moisture	Protein	Fat	Ash
Kaliningrad Bay	April	76–78	19–21	2.4–3.2	1.3–1.4
	May	76–77	18–20	3.4–4.7	1.3–1.5
Gulf of Riga	May–June	79–80	—	2.5–3	—
	August–January	73–75	—	9–10	—
Gulf of Finland	June	74.7	18	5.6	2.1
	September	72.4	17.0	8.6	1.8

from 2–10% and a little more depending upon how full of food the alimentary tract is. The head severed by a half circle cut constitutes 9–17%, depending upon the condition of the herring and the development of the gonads. In Fig. 59, the curve reflects the changes in the relative weight of the head which is an exact reflection of the curve for the trunk of dressed herring.

Weight proportions of the individual body parts of this herring in the period of feeding (June–August) as well as in the pre-spawning period (November–January) and post-spawning period (April–May) are given in Table 57.

During intensive feeding the alimentary canal of the Atlanto-Scandinavian herring contains highly active enzymes which destroy the abdominal cavity wall in salted herring. Therefore, at this time, semi-pressing or beheading of herring is carried out and they are rarely frozen. Studies in July–August showed that the output of the fish after beheading is 78.3%, after freezing 87.1% and the wastes are respectively 20.4 and 12.3%.

The chemical composition of herring meat is subject to large seasonal variations which are associated with changing conditions of feeding and spawning. Moreover, significant changes in age composition of herring have also been observed. Fat and moisture contents vary considerably; however, the periods of feeding and non-feeding influence protein content as well.

Figure 60 shows the changes in the average fat content of herring meat in different periods of the year. We note that fluctuations in the fat content of herring are highly variable. For example, in 30 specimens of this herring caught in July near the Faeroes Islands, the fat content varied from 4.7–24.0%.

The chemical composition of meat of two-year old herring from the Murmansk area and older fish from the Norwegian Sea are given in Table 58.

Simultaneous with the accumulation of fat in the fatty herring, the protein content increases in the nonfatty parts of the meat which makes the meat of herring denser; after boiling this makes the meat appear somewhat dry.

Accumulation of fat leads to an increase in the volume and weight of

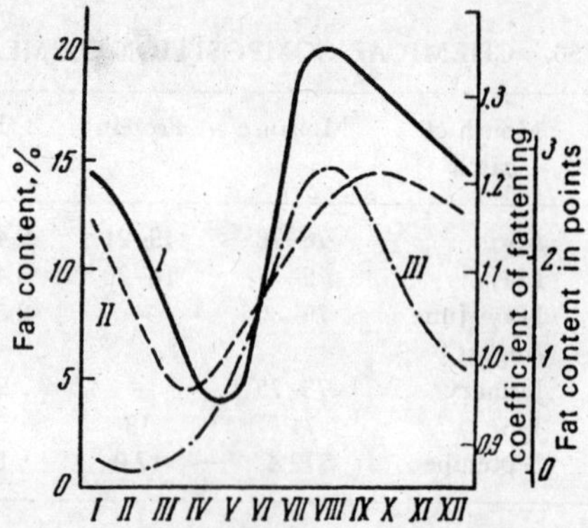

Fig. 60. Changes in average fat content of meat and the coefficient of condition of herring during the year: *I*—fat content in %; *II*—coefficient of condition; *III*—points of fat content.

TABLE 57. WEIGHT PROPORTIONS OF INDIVIDUAL BODY PARTS, %

Period of feeding	Stage of maturation	Head	Trunk	Meat with skin	Bones	Viscera		
						Entire	Gonads	Others
June–August	II–IV	12	75	67.5	7.3	13	4	9
November–January	IV–V	12	68	—	—	20	16	4
April–May	VI, VII–II and II	15–16	75	—	—	9	1	8

TABLE 58. CHEMICAL COMPOSITION OF MEAT, %

Age	Season	Moisture	Fat	Protein and ash
2 years	Spring	72.1–80.2 75.9	2.2–11.6 5.6	16.9–21.0 18.4
	Autumn	66.4–74.7 69.3	6.9–15.7 11.5	17.9–20.2 19.2
6 years	Spring	66.0–75.8 73.0	4.4–15.1 7.5	18.6–22.3 19.7
	Autumn	53.3–63.5 59.5	17.2–29.5 22.5	15.9–20.0 18.0

herring which is reflected in the increase of the coefficient of condition calculated by the formula $100/L^3G$, where G = weight of the fish in g, L = length in cm (cf. Fig. 60).

The index of fat content of the herring in Fig. 60 is shown in points determined on the basis of the occurrence of fat deposits in the viscera. As Fig. 60 indicates, this index can correctly indicate the fat content of the meat in the period of feeding. Comparison of the fat content of the meat in March–June by the point method gave the following results.

Fat content, points	0	1	2	3
Average fat content, %	6.0	8.6	13.4	17.5

Herring meat contains 0.1% calcium, 0.27% phosphorus, 0.06 mg% iodine. One g of herring meat contains 4 iu vitamin A, 15 iu vitamin D and 0.04–0.25 mg% vitamin B_1 and 0.24 mg% vitamin B_2.

The chemical composition of the head of herring changes relatively less (%): moisture 62–75, fat 7–13, protein 11–16, ash 4.2–7.5. These fluctuations depend upon how much meat is cut with the head.

The roe of herring contains little fat, from 0.5–3.5%. The protein content of the roe increases in the course of maturation, reaching 30–35% at stages IV–V of maturation. The quantity of phosphatides in the roe reaches 4.0–4.5%. The roe contains 10 mg% calcium, 200 mg% phosphorus and 0.015 mg% iodine.

The testes contain a much lower quantity of protein, from 17–24%: however, their fat content is higher, from 1.5–4.5%. A characteristic feature of their composition is the relatively high quantity of mineral matter, from 2.6–3.0%. The quantity of phosphatides in the testes is less than in the roe and amounts to 2.3–5%.

The composition of the viscera of herring undergoes more significant changes than the composition of the meat in the course of the year. For example, on an average the fat content changes from 4 to 35% (sometimes from 1 to 49%). The nitrogenous matter accounts for 9.5–19.0% (in the latter case, viscera and gonads were analyzed). The fat content of the viscera changes during different periods of the year in approximately the same manner as in the meat; however, accumulation and consumption of the fat in this case proceeds much more rapidly.

The liver contains 500–1,000 iu/g vitamin A and the oil from the trunk contains 150 iu/g. The roe contains 0.3–0.8 γ/g aneuria (B_1); the liver contains 5.9–7.4 γ/g riboflavin and the meat, 2.7–3.3 γ/g. The meat of this fish contains 39–63 γ/g nicotinic acid, roe 19–22, testes 26 and liver 50–65 γ/g. The meat also contains 95 γ/g pantothenic acid, in the liver 210, in the roe 140 and in testes 143γ/g. The canned products contain 16γ/g vitamin B_6. The meat has

0.14 γ/g vitamin B_{12}, the liver has 0.36, the roe 0.07–0.13, the testes 0.04–0.11 γ/g. The oil from the trunk contains 150 iu/g vitamin D.

Most of the herring caught is used for salting. The commercial quality of herring varies with the season. The best quality is that caught during July–October; it has larger quantities of fat and undergoes seasoning well after salting. When feeding ends (in November–January) enzyme activity decreases and such fish do not cure well in salting.

Fresh and frozen herring are in less demand for two reasons: herring fat is easily oxidized and imparts an unpleasant taste to the fish and, since herring meat contains relatively larger quantities of protein than common table fish its consistency after boiling or frying is dry. This defect is less expressed in lean herring.

Herring is a very valuable raw fish from which salted products, various delicious preserves and cold and hot smoked products can be prepared.

North Sea herring—*Clupea harengus harengus*. Herring fishing in the North Sea continues throughout the year. In early spring, February–April, herring are caught near the Viking Bank and south along the western slope of the Norwegian trough. In April–May, fishing starts in the northwestern parts of the sea. Gradually, the fishing effort shifts toward the south and comes to an end in December–January off the strait of La Manche.

In the northern part of the sea, the herring caught are usually 26–28 cm in length while in the southern part, their length is usually 22–26 cm. Food products made from this herring are high in quality.

The ratio between the length and weight of the North Sea herring caught in August was as follows:

Length, cm	16–18	20–24	26–28	29–30
Weight, g	120–168	124–172	164–180	190–215

TABLE 59. CHEMICAL COMPOSITION, %

Place of catch	Period of catch	Length, cm	Moisture	Fat	Protein	Ash
Northwestern part of the sea	May	23–24	59–63	18–22	17–19	0.8–1.5
	June	22–23	60–70	12–20	17–18	1–2
Southwestern part of the sea	July	22–24	56–58	15–26	16–19	1–1.5
	August	25	66	15	18	1.3

RATIONAL METHODS OF PROCESSING HERRING

Fatty herring

Mild salted herring in crates lined with polymer sheets; mild salted herring in cans of 1.5–3.0 kg capacity; sun cured products packed in small barrels; preserves with different kinds of sauces: apple, lemon, dill, wine, white and pink sauces; the medium salted herring in small barrels; products of cold and hot smoking.

Lean herring

Canned products: autolysates† and pastes: preserves with ferments and sauces; smoked dry herring; fish flour for food purposes; fish meal for animal feed.

Pacific herring—*Clupea harengus Palassi**. A few shoals of Pacific herring are known to extend to the northwestern Pacific Ocean; the Sakhalino-Hokkaido, Okhotsk, Gizhiginsk-Kamchatka and the Olyutorsk-Navarinsk shoals which form prespawning and feeding concentrations. Moreover, in the Barents Sea there are winter accumulations of Pacific herring.

This herring attains commercial size (length more than 21 cm) in the three to four years of its life. It grows significantly around eight to ten years, sometimes reaching 35–40 cm in length. However, because of large scale fishing such large size herring are rare. In the catches four to five year old herring 25–28 cm in length, weighing 200–300 g are more common.

The output of fillets with the skin may constitute 50–60% of the weight of the fish.

The proportion of nitrogenous compounds including proteins in herring meat reaches 15–19% and that of mineral matter 1.0–1.5%. The fat content is subject to the greatest changes and consequently the moisture content also. In lean spent herring the moisture content may reach 81% and the fat content 2%. In the period of feeding the fat content in the meat of more fatty herring may reach 33% and moisture about 50%.

Fat content in meat (%) of herring in different phases of its annual cycle

Spawning	5–12
Recently spent	2–7
Beginning of feeding	7–20
Feeding of fatty fish	14–35
Resting	20–7

The larger the size of the herring and the better the conditions of feeding, the earlier the fat content of its meat increases. During feeding, the fat first

† Hydrolysates—Editor.

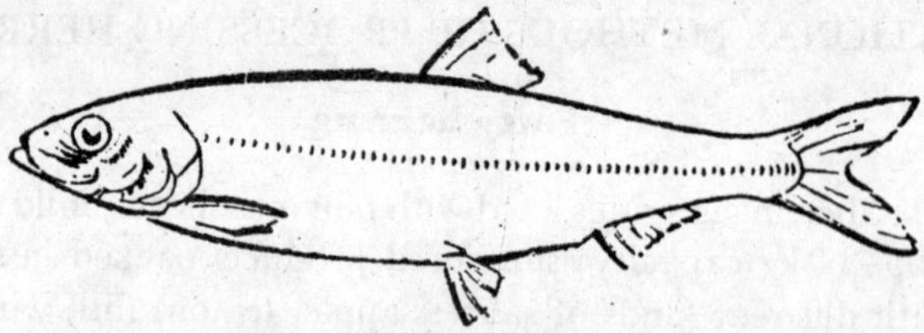

Fig. 61. Pacific herring—*Clupea harengus Palassi.**

TABLE 60. WEIGHT PROPORTIONS OF INDIVIDUAL BODY PARTS, %

Condition of herring	Head	Trunk	Gonads	Viscera
Spawning	12	64	21	3
Spent	15	75	2	8
Feeding	13	76	3	8
Resting	12	77	9	2

accumulates in the viscera and then in the meat. In spawners there is more fat in the head than in the meat; in feeding fish, fat content is low.

Upon catching, herring are sent for salting in barrels or in cans onboard ships at sea or in vats at the shore-based plants. In salting the whole herring, up to 16% of the weight is lost, but with intermittent salt curing with a salt content of 6–10% in the meat and subsequent storage in isotonic brine this weight loss is significantly reduced since the fish swell.

In salting in cans with 8.5% salt and 1.5% sugar, the weight of the finished mildly salted herring is equal to the weight of the raw material. Preserves are made from fillets of mildly salted herring with the addition of oil, different sauces and condiments.

Canned herrring products age quickly; for this reason and because of the tender consistency of the sterilized products their transportability is low. Canned preparations from dry or blanched fillets of frozen herring, packed in flat fancy shaped cans, have longer shelf life.

New Zealand sprat—*Clupea antipodum*. Distributed near New Zealand.

Weight of some specimens caught, 4–10 g, length 4–9 cm and depth 1.5–3.9 cm.

While dressing the fish, the proportion of the trunk is 64–70% of the total weight of the fish, head 25% and viscera 8%. Chemical composition of meat (%): moisture 75.9–78.4, fat 0.7–2.1, protein 18.5–20.3, ash 1.8–2.1.

Sprat can be used for frozen products with careful glazing. Fresh and frozen sprat can be processed by sun curing and salting in the tins. The finished product has a bitter taste.

Fig. 62. Van Hasselt's sprat—*Dussumieria hasseltii.**

Genus *Dussumieria*

Van Hasselt's sprat—*Dussumieria hasseltii**. Caught in the Gulfs of Tonkin and Siam. A small non-commercial fish 10–18.5 cm in length, weighing 6.2–9.0 g.

Chemical composition of meat of this fish caught in May–June (%): moisture 75.6–78.1, fat 0.7–1.4, protein 19.4–23.2, ash 1.2–2.3.

The canned products may be natural, blanched in oil and others characterized by unsatisfactory quality. Culinary preparations (boiled, fried, cutlets) are moderate in taste. This fish should be included in those raw fish more appropriately used for fish meal.

Genus *Etrumeus*

Round herring—*Etrumeus teres*. A small fish caught in small numbers along with other commercial fish from the western Central Atlantic.

TABLE 61. WEIGHT PROPORTIONS OF INDIVIDUAL BODY PARTS OF ROUND HERRING OF DIFFERENT REGIONS, %

Place of catch	Head	Trunk	Viscera	Fins
Gulf of Mexico	19	70	8	3
Georges Bank	15	66	16	2

TABLE 62. CHEMICAL COMPOSITION OF FISH FROM DIFFERENT REGIONS, %

Place of catch and body parts	Moisture	Protein	Fat	Ash
Gulf of Mexico				
meat	74.4	21.3	2.8	1.5
whole fish	74.6	20.5	2.8	2.1
Georges Bank				
meat	52–64	17.4	14–29	1.2
viscera	71.3	17.8	5.4	5.6

Size characteristics of round herring caught in June in the Gulf of Mexico: length 11–13 cm, maximum depth 2.0–2.5 cm, maximum thickness 1.5–2.0 cm, weight 20–30 g. Length of round herring caught in September at Georges Bank 15–19 cm and weight 60–110 g.

The meat of boiled fish is white, soft, juicy, tasty with a mild flavor. This fish is recommended for canning as "sardines in oil" and for various types of preserves.

Pacific round herring—*Etrumeus micropus*. Caught in an admixture of fish. Distributed in the Indian and Pacific Oceans and adjoining seas, Sea of Japan, East China Sea, near the Hawaiian Islands and sometimes reaching the coast of California. Usual length 18–20 cm, but sometimes 40 cm.

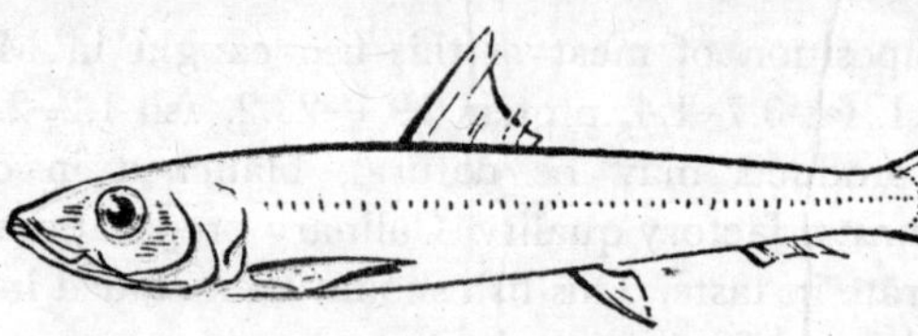

Fig. 63. Pacific round herring—*Etrumeus micropus*.

Length of fish caught in February–May 15.5–18.5 cm and average weight 65.0–85.8 g.

Weight proportions of individual body parts (%): head 12.0–18.4, bones and fins 7.6– 10.0, viscera 4.1–6.6, gonads 7.0. Output of fillets 60.7–66.0%.

Chemical composition of meat (%): moisture 72.3–76.5, fat 3.9–4.5, protein 18.5–22.0, ash 1.1–1.2.

This meat has satisfactory taste when cooked. It can be used for preparing culinary products.

Genus *Harengula*

Harengula sardine—*Harengula kenigsbergeri*. Found in the shelf region of the Indian Ocean. Length of fish caught in February–July 10.2–17.3 cm and weight 32.5–71.8 g.

Weight proportions of individual body parts (%): head 22.1, trunk with fins 68.5, viscera 7.6.

Chemical composition of meat (%): moisture 75.8–77.8, protein 19.0–20.6, fat 0.3–1.2, ash 1.7.

This is a very fleshy fish. At temperatures of – 30° C frozen sardine can be stored for a long time. It can be used for canned products such as "blanched fish in oil".

Genus *Ilisha*

Chinese herring—*Ilisha elongata*. Caught in small numbers with other fish. Length of fish caught in September in the Indian Ocean 19–23 cm, weight 90–190 g.

Weight proportions of individual body parts (%): head 20.4–21.8, trunk with fins 70.0–71.3, viscera 5.6–9.6, skin 2.7, bones 6.4. Output of fillets (without skin) 5[illegible].6%.

Chemical composition of meat (%): moisture 77.2–78.6, fat 0.9–2.0, protein 19.5–22.0, ash 1.7.

When boiled, this meat is good in taste. The fish is an excellent raw material for preparing cooked products. Reports in literature indicate that in certain periods the gonads and the membranes lining the abdominal cavity can be poisonous.

Indian herring—*Ilisha indica*. Caught in small numbers with other fish. In catches in August from the Persian Gulf, most fish are about 19 cm (17.8–19.5) in length and weigh on an average 115 g (100–125) with maximum depth 6 cm (5.5–6.4) and maximum thickness 2 cm (1.9–2.2).

Evisceration of this fish upon catching is essential since its roe is poisonous and the guts are infested with helminths.

Weight proportions of individual body parts (%): head on an average 17.8, trunk 68.7 (62.5–72.7), including meat with skin 57.4 and bones 11.3, scales 0.4, fins 2.6, entire viscera 6.9.

After cooking, the meat becomes white, soft and tasty. There are many fine, pointed bones in the meat. The soup is rich, fatty, tasty with a large quantity of extracted substances. Fried meat is light in color, soft and good in taste.

This fish should be canned in oil.

TABLE 63. CHEMICAL COMPOSITION, %

Body parts	Moisture	Protein	Fat	Ash
Meat	75.9	20.2	2.3	1.6
Head, fins, guts	68.1	17.0	7.7	6.6

Genus *Opisthonema**

Atlantic thread herring—*Opisthonema** *oglinum*. Caught with commercial fish in the western Central Atlantic. Usual sizes: length 22–28 cm, weight 150–220 g.

Weight proportions of individual body parts (%): trunk 68.6–70.9, head 18.8–21.0, viscera 4.3–9.0, scales 1.0, fins 1.4.

Chemical composition of meat of fish caught in January (%): moisture 70.3, fat 5.7, protein 22.5, ash 1.5.

After cooking, the meat has a pleasant but poorly expressed taste and somewhat loose consistency. There are many fine bones which make it difficult to eat. This species is characterized by compactly arranged scales. A satisfactory product is obtained after salting in which case the bones are less noticeable.

After removing the scales, this fish should be used for canned preparations in oil.

Genus *Pellona*

Pellona herring—*Pellona ditchella.* A commercial fish. Distributed in the central Indian and Pacific Oceans.

In catches off the coast of Pakistan, most fish are about 30 cm (24–32) in length and weigh around 300 g (160–385). In catches from the western shelf of India, most fish are about 20 cm (19–28) in length and weigh about 200 g (126–263), with maximum depth 8–10 cm and maximum thickness 2.4–3.0 cm.

Weight proportions of individual body parts of this herring (%) fluctuate significantly depending upon the area of catch.

In fish caught in October off the coast of Pakistan the head, separated by an oblique cut, on an average constituted 20.5% (19.3–22.1), trunk 70.2%

Fig. 64. Pellona herring—*Pellona ditchella.*

TABLE 64. CHEMICAL COMPOSITION, %

Body parts	Moisture	Fat	Protein	Ash
Meat of fish from Pakistan coast	76.7	1.3	20.3	1.5
Meat of fish from western shelf of India	68.5	6.9	22.4	1.7
Inedible parts of body (offal) of fish from Pakistan coast	66.9	5.6	19.1	7.9

(66.2–71.9), meat on an average 52.2%, skin 2.9%, bones 15.1% (11.3–21.6), scales 1.9%, fins 1.3%, entire viscera 5.1% (4.4–6.1).

The West Indies shelf catch of herrings in December shows that the head if separated with an inclined cut on the average constitutes 14.9% (11.3–18.7) of the body weight. The figures for other organs are: trunk 68.5% (61.7–74.0), scales 2.6%, fins 1.4% and viscera 11.8% (5.3–21.5).

The meat has many fine pointed bones. After cooking, it is loose and tasteless and on chewing gives a sound under the teeth like crunching plastic. The soup prepared from this fish is colorless and tasteless. In salted form also the meat has low taste qualities. This fish should be used primarily for making fish meal.

Genus *Pomolobus*

Alewife—*Pomolobus mediocris*. Distributed in coastal regions of the northwestern Atlantic, at Georges Bank and southward. Usual length of the fish is 26–33 cm.

Size characteristics: length 18–28 cm, maximum depth 4.5–7.0 cm, maximum thickness 2–3 cm, weight 76–244 g.

Weight proportions of individual body parts (%): trunk 73–76, head 13–14, viscera 10–13, bones 4.7–5.2, fins 1.2–1.5, scales 2.1–2.4. Output of fillets 63–64%.

The taste qualities of this fish, unlike the Atlantic herring, are poor. The salted alewife has a specific taste which resembles the smell of fried sunflower seeds; this lowers its quality. This fish should be used for salting or freezing

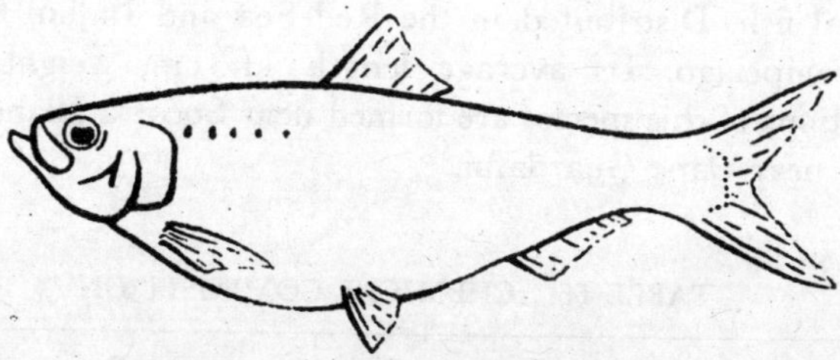

Fig. 65. Alewife—*Pomolobus mediocris.*

TABLE 65. CHEMICAL COMPOSITION, %

Body parts	Moisture	Fat	Protein	Ash
Meat	60.0–72.4	7.8–21.0	17.2–20.0	1.2–1.6
Head	68.7–69.2	9–9.2	13.8–14.2	5.8–6.4
Bones	61–61.8	10–11.1	17.5–18.0	8.3–8.5
Viscera	61.5–65.8	17.5–22.5	13.5–14.1	1.3–1.5

followed by canning and preparation of smoked products. The alewife with low fat content can be better utilized for fish meal.

Genus *Sardinella*

This genus includes 16 species living in tropical waters, rarely in the subtropical Atlantic and Indian Oceans, in the western Pacific Ocean and in the adjoining seas.

Oil sardine—*Sardinella longiceps*. This is a large sardine, an important constituent of fishing. It is found in large concentrations in the northwestern Indian Ocean. It is mainly distributed along the western coasts of the Indian Ocean, rarely along the eastern coast near Sri Lanka from the Andaman Islands to the Malay archipelago. The average length of the oil sardine is 18–20 cm.

In catches in October in the Gulf of Aden, there was a predominance of individuals about 12 cm (10.6–12.2) in length, weighing about 20 g (16–27), with maximum depth on an average 2.3 cm (2.1–2.5) and maximum thickness 1.3 cm (1.0–1.5).

Weight proportions of individual body parts (%): head 23.7 (18.7–27.4), trunk 63.0 (58.5–68.0), fins 0.9 (0.7–1.1), entire viscera 11.6 (7.2–15.9).

Chemical composition of meat (%): moisture 70.7, fat 7.0, protein 19.0, ash 1.3.

The meat of the oil sardine has a sour taste, but to a lesser extent than in gilt sardine, *Sardinella aurita*. This fish should be used for processing canned products in oil and smoked preparations.

Fringe scale sardine—*Sardinella fimbriata*. Also constitutes an important commercial fish. Distributed in the Red Sea and Indian Ocean up to the Malay archipelago. Its average length 11.3 cm, weight 25.4 g. Largest concentrations of this species are formed near Socotra Islands, in the Gulf of Aden and near Cape Guardafui.

TABLE 66. CHEMICAL COMPOSITION, %

Body parts	Moisture	Protein	Fat	Ash
Meat	$\frac{76.2–77.2}{76.7}$	$\frac{0.2–0.4}{0.3}$	$\frac{20.1–20.6}{20.4}$	$\frac{1.4–1.5}{1.5}$
Inedible parts (offal)	74.0	0.9	16.0	8.0

In catches in March from Socotra Islands, most fish are 11.5 cm (11.0–12.9) in length, weighing about 20 g, with maximum depth 2.5 cm (2.1–3.0) and maximum thickness about 1.2 cm (1.0–1.5).

In September in the Gulf of Aden the fish caught were 13 cm (12–14) in length and weighed about 30 g (24–36).

Weight proportions of individual body parts (%): head 18.0 (16.5–19.6), trunk 68.2 (67.7–68.6), including meat and skin 57.9 (57.0–58.8), bones 9.8, scales 1.2, fins 1.1, entire viscera 10.5 (7.8–13.2).

In cooked form, the meat is gray, somewhat dry; after frying, it is soft and juicy. It has the mild sourish taste typical of sardines. This fish should be canned in oil.

Allex sardine—*Sardinella allexia.* Not an important commercial fish. Found in the Gulfs of Tonkin and Siam. Length 13.0–14.5 cm, weight 32–47 g. Trunk 64.2% of the body weight. Chemical composition of meat (%): moisture 74.2, fat 2.2, protein and mineral matter 23.6.

Tembang—*Sardinella jusseiu**. Found in the Gulfs of Siam and Tonkin. Not important in fishing. Generally encountered in localized shoals in a restricted area of distribution. Length of fish caught in May–December 9.0–19.7 cm, weight 10–97 g.

Weight proportions of individual body parts (%): head 20.0–20.1, trunk 64.1–67.3, bones 12.6–13.3, fins 2.7–3.2, gonads 2.6–4.8, viscera 4.3–4.7. Output of fillets with skin 50.8–54.7%.

Chemical composition of meat (%): moisture 74.4–76.9, fat 0.4–0.8, protein and mineral matter (variable) 22.7–24.9.

This is a valuable raw material for canning.

Sardinella clupeoides. No commercial significance. Found in the Gulf of Tonkin. Length 16.0–18.5 cm and weight 75–107 g.

Weight proportions of individual body parts (%): head 18.3, trunk 70.0, viscera 5.4, bones and fins 17.7, liver 1.7. Output of fillets with skin 54.9%.

Chemical composition of meat (%): moisture 74.5, fat 0.1, protein and mineral matter (variable) 23.4.

Gilt sardine—*Sardinella aurita.* A shoaling pelagic fish. Found all along the western coast of Africa from the Straits of Gibraltar to the southern tip of the continent.

Usual length of the fish 15–27 cm, but individuals up to 35 cm have been caught. Size characteristics of the fish: length 11–27 cm, maximum depth 2.5–6 cm, maximum thickness 1.0–3.5 cm, weight 20–440 g.

Weight proportions of individual body parts (%): trunk 62–79, head 15–28, viscera 6–19, meat 43–64, skin 1–3, bones 4–8, fins 0.5–2.0, scales 1.4.

Chemical composition of meat (%): moisture 58–77, fat 1.2–19.7, protein 17.1–22.2, ash 0.7–2.6.

The gilt sardine from the Dakar region fluctuates greatly in fat content not only in different periods of catch but also in the same catch, since different shoals are represented in the catch. The maximum fat content in this fish has been recorded in June–July and the minimum in March.

Mineral content of the meat of this fish is (mg%): potassium 350–420, calcium 75–85, magnesium 55–56, phosphorus 265–285, iron 4.1.

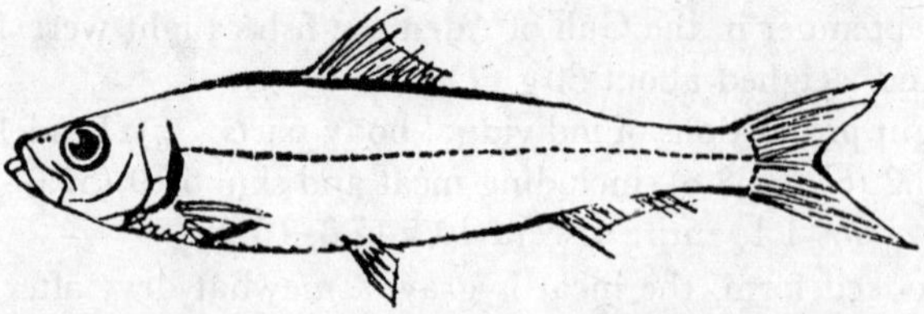

Fig. 66. Gilt sardine—*Sardinella aurita.*

TABLE 67. CHEMICAL COMPOSITION, %

Body parts	Moisture	Fat	Protein	Ash
Heads	66–73	1.1–3.8	11–14.5	8–12
Bones	42–72	2.1–9.0	19–25	13–32
Fins	23–48	0.2–0.9	20–23	30–50
Skin	72–73	1.3–2.8	18–24	1.5–2
Scales	28–33	0.9–1.6	31–33	34–36
Viscera	74–80	1.3–5.8	13–14	2–2.5
Whole fish	60–73	2.4–18.3	16.6–20	2–4.2

The meat of this fish is soft and juicy with a mild acid taste. This is a good raw material for preparing high quality canned products such as "Atlantic sardines in oil".

Spanish sardine—*Sardinella anchovia.* Distributed in the western Atlantic Ocean from Cape Cod to Rio de Janeiro. Commercial importance in the western Central Atlantic. Fish usually not longer than 15–20 cm.

Size characteristics of fish caught in June in the Gulf of Mexico near the Yucatan peninsula: standard length 15.5–18.0 cm, thickness 2.0–2.5 cm, weight 55–80 g.

TABLE 68. CHEMICAL COMPOSITION, %

Body parts	Moisture	Fat	Protein	Ash
Meat	73.7	3.4	21.8	1.1
Whole fish	69.4	8.5	18.3	3.8

Weight proportions of individual body parts (%): trunk 68, head 18, viscera 12, fins 2.

After cooking, the meat is soft and tasty. Delicious canned products, "sardines in oil", are prepared from this fish.

Flat sardine—*Sardinella eba.* Distinguished from the Spanish sardine by a deeper and flatter body. Little commercial value. Flat sardine caught in May: total length 32 cm, standard length 27 cm and weight 360 g.

Weight proportions of individual body parts (%): head 19, trunk with fins 70, viscera 11. Chemical composition of meat (%): moisture 75, fat 0.7, protein 22.7, ash 1.6. The flat sardine should be used for canning.

Genus *Sardinops*

This genus includes fish of one species, with a few subspecies which live in the subtropical Pacific and Indian Oceans and some parts of the Atlantic Ocean, along the coasts of North and South America, East Asia, New Zealand, Australia and the southern tip of Africa.

Australian sardine—*Sardinops neopilchardus.* Distributed around New Zealand, near the southeastern and western coasts of Australia and northern coasts of Tasmania. Attains length of 28 cm but in commercial catches most fish are 9–17 cm in length and weigh 7–40 g. Length of fish caught in May–January: 9.5–17 cm and weigh 7.5–30 g.

Weight proportions of individual body parts (%): trunk 65.0–73.3, head 19.2–24.3, viscera 3.9–6.4, fins 1.5–1.7.

Chemical composition of meat (%): moisture 75.4–80.6, protein 17.5–21.7, fat 0.3–1.4, ash 1.5–2.4.

Fresh and frozen sardines are best used in canned products labelled "blanched fish in oil" and "smoked fish in oil". Canned preparations from frozen sardines have somewhat dry meat. Preserved products mature rapidly (one month after preparation) and after two months they are overmature.

The Australian sardine should be supplied from the place of catch for

Fig. 67. Australian sardine—*Sardinops neopilchardus.*

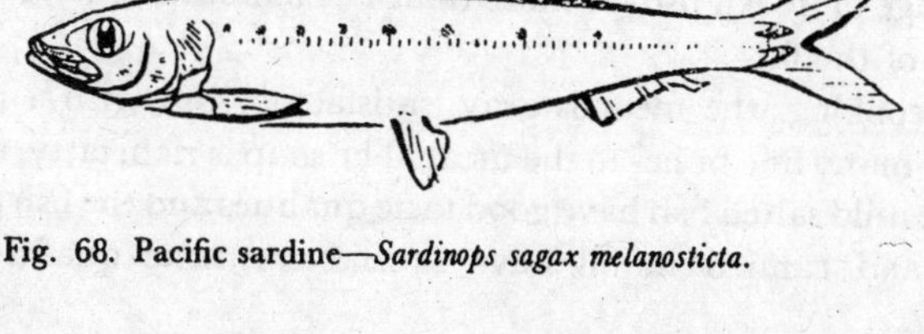

Fig. 68. Pacific sardine—*Sardinops sagax melanosticta.*

canning or freezing with essential glazing. The storage period for frozen, glazed sardines before canning is 1–1.5 months at −18°C.

Pacific sardine—*Sardinops sagax melanosticta*. Found in the western Pacific Ocean and in the East China Sea and Sea of Japan. About 25 years ago the Pacific sardine was seen in the Okhotsk and Bering Seas up to the coasts of Kamchatka and Komander Islands. Now it rarely enters from the south into the Primorye region or north of it; therefore, it has no commercial significance in this area.

Length of sexually mature sardine 17–22 cm, weight 65–105 g. Average output of trunks 57%. At the Primorye coast the fat content in the meat of this sardine was 15–20%, sometimes even 33%. The quantity of protein in the meat was 18–20%.

This is a good raw material for canning in oil as well as for preparing mild salted products.

California sardine—*Sardinops caerulea*. Found from the coast of southern Alaska to the Gulf of California. This species has no commercial significance presently. Weight proportions of individual body parts on the basis of 20 specimens caught in February–May (%): trunk 65.1, head 19.7, viscera 13.0, bones 8.1, fins 2.2, skin 3.4.

Chemical composition of meat (%): moisture 75.8–80.6, protein 17.5–21.7, fat 0.3–0.8, ash 1.6–1.9.

Culinary preparations made from the frozen fish are low grade. Salted fish had mediocre flavor because of a bitter taste. The products of hot smoking from the whole frozen California sardine, in spite of marked bitterness of the viscera, received satisfactory grading.

The technique for preparing conserves using fresh California sardine should be developed.

South African sardine—*Sardinops ocellata*. An important constituent in commercial fishing. Distributed in the Atlantic Ocean south of Angola, along the coasts of southwest Africa and the Republic of South Africa.

Usual length of the fish 17–24 cm, weight 90–165 g, maximum depth 3–4 cm, maximum thickness 2.0–2.5 cm.

Weight proportions of individual body parts (%): trunk 61–66, head 19–26, viscera 8–16, fins 1–2, scales 0.6.

Subcutaneous yellowing is observed in the frozen sardines after five days of storage as a result of the release of carotenoids. Frozen sardines with subcutaneous yellowing after storage for 1–1.5 months are included in grade 1. After storage of frozen fish for more than 1–1.5 months, the yellowing is due to oxidation of the fat.

After cooking, the meat is gray, satisfactory but with a mild acid taste. There are many fine bones in the meat. The soup is rich, fatty, transparent and tasty. The mild salted fish have good taste qualities and the fish prepared by hot smoking and canned in oil have satisfactory taste qualities. This fish is

TABLE 69. CHEMICAL COMPOSITION, %

Body parts	Moisture	Protein	Fat	Ash
Meat	66.7–72.6	18.6–21.4	4.2–10.3	1.5–1.8
Heads, bones, fins	66–70	12.7–14.0	9.6–11.6	5.1–7.2
Viscera	69.8–73.2	11.4–15.4	5.8–11.5	2.4–2.7
Whole fish	63.6	21.2	6	9.2

recommended for preparing mild salted, smoked and canned products.

Genus *Sardina*

This genus includes one species with two subspecies distributed in the subtropical Atlantic Ocean, along the coasts of Europe and North Africa and in the Mediterranean and its adjoining seas.

Morocco sardine—*Sardina pilchardus*. Great commercial significance in the eastern Central Atlantic from the Canary Islands to the La Manche Strait. Usual length of fish in catches 13–18 cm, sometimes 25 cm.

Weight proportions of individual body parts (%): trunk with fins 66, head 20, viscera 12, fins 2.

Delicious canned products are prepared from pilchards.

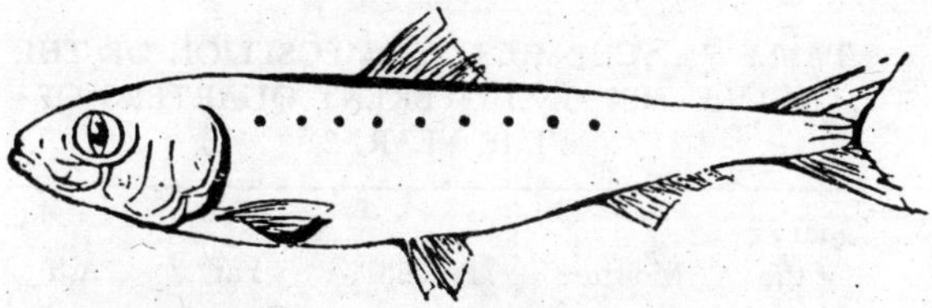

Fig. 69. Morocco sardine—*Sardina pilchardus*.

TABLE 70. CHEMICAL COMPOSITION, %

Body parts	Moisture	Fat	Protein	Ash
Meat	67.9	8.9	20.2	2.4
Head, bones	72.0	7.5	12.7	6.4
Viscera	70.2	13.3	12.5	2.2

Flat sardine—*Sardina pastera*. Found in the Great Australian Bight. No commercial significance at present. Fish with large compactly set scales; length 14.0–17.5 cm and weight 32.5–75.0 g.

Weight proportions of individual body parts (%): head 16.9–22.0, scales 3.2–6.3, trunk 67.0–72.8, viscera 7.6, fins 1.7.

Chemical composition of meat (%): moisture 69.7–76.4, fat 0.4–3.1, protein 21.7–22.2, ash 1.9–2.3.

This is a small and fairly fleshy fish. It can be used for canned products "blanched in oil"; however, in this case the process of removing the scales from the fish is quite difficult. During sun curing and salting flat sardine mature slowly, but after maturation they have soft, tender and fatty consistency and good taste. This fish should be frozen and later used for canning or preparing sun cured products.

Genus *Sprattus*

Baltic sprat—*Sprattus sprattus balticus.* Distributed in the Baltic Sea and its gulfs. Usual length 9–16 cm.

Weight proportions of individual body parts (%): trunk with fins 76–80, head 12–15, viscera 6–10.

The fattiest sprat are found in the autumn–winter period.

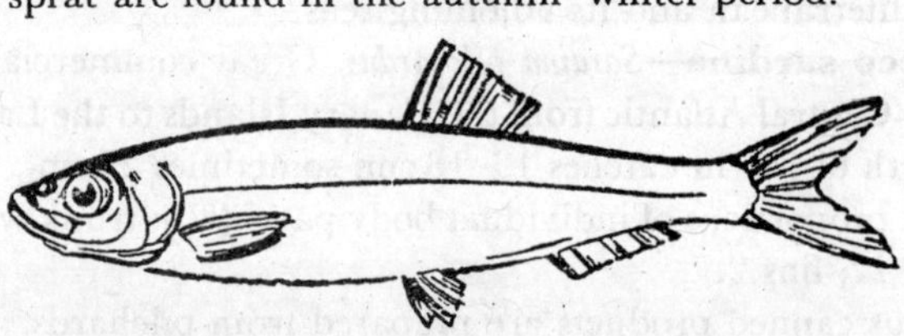

Fig. 70. Baltic sprat—*Sprattus sprattus balticus.*

TABLE 71. CHEMICAL COMPOSITION OF THE WHOLE FISH IN DIFFERENT QUARTERS OF THE YEAR, %

Quarter of the year	Moisture	Protein	Fat	Ash
I	71–73	13–14	10–13	1.9–2
II	75–79	13–14	4.5–8	1.9–2.1
III	70–78	13.8–15.2	5.8–12.5	1.9–2.2
IV	70–74	13–14.6	12–13.7	2–2.2

The Baltic sprat is used for high quality canned products, "sprat in oil" and for delicious sun cured preserves.

Family CONGRIDAE—CONGER EELS

The conger eels live in the tropical and temperate waters of the Atlantic, Indian and Pacific Oceans. They have a long body with no scales. They attain large size.

Three genera of eel belonging to this family are known from the Indian and Atlantic Oceans: *Conger, Cerisona, Uroconger*. The most widespread eel belong to the genus *Conger*.

Conger (eel)—*Conger conger*. Distributed in the Indian and North Atlantic Oceans; also found in the Mediterranean, Black and Baltic Seas. A large fish, more than 3 m in length and weighing 65 kg.

After cooking, the meat is white and soft with many large and small bones. Its taste is typical of table fish. In taste and food value it is greatly inferior to the fresh water eel. The soup has a pleasant taste. After frying, the meat is very tasty. In natural canned condition and after canning in oil the meat is loose, slippery and watery. Canned in tomato sauce, the meat and sauce are dark in color.

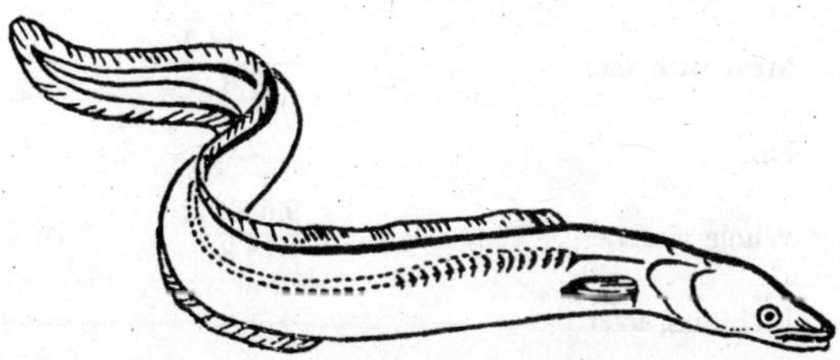

Fig. 71. Conger (eel)—*Conger conger*.

TABLE 72. SIZE CHARACTERISTICS OF CONGER (EEL)

Parameters	Indian Ocean off Diu	Strait of La Manche
Length, cm	109–141 / 125	120
Maximum depth, cm	8–14 / 11.5	14
Maximum body thickness, cm	6–10 / 8	10
Weight, g	2,290–6,239 / 4,300	5,300

Eel meat is very popular in the coastal regions of South America. It is a table fish; also recommended for hot smoked products.

We note that the blood and mucus (slime) of eel are poisonous; therefore, it is essential that upon catching the fish are thoroughly washed and bled.

TABLE 73. WEIGHT PROPORTIONS OF INDIVIDUAL BODY PARTS, %

Body parts	Indian Ocean off Diu	Strait of La Manche
Head	10.5–14.8 / 11.9	11.0
including meat of occiput and nape	2.4	—
Trunk	68.4–72.1 / 69.3	73.0
Bones	11.1–13.3 / 11.5	—
Fillet	—	58
Meat with skin	57.3–58.7 / 57.8	—
Fins	2.6	—
Whole viscera	9.6–17.2 / 15.8	16
Including liver	1.0	—

TABLE 74. CHEMICAL COMPOSITION OF MEAT, %

Parameters	Indian Ocean off Diu	Strait of La Manche
Moisture	78.6	76.4
Fat	0.1	3.8
Ash	1.7	1.2

Family CYCLOPTERIDAE—LUMPFISHES

These are very widely distributed marine fish living in the Pacific as well as Atlantic Oceans. The members of this family are also found in the Arctic and Antarctic waters. Lumpfish is a commercially important member of this family.

Lumpsucker—*Cyclopterus lumpus*. Distributed all over the northern Atlantic Ocean and in the Barents and White Seas.

Attains length of 60 cm and weight 5.5 kg. Female fish are larger than males. There is generally a predominance of females 30–35 cm in length and males 15–25 cm in length in the catches.

Weight proportions of individual body parts of the lumpsucker are

TABLE 75. WEIGHT PROPORTION OF INDIVIDUAL BODY PARTS, %

Date of catch	Weight, g	Length, cm	Stage of maturation	Head	Trunk	Meat	Skin	Bones	Fins	Crest	Viscera			
											whole	including		
												liver	roe	other
April 15	4132	42	III	20.9	50.1	24.2	21.4	4.5	1.3	9.6	16.9	4.0	3.1	9.8
April 15	3100	40	III–IV	14.5	40.1	20.2	14.8	5.2	—	7.3	34.6	2.8	22.7	9.1
November 18	2355	36	?	13.8	56.7	31.6	19.3	5.8	1.4	6.4	20.0	5.4	3.4	11.1

Fig. 72. Lumpsucker—*Cyclopterus lumpus*.

characterized by low content of meat (25%) and larger quantity of thick skin 18.5% of the whole fish. The head accounts for 14–21%. The roe in stages III and IV of maturation reaches 35% of the weight of whole fish. The lumpsucker has a fairly large liver, 3–5%. The results of the three analyses of lumpsucker females caught in the Barents Sea are presented in Table 75.

The edible parts of the lumpsucker account for 31–46%. The chemical

TABLE 76. CHEMICAL COMPOSITION, %

Body parts	Moisture	Protein	Fat	Ash
Meat	75.4–88.2 / 81.8	5.1–9.0 / 8.1	4.6–14.9 / 9.5	0.3–1.0 / 0.5
Meat, bones	71.8–81.9 / 76.9	7.4–10.7 / 8.6	9.6–17.7 / 13.6	0.7–1.1 / 0.8
Liver	52.4–69.8 / 62.2	7.6–12.5 / 9.9	14.4–37.3 / 24.6	1.2–1.5 / 1.3
Roe	82.5–84.2 / 83.2	5.9–9.0 / 8.1	5.5–7.8 / 5.9	0.7–1.0 / 0.8
Testes	— / 88.4	— / 8.1	— / 1.3	— / 1.4
Skin	88.6–91.7 / 90.2	7.6–9.4 / 8.3	0.3–0.5 / 0.4	0.4–1.2 / 0.8
Crest	84.8–92.3 / 93.6	5.5–6.1 / 5.8	0.0–0.2 / 0.1	0.3–1.0 / 0.7
Bones	88.3–90.1 / 89.2	— / 5.9	3.0–3.6 / 3.3	— / 1.5
Head	86.9–90.0 / 88.5	5.8–8.4 / 7.9	0.4–2.3 / 1.7	1.1–1.4 / 1.2
Fins	84.7–89.2 / 86.9	7.4–9.4 / 8.6	0.2–0.5 / 0.4	1.3–4.0 / 2.6
Viscera	90.6–92.8 / 91.5	3.0–4.8 / 3.9	0.6–1.4 / 1.0	0.6–1.9 / 1.3

composition of the lumpsucker is unique: all the parts of its body contain a lot of water and large quantities of fat. Males are a little fattier than females and have denser meat. The most significant feature of the lumpsucker is the low protein content: even the roe contains only a total of 6–9% protein. Also noteworthy is the low mineral content of the bones due to which it can be easily cut by a knife.

The lumpsucker has low commercial value. To remove the excessive water and to compress the meat before preparing it as food, it is salted and dried. The roe is the most valuable and tasty part of the lumpsucker.

Family CORYPHAENIDAE—DOLPHINS

Dorado—*Coryphaena hippurus*. Lives in the tropical zones of the Atlantic, Indian and Pacific Oceans. In the Atlantic Ocean, found in large numbers in the Gulf Stream zone. A large fish attaining 72–154 cm in length and weighing 1.8–20.1 kg.

The dorados caught in different zones of the world's oceans differ in weight of individual body parts and chemical composition.

Length of dorados caught in September–December in the Pacific Ocean 100–105 cm, weight 8.9–9.5 kg. Weight proportions of individual body parts (%): head 15.2–23.6, trunk 63.8–70.0, fillets without skin 54.4–56.0, viscera 5.9–10.0, bones 5.4–10.0, fins 0.4–4.0. Chemical composition of dorado meat (%): moisture 70.7–76.2, protein 18.4–24.5, fat 0.2–3.2, ash 1.4–1.6.

Length of dorados caught in December–March in the Indian Ocean 54 cm (average of the lengths of five dorados), weight 1.8 kg. Weight proportions of individual body parts (%): head 24.3, trunk 63.6, scales 0.6, fins 2.6, viscera 8.4. Chemical composition of meat (%): moisture 74.6–76.1, protein 18.8–19.8, fat 1.0–3.2, ash 1.5.

Length of dorados caught in May–June in the western Central Atlantic 72–154 cm, weight 1.8–20.1 kg.

Weight proportions of individual body parts (%): head 12.9–22.2, trunk with fins 66.7–77.1, viscera 8.5–13.9.

Chemical composition of meat (%): moisture 75.0–76.0, protein 21.7–22.3, fat 0.5, ash 1.5.

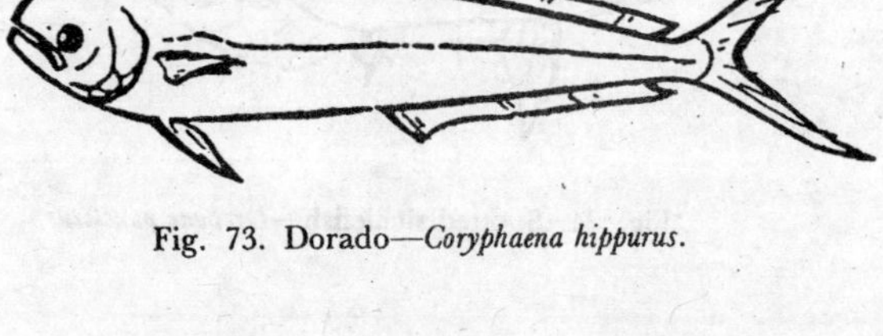

Fig. 73. Dorado—*Coryphaena hippurus*.

The meat of the dorado is characterized by specific taste qualities. The dorado belongs to the group of table fish.

After catching, it should be frozen for subsequent use for culinary purposes.

Family DREPANIDAE—SICKLEFISHES OR DREPANES

These fish live in the tropical parts of the Indian and Pacific Oceans and they have extended far south; also in the Atlantic Ocean along the western coast of Africa. These are considered table fish. Drepanidae includes one genus with two species.

African sicklefish—*Drepane africana.* Caught in small numbers along with other fish in commercial catches in the southeast Atlantic.

A demersal fish. Length of the African sicklefish caught in April near Cape Frio 21–25 cm, maximum depth 21–23 cm, maximum thickness 4–5 cm and weight 0.6–0.9 kg.

Weight proportions of individual body parts (%): trunk with fins 63, head 30, viscera 7.

Chemical composition of meat (%): moisture 76.7–79.5, fat 0.3–1.8, protein 18.1–19.1, ash 1.4–1.9.

The meat is light in color. The meat of fried fish has a typical and specific taste. The taste qualities of the fish are satisfactory. After culinary processing the meat becomes somewhat dry. This fish is recommended for marketing after freezing.

Spotted sicklefish—*Drepane punctatus.* A fish of some commercial significance. Attains length of 40 cm. In catches from the Gulf of Aden (in May, June, August, September), near Cape Fartak (in May) and in the Arabian

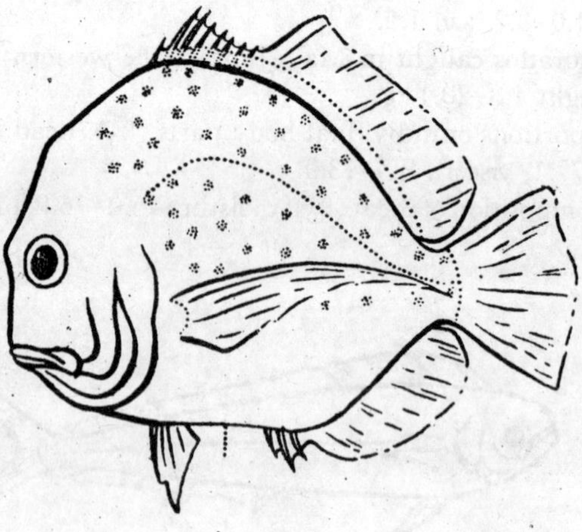

Fig. 74. Spotted sicklefish—*Drepane punctatus.*

TABLE 77. CHEMICAL COMPOSITION, %

Body parts	Moisture	Fat	Protein	Ash
Meat	74.0–79.5 / 77.5	0.4–2.2 / 1.1	18.2–22.5 / 19.4	1.3
Inedible parts (offal)	69.0	4.4	18.0	8.7

Sea (in January), most spotted sicklefish were 18–32 cm long and weighed 280–1,400 g.

Weight proportions of individual body parts showing considerable fluctuations (%): head separated by oblique cut 25.2 (18.5–30.6), trunk 62.2 (57.5–70.0), meat 44.0 (35.7–58.8), skin 4.5 (3.8–5.2), bones 15.4 (11.2–17.9), scales 2.1 (1.3–2.5), fins 2.7 (2.4–3.7), whole viscera 5.4 (3.7–6.1).

After boiling and frying the meat is white, varying in consistency from soft to dense. The taste is pleasant, typical of a good table fish. The soup is mediocre in taste. The meat of fish caught along the coast of Australia, in spite of higher fat content, was characterized by low taste qualities and unpleasant smell.

The fish caught along the coast of Australia are recommended for fish meal and those from the Gulf of Aden and from near Cape Fartak should be used as table fish for second courses and for hot smoked products.

Family DACTYLOPTERIDAE—FLYING GURNARDS

Common flying gurnard—*Dactylopterus volitans*. Attains length of 90 cm. Weight proportions of individual body parts of the common flying gurnard 23–30 cm in length, with maximum depth 3.5–5 cm, maximum thickness 4.5–6 cm, weight 200–500 g (%): trunk with fins 59, head 31, viscera 10. Output of fillets is 40%.

The meat of fried fish is hard, fibrous, with a noticeable specific taste. The meat is edible but its taste qualities are low. The fish should be used for fish meal.

TABLE 78. CHEMICAL COMPOSITION, %

Body parts	Moisture	Fat	Protein	Ash
Meat	72.2	5.5	20.9	1.4
Head, bones, viscera	56.1	11.2	19.5	13.2

Family ENGRAULIDAE—ANCHOVIES

This family includes 17 genera and about 40 species of fish which are widely distributed in the tropical and temperate seas of the northern and southern hemispheres, mostly in the southern parts of the Pacific Ocean.

Genus *Engraulis*

European anchovy—*Engraulis encrasicholus*. Lives in the waters of the Atlantic Ocean, widely distributed in the Black and Azov Seas. Large concentrations of anchovy found near Senegal, Cape Palmas, in the Gulf of Guinea and near southwest Africa.

Fig. 75. European anchovy—*Engraulis encrasicholus*.

Usual length of this anchovy 12–15 cm, sometimes 20 cm.

Weight proportions of individual body parts of this anchovy caught in January–February near South America (%): head 19.0, meat 61.2, trunk 66.0, skin, bones and fins 5.2 and viscera 12.2.

TABLE 79. CHEMICAL COMPOSITION, %

Body parts	Moisture	Fat	Protein	Ash
Meat	72.7	3.3	23.0	2.0
Whole fish	73.7	2.6	20.2	3.5

The meat of this anchovy has good taste. It can be used for salt cured products as well as conserves.

Argentine anchovy—*Engraulis anchoita*. Found along the Atlantic coast of South America. Length of fish caught in April on the shelf off Uruguay 15 cm and weight 25 g.

Weight proportions of individual body parts (%): head 17.8, meat 63.8, skin and fins 0.2, bones 5.2, viscera 12.3.

Chemical composition of meat (%): moisture 64.1, fat 15.3, protein 18.7, ash 1.1.

Salted anchovy undergoes seasoning well; it should therefore be processed for salt curing.

Northern anchovy—*Engraulis mordax*. Main distribution along the Pacific coast of America, from Vancouver to the southern tip of the Californian peninsula. Weight 10–25 g, length 9–12 cm.

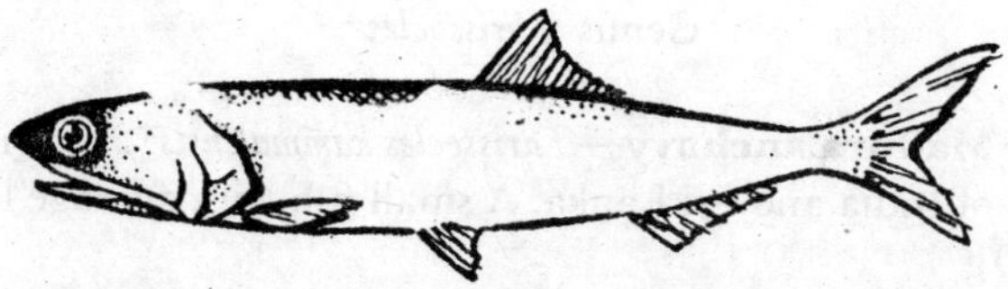

Fig. 76. Northern anchovy—*Engraulis mordax.*

Weight proportions of individual body parts (%): head 24.5–28.2, trunk 58.7–63.0, viscera 8.5–8.8, bones 5.2, fins 2.3, skin 2.5. Output of fillets 51%.

Chemical composition of meat (%): 71.2–78.7, protein 17.3–19.1, fat 0.7–9.2, ash 1.5–2.3. Chemical composition of inedible parts of this anchovy (%): moisture 72.7, protein 10.7, fat 6.8, ash 7.3.

The products made by spicy salting have received satisfactory grading. It is good for small sized packing, even without gutting.

Japanese anchovy—*Engraulis japonicus*. Lives in the Sea of Japan up to Alexandrovsk-Sakhalinsk along the Primorye coast and the coasts of Japan and Korea. Length 11.0–18.5 cm, weight 15–45 g. Weight proportions of individual body parts (%): trunk 57–64, viscera 18.8. Chemical composition of meat (%): moisture 62.3–74.7, protein 13.3–21.0, fat 3.3–18.0, ash 1.1–1.8.

Fig. 77. Japanese anchovy—*Engraulis japonicus.*

This is a very tender fish requiring careful handling during processing. The Japanese anchovy is mainly used for products of spicy salting and canned products, "blanched fish in oil" or "smoked fish in oil".

Genus *Diaphus*

Headlight anchovy—*Diaphus antiorbitalis*. Caught near the Hawaiian Islands. Length of the fish caught in March on an average 15 cm (11–18) and weighing 40 g.

Weight proportions of individual body parts (%): head 25.3, trunk 61.2, viscera 10.6, bones 8.0, fins 2.6. Output of fillets 53.3%.

Chemical composition of meat (%): moisture 71.1, protein 17.2, fat 10.0, ash 1.4.

This species should be used for conserves and preserves.

Genus *Thrissocles**

Indian or Madura anchovy—*Thrissocles kammalensis**. Distributed in the coastal waters of India and Sri Lanka. A small fish with average length 15 cm and weight 50 g.

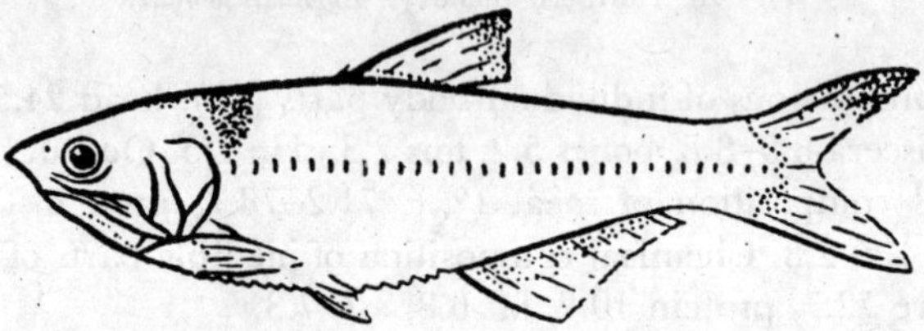

Fig. 78. Indian or Madura anchovy—*Thrissocles kammalensis* *.

Weight proportions of individual body parts (%): head 18.5, trunk 70.5, fins 2.2, intestines 7.5. Chemical composition of meat (%): moisture 72.4, protein 16.3, fat 8.2.

The meat of cooked fish has a pleasant smell, somewhat dry consistency and good taste. It has many fine bones. After salting, the fish does not mature and acquires a hard consistency.

The Indian anchovy can be used for conserves in oil.

Genus *Setipinna*

Setipinna anchovy—*Setipinna taly*. Lives in the western Pacific Ocean and in the Great Australian Bight. Length of fish 8.9–14.1 cm, weight 10–35 g. Weight proportions of individual body parts (%): head 15.3, trunk 76.9, viscera 3.8, fins 3.8.

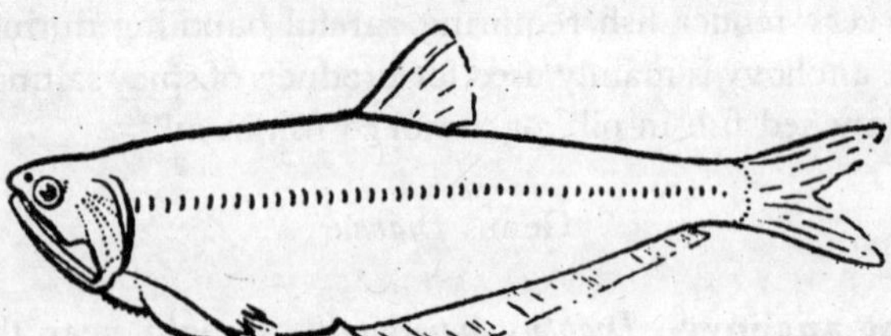

Fig. 79. Setipinna anchovy—*Setipinna taly*.

The meat is white, pleasant in taste, but very sour and not suitable for canning or culinary processing. It should be used only for fish meal.

Family EMMELICHTHYIDAE—BONNETMOUTHS OR REDEYES

These fish live in the deep waters of the Indian and Atlantic Oceans. Two genera are known from the coastal regions of South Africa.

Genus *Plagiogeneion**

Rosy redeye—*Plagiogeneion rubiginosus**. Caught in small quantities with other fish. Attains length of 60 cm. In catches from the shelf off Cape Agulhas in the Indian Ocean in February, most fish are about 40 cm (38–45) in length and weigh around 1,400 g (1,010–1,710).

After cooking, the meat is light colored, friable and pleasant in taste. The soup is transparent, yellow-green, very tasty, with a meat-like (non-fish) taste.

TABLE 80. WEIGHT PROPORTIONS OF INDIVIDUAL BODY PARTS, %

Head	Trunk		Skin	Bones	Scales	Fins	Viscera	
	whole	including meat					whole	including liver
29.0–34.5 / 32.0	51.2–57.3 / 54.3	35.6–43.1 / 40.0	3.4	8.0	1.3	2.0	7.4–10.5 / 9.2	1.2

TABLE 81. CHEMICAL COMPOSITION, %

Body parts	Moisture	Protein	Fat	Ash
Head	78.7	19.4	0.5	1.3
Inedible parts (offal)	70.1	17.5	3.8	8.6

This fish can be used as a table fish for first and second courses and hot smoked products.

Australian perch—*Plagiogeneion macrolepis**. Found on the shelf of the Indian Ocean, little commercial value. Length of fish caught 19.5–26.0 cm, weight 195–400 g. Weight proportions of individual body parts (%): head 11.2, trunk 75.1, bones 12.8, viscera 8.3, scales 1.8. Output of fillet with skin is 62.3%. Chemical composition of meat (%): moisture 77.2, protein 19.1, fat 1.4, ash 1.0.

When Australian perch is found in catches, it should be frozen in blocks for subsequent frying in fillets and steaks or prepared for hot smoked products as cured balyks and for dressed and beheaded fish.

Schlegeli perch—*Erythrichthys schlegeli.* Not a commercial fish. Found in the South China Sea. Length 10–12 cm, weight 20–42 g. Weight proportions of individual body parts of dressed fish (%): head 26.8, trunk with fins 73.0.

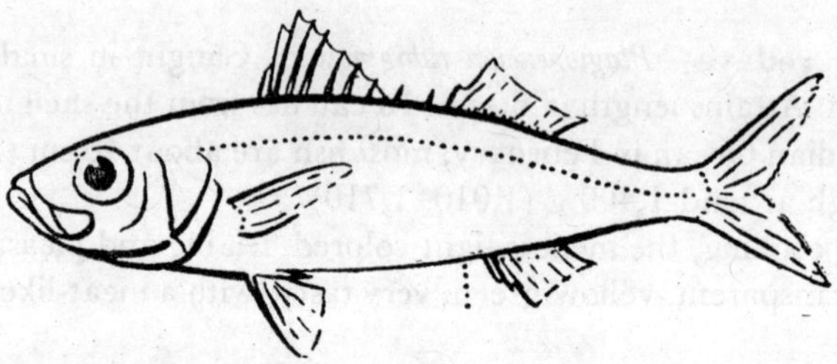

Fig. 80. Schlegeli perch—*Erythrichthys schlegeli.*

Family ELOPIDAE—TENPOUNDERS OR TARPONS

Banana fish—*Elops senegalensis.* Found along with other fish in catches off the African coast of the Atlantic Ocean where rivers drain into the ocean. Usual length of fish in catches 30–45 cm, some specimens up to 60 cm.

Weight proportions of individual body parts of banana fish 33–38 cm long and weighing 0.3–0.6 kg, %: head 14, trunk with fins 79, viscera 7. Output of fillets 58%.

TABLE 82. CHEMICAL COMPOSITION OF BANANA FISH CAUGHT IN APRIL, %

Body parts	Moisture	Protein	Fat	Ash
Meat	75	23	0.6	1.4
Head, bones and fins	63	21	7.6	8.4
Viscera	71	18	10.0	2.0

The banana fish should be processed for freezing and for preparation of smoked products.

Family EPHIPPIDAE—SPADEFISHES

Leaffish or round batfish—*Platax orbicularis*. A non-commercial fish. Lives in the waters off northwest Australia. Length 45–47 cm, weight 2,900–3,100 g.

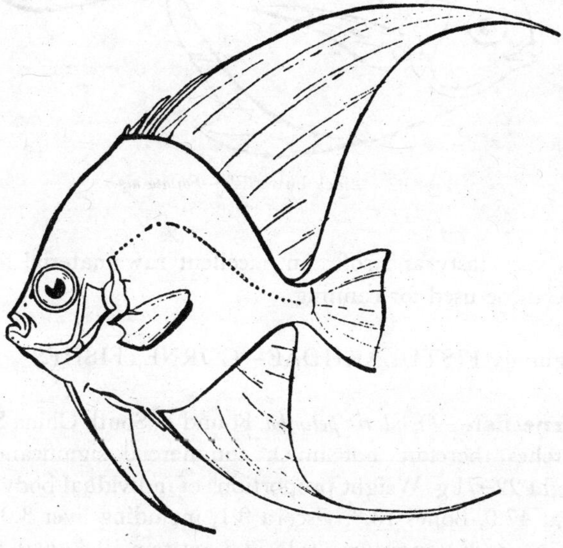

Fig. 81. Leaffish—*Platax orbicularis*.

Weight proportions of individual body parts of fish caught in December–May, (%): head 23.4, trunk 61.9, scales 1.6, fins 3.5, bones 16.0, skin 5.8, viscera 8.7. Output of fillets 40%.

Chemical composition of meat with skin (%): moisture 67.3–76.2, protein 19.9–22.5, fat 1.6–8.2, ash 1.4–2.1.

This is a very fleshy fish. The raw meat is gray. After heating it becomes lighter in color but acquires a greenish tinge. It is tasty but has an unpleasant external appearance. These fish should be used for fish meal.

Family FORMIONIDAE—BLACK BUTTERFISHES

Black butterfish—*Formio niger*. Not very important commercially in the South China Sea. Length 23–28 cm, weight 425–670 g. Weight proportions of individual body parts of fish caught in October (%): head 16.6, trunk with fins 76.5, viscera 6.9.

Chemical composition of meat (%): moisture 73.6, protein 21.2, fat 2.3 and ash 2.2.

Fig. 82. Black butterfish—*Formio niger*.

Its meat is very tasty and soft; an excellent raw material for culinary preparations. Can be used for canning.

Family FISTULARIIDAE—CORNETFISHES

Pacific cornetfish—*Fistularia petimba*. Found in South China Sea. Rarely present in catches therefore not much commercial significance. Length 40–47 cm, weight 22–71 g. Weight proportions of individual body parts (%): head 25.6, meat 47.0, bones 18.3, viscera 9.1, including liver 3.0. Chemical composition of meat (%): moisture 76.1–77.3, protein 19.6–20.0, fat 1.0–1.7, ash 1.4.

Fig. 83. Pacific cornetfish—*Fistularia petimba*.

The canned meat of this cornetfish (natural) is quite tasty and resembles the meat of prawns. It can be used for preparing natural canned products.

Red cornetfish—*Fistularia villosa*. Caught along with other fish in the eastern Central Atlantic. Fish caught in April were 73 cm long and weighed 0.3 kg. Weight proportions of individual body parts (%): head 30, trunk 52, viscera 17, fins 1.

Chemical composition of meat (%): moisture 75.0, protein 23, fat 0.4, ash 1.6.

The taste is good. Sun cured products can be made from this fish.

Family GADIDAE—CODS

This family contains 22 genera with 58 species. Cod are distributed in the northern, temperate, tropical and subtropical waters of the Atlantic and Pacific Oceans and in the Arctic Ocean. Several species of cod live in the southern hemisphere.

Almost all cod are marine fish except burbot which is a permanent inhabitant of fresh water. Among the most important commercial fish from this family are cod, hake, haddock, walleye pollock and saithe.

Genus *Gadus*

This genus has one species and several subspecies distributed in the northern Atlantic and Pacific Oceans and in the adjoining seas. It is divided into six subspecies with the common names Atlantic cod, Baltic cod, Kildin cod, White Sea cod, Greenland cod and Pacific cod. They are marine shoaling pelagic fish. Of these, the Atlantic cod has the greatest commercial significance.

Atlantic cod—*Gadus morhua morhua* (L.) It is distributed from Novaya Zemlya in the east to the American coast in the west and from the latitude of Bay of Biscay to 80°N near Spitsbergen to 70°N in the Davis Strait. Its spawning takes place from February to April.

The relative weight of the liver fluctuates depending upon the fattiness of the fish, that is, conditions of feeding as well as age: the larger the cod the greater is the weight of its liver. The minimum relative weight of the liver, 1.9%, was recorded in a cod weighing 0.6 kg and the maximum relative weight 22.1%, was found in a cod weighing 3.5 kg.

By severing the head with straight cut in a line perpendicular to the body, the weight of the trunk comes to 52.5% and the head along with the pectoral fins 31.5%. In such a dressing part of the liver is also cut (about 20%). If the head is removed by a straight cut without the pectoral fins about 7% of the liver is removed along with the head. In this case the output of the trunk is 61.6%, head 27.5%. When the head is separated by a cut at an angle of 60°, the trunk again constitutes 61.6% and the head 26.4%. In this method of dressing the liver remains intact. Separation of the head by two cuts slanting in relation to the plane of symmetry, starting behind the pectoral fins and terminating at the parietal bone, also does not touch the liver. In this case the trunk constitutes 60.2% and the head 25.7%.

The weight of the roe increases as it matures: in sexually mature females it accounts for 3–4% toward November, in November–December its proportion is 5% and in February 10%. Before spawning, the roe develops considerably and its food value decreases.

Chemical composition of cod meat changes little in different seasons and at different ages.

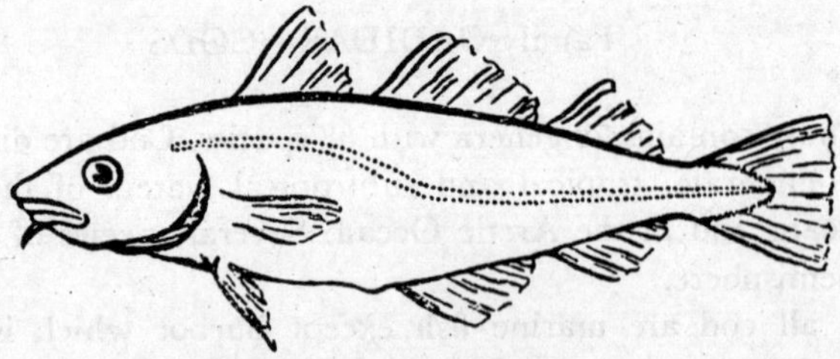

Fig. 84. Atlantic cod—*Gadus morhua morhua* (L.).

TABLE 83. WEIGHT PROPORTIONS OF INDIVIDUAL BODY PARTS OF THE ATLANTIC COD (DRESSED BY COMMERCIAL METHODS, THAT IS, BY SEPARATING HEAD ALONG THE CRANIAL BONES), %

Size of cod	Head	Trunk	Meat	Skin	Bones	Fins	Viscera	
							whole	including liver
Large, more than 1.5 kg	20.3	64.5	52.2	3.5	6.9	1.9	13.1	5.0
Small, less than 1.5 kg	20.9	65.4	53.9	3.5	8.0	1.8	11.6	3.6

The average chemical composition of the liver (Table 84) is somewhat lower in value since the analytical material was obtained in years when predominantly small fish were caught.

Chemical composition of the head, bones and guts undergoes some

TABLE 84. AVERAGE CHEMICAL COMPOSITION, %

Body parts	Moisture	Protein	Fat	Ash
Meat	80.0–82.5 80.4	16.9–19.4 17.8	0.0–0.3 0.2	1.0–1.3 1.2
Liver	20.9–64.4 41.0	5.2–21.6 11.0	15.0–73.2 47	—
Head	79.5	13.9	0.3	5.6
Bones	75.0	15.2	0.4	9.0
Fins	74.6	14.0	1.0	8.7
Guts	81.3	13.9	1.5	1.6
Roe	73.1–93.0	5.2–21.5	0.1–1.2	0.5–2.0
Testes	80.3–86.7	11.1–14.7	0.4–2.0	1.0–1.8

variation; Table 84 presents only average values. The average compositions of fish roe and testes are, on the contrary, given in a range, since their average composition is of no interest.

Cod meat contains the following mineral substances (mg %): potassium 361, calcium 57, magnesium 24, phosphorus 222, iron 0.7.

The liver of Atlantic cod has a potency of 270–10,300 iu/g of vitamin A; in hot-rendered oil supplied by trawlers at ports its potency is 150–800 iu/g. The quantity of aneurin (vitamin B) in cod liver is 2.5–3.5 γ/g, in the roe 6–12 γ/g; riboflavin in the meat is 1.4 γ/g, in the liver 5–10, in the roe 6–15 and in the testes 3–4 γ/g, nicotinic acid in the meat is 22 and in the liver 23 γ/g and pantothenic acid in roe 330–380 γ/g.

The cod is a typical table fish. It is used mainly for first and second courses. Therefore, the best method of processing is cooling and freezing including filleting. In salt curing of cod there is a large loss of nutritional substances and deterioration in taste; however, this method is still fairly widely used to supply cured fish to regions which do not have refrigeration facilities.

There are many second course dishes made from cod: mixed preparations with vegetables, boiled in the Polish manner, cooked in sauce, fried with different sauces, cutlets, fish balls, pudding and so on; all of these have good taste.

Canned preparations from cod liver are very nutritious and tasty. The medicinal importance of cod liver oil, extracted from the cod liver by hot-rendering, has decreased since the pharmaceutical industry has started making synthetic vitamin A in the form of tablets. The value of the medicinal cod liver oil as a source of unsaturated fatty acids is greater if users consume it in the form of canned liver.

Pacific cod—*Gadus morhua macrocephalus*. An important commercial fish. Lives in the Bering and Okhotsk Seas, in the Sea of Japan and in the eastern Pacific Ocean.

Length of Pacific cod 45–93 cm, weight 0.7–14.0 kg. In catches, a predominance of fish weigh 2–14 kg and only 10% of the catches account for fish weighing less than 2 kg.

Weight proportion of the individual body parts (%): trunk 40.5–45.5,

Fig. 85. Pacific cod—*Gadus morhua macrocephalus*.

head 33.6–34.2, bones 9.4–12.1, viscera and fins 20.8–25.2, including liver 5.6–6.3. Output of fillets 28.4–36.8%.

Average chemical composition of meat (%): moisture 79.5, protein 17.1, fat 0.9, ash 1.2.

This cod is a typical table fish. It is pleasant in dressed and beheaded form as well as in the form of fillets.

The liver of this cod is very important as it contains large quantities of oil. The oil extracted from the liver of this species is rich in vitamins A and D. The vitamin A content varies from 85–500 iu for 1 g of liver.

Genus *Eleginus*

Navaga—*Eleginus navaga.* A cold loving fish living in the coastal waters of the Barents, Kara and White Seas and from the Kola Bay to the Ob inlet. Attains a length of 40 cm. Length of navaga in catches ranges from 16–24 and 12–20 cm and weight 16–59 g.

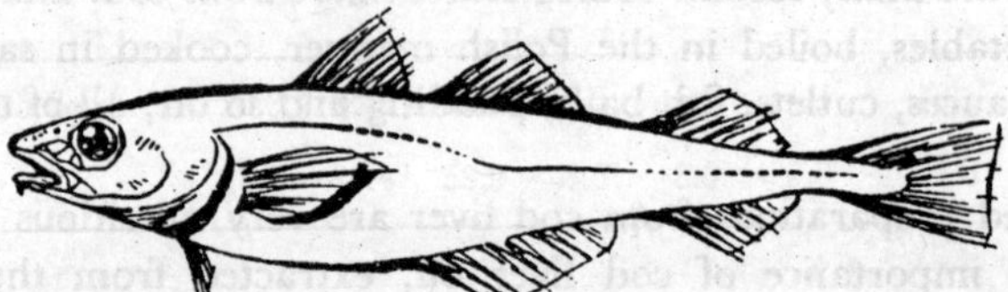

Fig. 86. Navaga—*Eleginus navaga.*

Weight proportions of the roe reach 18% and of the testes 15%. The relative weight of the liver increases with increase in the age of the fish. In the large navaga it is about 5%.

In its chemical composition navaga differs slightly from cod by higher fat content in the meat.

TABLE 85. WEIGHT PROPORTIONS OF INDIVIDUAL BODY PARTS, %

Head	Trunk	Meat with skin	Bones	Fins	Viscera whole	Viscera including liver
26.7	50.5	—	—	—	16.9	3.8
21.1	55.5	48.3	7.2	1.3	15.9	—

Note: The data presented in Table 85 were obtained by different methods of separating the head.

TABLE 86. CHEMICAL COMPOSITION, %

Body parts	Moisture	Protein	Fat	Ash
Trunk	77.9–80.3 / 79.3	15.0–17.4 / 16.2	0.3–2.8 / 1.4	1.9–2.7 / 2.4
Liver	59.9–74.0 / 65.1	12.2–13.0 / —	7.5–23.6 / 13.9	1.0–1.4 / —

In taste qualities navaga is superior to all cod and it is especially tasty when fried with breadcrumbs and sprinkled with lemon juice.

Saffron cod or Pacific navaga—*Eleginus gracilis*. Lives in the Sea of Japan, the Okhotsk and Bering Seas and in the eastern Pacific Ocean.

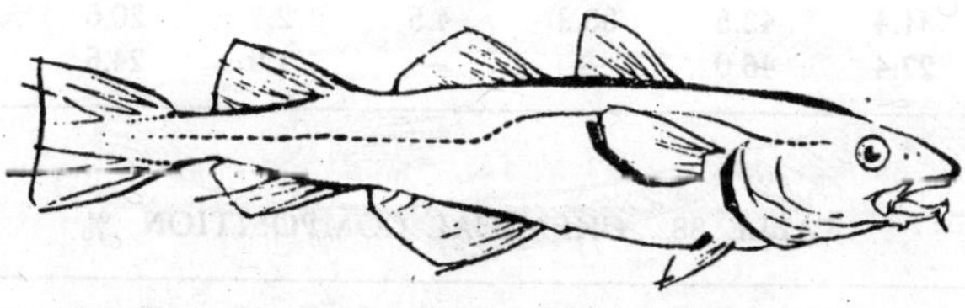

Fig. 87. Saffron cod or Pacific navaga—*Eleginus gracilis*.

Length of the saffron cod 38–47 cm, weight 0.15–1.3 kg. The smallest fish (0.15–0.25) was caught in Primorye and the largest (weighing 0.7–1.3 kg) was caught in the Bering Sea.

Weight proportions of individual body parts (%): head 16–22, trunk 60.0–70.8, fins 2.3–2.7, viscera 8.4–14.6, including liver 1.6–4.2.

Chemical composition of meat (%): moisture 80.7–81.3, protein 13.7–16.5, fat 0.7–1.1, ash 1.6–2.0.

The meat of this fish is lean and has a pleasant taste. It is used for canning and for culinary products.

Genus *Boreogadus*

Arctic cod or polar cod—*Boreogadus saida*. Distributed in the Barents and Karsk Seas and in regions covered by ice in winter. Spawning takes place in January–February.

Length of polar cod caught in spring 18 cm, maximum depth 1.7 cm, maximum thickness 1.6 cm.

The food value of the polar cod is low because of the small quantity of edible parts and low protein and fat contents. The polar cod should be supplied

Fig. 88. Arctic or polar cod—*Boreogadus saida*.

TABLE 87. WEIGHT PROPORTIONS OF INDIVIDUAL BODY PARTS, %

Weight, g	Head	Trunk	Meat with skin	Bones	Fins	Viscera		
						whole	including	
							gonads	liver
516–1,020	31.4	43.5	38.3	4.5	2.7	20.6	—	11.2
—	27.4	46.0	—	—	1.9	24.6	14.6	5.0

TABLE 88. CHEMICAL COMPOSITION, %

Body parts	Moisture	Protein	Fat	Ash
Meat with skin*	76.7–81.4 / 79.1	16.3–19.1 / 17.7	0.4–0.8 / 0.6	1.1–1.6 / 1.4
Trunk**	81.5–84.4 / 82.5	10.5–14.7 / 12.7	0.5–1.0 / 0.8	2.0–3.4 / 2.8
Liver	27.3–42.5 / 36.0	6.2–7.9 / 7.3	47.5–60.0 / 53.8	0.6–0.9 / 0.8
Head	79.3–83.0 / 81.6	11.7–13.6 / 12.4	0.7–1.2 / 1.1	3.2–3.6 / 3.3
Viscera	72.3–79.6 / 77.3	12.1–13.7 / 13.1	3.1–10.2 / 6.0	1.3–1.8 / 1.6

* Fish caught in spring.
** Fish caught in winter

for fish meal although the possibility of using this fish for food purposes cannot be totally ruled out.

Genus *Macruronus*

Magellan hake—*Macruronus magellanicus*. Lives in the northwestern Atlantic Ocean. The length of the fish examined 77 cm and weight 1,440 g.

Weight proportions of individual body parts (%): head 18.3, meat 52.7, fins and skins 4.2, bones 8.6, viscera 14.5.

Chemical composition of meat (%): moisture 78.4, protein 16.9, fat 2.9, ash 1.1.

The meat of this hake should be used for canning.

Genus *Merluccius*

These fish are distributed in the Atlantic and Pacific Oceans. In the Indian Ocean they are found only along the southern coast of Africa. This genus, comprising hakes, is represented by nine species.

Cape hake—*Merluccius capensis*. Distributed in the Atlantic Ocean, south from Angola, in the Indian Ocean, along the coast of Africa. Most favorable season for fishing April–May. Attains a length of 1.2 m; usual length 50–90 cm.

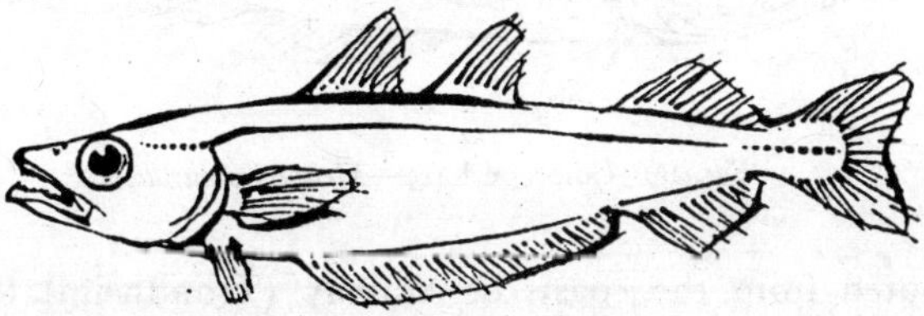

Fig. 89. Cape hake—*Merluccius capensis*.

TABLE 89. WEIGHT PROPORTIONS OF INDIVIDUAL BODY PARTS, %

Head	Trunk	Meat with skin	Bones	Scales	Fins	Viscera		
						whole	including	
							roe	liver
26.0	59.2	46.9	6.1	1.2	1.8	1.8	—	—
22.5	63.9	48.9	—	—	0.6	7.7	4.2	2.7

TABLE 90. CHEMICAL COMPOSITION, %

Place of catch	Body parts	Moisture	Protein	Fat	Ash
Indian Ocean	Meat	77.5	19.3	1.1	1.3
	Offal	78.1	15.6	1.3	5.3
Southeastern Atlantic	Meat	81.7	16.4	0.3	1.3

Length of fish caught 78 and 51.5 cm and weight 5,117 and 1,380 g respectively; maximum depth 15.2 cm, maximum thickness 9.2 cm.

The liver in the fish caught off the coast of South Africa contained 0.6–0.8 γ/g vitamin B_1, 0.8–1.0 γ/g riboflavin, 21–26 γ/g nicotinic acid, 24–30 γ/g pantothenic acid, 8–10 γ/g vitamin B_6, 0.5–0.6 γ/g vitamin B_{12} and 1.0–1.2 γ/g folic acid.

This hake is a table fish with mediocre taste qualities. Its meat is not suitable for natural canned products (it becomes over-boiled).

Common hake—*Merluccius merluccius.* A good commercial fish. A fairly large demersal fish attaining length of 1.2 m. Length commonly encountered in the fishery 40–80 cm.

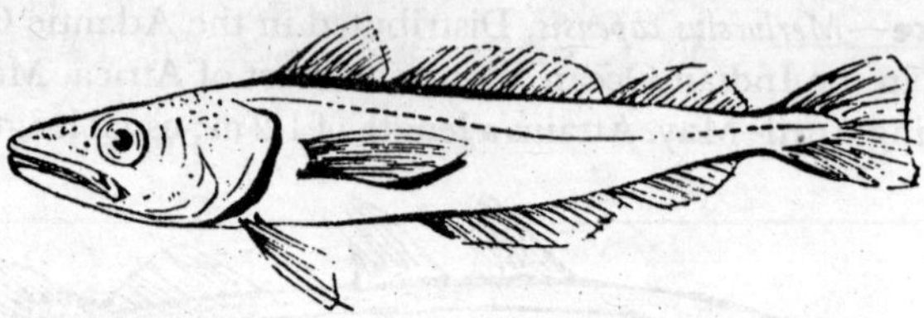

Fig. 90. Common hake—*Merluccius merluccius.*

Distributed from the coasts of Norway (Trondheim) to the coasts of Senegal, as well as in the Mediterranean and Black Seas. Length of common hake 27–80 cm, weight 180–3,700 g, maximum depth 6–15 cm, maximum thickness 3.0–6.5 cm.

Weight proportions of individual body parts (%): head 25.5, trunk 61.1, meat 52.6, bones and scales 8.5, fins 2.7, viscera 8.0, including liver 3.9.

The weight proportions of individual body parts of the common hake differ, depending upon the method of separation of the head. When it is separated by cutting along the cranial bones it accounted for 18.6%, by cutting with a straight cut including pectoral fins; when together with the head a part of the liver is separated, its weight accounts for 34.9%. The proportions of the remaining parts of the body change accordingly.

TABLE 91. CHEMICAL COMPOSITION, %

Body parts	Moisture	Protein	Fat	Ash
Head	81.7	12.3	0.3	5.0
Meat	80.1	17.5	0.9	1.5
Viscera	80.9	13.9	2.5	1.5
Liver	58.4	9.0	34.6	1.0

The meat of the hake contains the following mineral substances (mg%): potassium 352, calcium 33.2, magnesium 37.7, phosphorus 421 and iron 4.4.

The liver oil contains 1,000–51,000 iu/g vitamin A, 23 γ/g vitamin E (single analysis).

The common hake is a good table fish of mediocre value. It is suitable for first and second courses; it is especially good when fried and sprinkled with lemon juice and for preparing products with different sweet and sour sauces.

Pacific hake—*Merluccius productus*. A commercial fish, caught in the Pacific Ocean from southern California to northeast Alaska.

Length of the pacific hake 32.5–63.5 cm, sometimes reaching 91.5 cm, weight 550–1,830 g, sometimes growing to 3–4 kg.

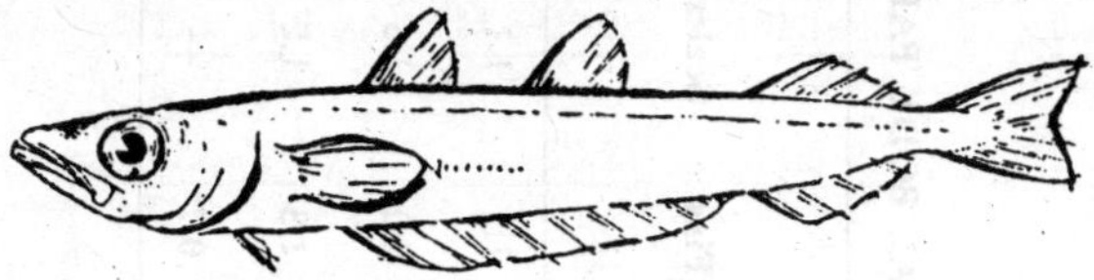

Fig. 91. Pacific hake—*Merluccius productus*.

Weight proportions of individual body parts (%): head 15.3–25.2, trunk 57.4–70.8, viscera 8.0–14.5, including liver 2.3–7.6, fins 1.5–4.1. Output of fillets 47.8–57.7%.

Chemical composition of meat (%): moisture 76.9–84.3, protein 13.8–19.2, fat 0.7–2.7, ash 1.2–1.7.

After cooking, the meat of this hake is white and has soft and juicy consistency; culinary preparations are made from it. Canned products in tomato sauce and smoked products in oil can also be prepared from frozen hake.

The Pacific hake is frozen intact immediately after catching, in dressed and beheaded form and in fillet form. If raw fish are retained for a long time before freezing, the meat acquires an unpleasant smell and has poor consistency. The liver should be used as a raw source of vitamins and as an additional material for preparing fish containing vitamins and fish-vegetable conserves.

Silver hake—*Merluccius bilinearis*. An important commercial fish, occurring in a relatively small area of the northwest Atlantic—Georges Bank, New Shetland shelf and on the shelf of the USA.

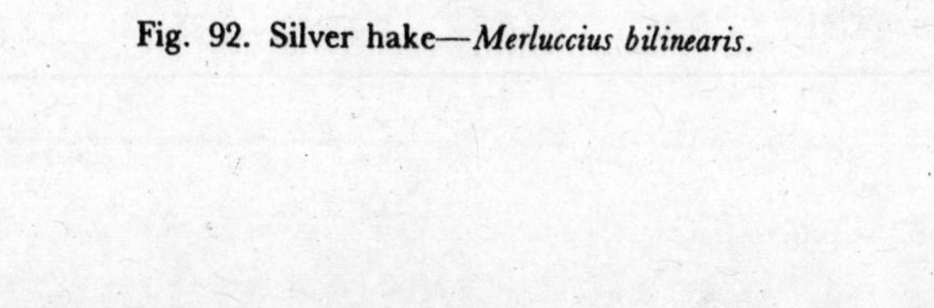

Fig. 92. Silver hake—*Merluccius bilinearis*.

TABLE 92. WEIGHT PROPORTIONS OF INDIVIDUAL BODY PARTS, %

Stage of maturation of gonads	Average length, cm	Average weight, g	Head	Trunk	Meat with skin	Bones	Fins	Scales	Viscera		
									whole	including	
										gonads	liver
I	29.7	254	16.2	70.3	62.3	6.8	1.2	1.1	10.3	3.1	3.7
II	25.6	203	17.0	81.8	54.7	6.1	2.0	1.6	17.4	6.7	5.4
III	33.8	415	15.6	82.8	57.8	5.7	1.3	1.2	16.2	5.0	7.8
IV	26.1	123	21.8	68.6	61.1	6.6	0.7	—	6.7	—	2.2

Length of silver hake 25–35 cm, maximum length 80 cm.

In its chemical composition the meat of the silver hake is closer to the White Sea navaga. It contains a much greater quantity of fat compared to most members of the cod family.

The liver oil contains a little more than 1,000 iu/g of vitamin A and about 80 γ/g of vitamin E.

The silver hake, if properly frozen, is close to the White Sea navaga in taste. It makes delicious dishes [illegible] fried.

TABLE 93. CHEMICAL COMPOSITION, %

Body parts	Moisture	Protein	Fat	Ash
Meat	77.1–81.2 / 79.1	15.6–17.9 / 16.7	1.1–4.1 / 2.7	0.9–1.6 / 1.3
Roe in stages III–IV	67.8–72.5 / 70.4	18.0–20.2 / 19.0	4.8–8.8 / 7.2	0.4–1.6 / 1.1
Testes	80.0–86.3 / 84.0	10.5–12.3 / 11.4	0–1.5 / 0.7	1.5–2.4 / 1.9
Liver	31.8–63.0 / 42.9	7.3–11.6 / 9.6	22.5–59.5 / 44.3	0.4–1.2 / 0.9
Head	77.9–82.7 / 81.2	9.7–13.0 / 11.2	0.1–4.1 / 1.3	4.5–5.7 / 5.3
Bones	74.4–76.4 / 75.7	14.0–14.7 / 14.2	0.4–2.0 / 1.1	8.2–9.0 / 8.5
Viscera	77.7–81.6 / 79.9	13.3–14.6 / 13.9	3.0–5.7 / 3.6	1.2–1.5 / 1.3

Argentine hake—*Merluccius hubbsi*. Caught in the southwest Atlantic, from Cape Frio (Brazil) to Patagonia. Length 40–50 cm, sometimes 90 cm. Length of Argentinian hake caught 71 and 34–42 cm, weight 3,750 and 340–875 g respectively; maximum depth 14.5 and 6–7 cm, maximum thickness 9 and 4–6 cm.

TABLE 94. WEIGHT PROPORTIONS OF INDIVIDUAL BODY PARTS, %

Head	Trunk	Meat	Fins	Skin, scales, bones	Viscera	
					whole	including liver
19.5	52.0	44.2	3.0	7.8	24.7	6.5
22.3	56.4	44.7	4.2	10.9	16.2	—

Chemical composition of meat of Argentinian hake (%): moisture 79.1–79.9, protein 18.8–17.1 [*sic*], fat 0.9–1.6, ash 1.2–1.4.

Genus *Micromesistius*

This genus is represented by two species, one distributed along the coast of Europe, the other along the southeastern coast of South America.

Southern putassu—*Micromesistius australis*. A commercial fish, distributed along the coast of South America, from the Gulf of Rio de la Plata to the Antarctic waters of Scotia Sea. Average length 45–47 cm sometimes reaching 60 cm, average weight 800–900 g.

Weight proportions of individual body parts (%): trunk 61.5, meat with skin 51.5, head 16.7, bones 6.4, fins 2.6, viscera 20.8, including liver 9.2, gonads 1.4.

The southern putassu is a good table fish. It should be marketed in frozen form. Its meat can also be used for preparing culinary products and conserves in oil. The liver is a good raw material for natural canned products.

TABLE 95. CHEMICAL COMPOSITION, %

Body parts	Moisture	Protein	Fat	Ash
Head	79.4	10.0	0.8	—
Meat	79.7–81.6 / 80.6	14.0–16.8 / 15.2	0.5–3.2 / 1.5	0.9–1.2 / 1.1
Viscera	81.6	9.7	3.7	—
Liver	28.8–40.2	4.8	48.2–62.6	—
Bones	73.4	11.3	1.3	—
Skin	73.8	31.2	1.2	—

Putassu (Blue whiting?)—*Micromesistius poutassou*. Lives in the Atlantic Ocean along the coast of Europe, from Gibraltar to the Nordkapp. Caught near the Bear Islands, often found west of Spitsbergen. Usually occurs off the southern and western coasts of Iceland, sometimes caught near the southwest coast of Greenland. In the south distributed to Gibraltar and along the western Mediterranean Sea; also found off the coasts of Italy and Greece.

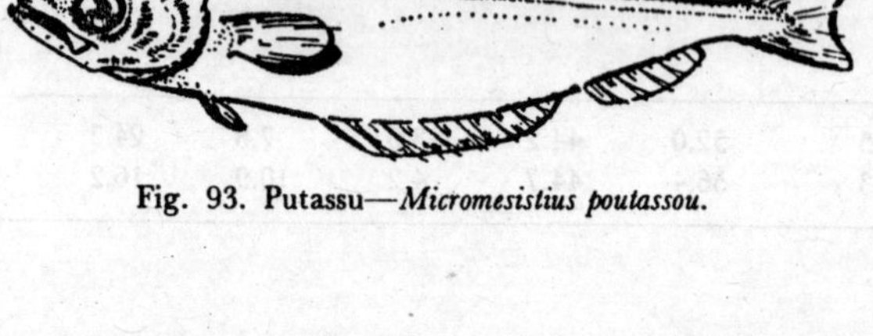

Fig. 93. Putassu—*Micromesistius poutassou*.

Average length 26–28 cm, maximum 35–40 cm, weight 250–300 g.

Length of fish caught 20–25, 31.0–36.5 and 25.9 cm. Respective weights 70–119, 96–249 and 146 g, maximum depth 3.5–5.0 cm and maximum thickness 2.0–2.5 cm.

TABLE 96. WEIGHT PROPORTIONS OF INDIVIDUAL BODY PARTS, %

Head	Trunk	Meat with skin	Bones	Fins	Viscera whole	Viscera including liver	Viscera including gonads
23.7*	60	57.9	4.5	1.9	12.0	5.1	—
17.8**	65.9	58.4	7.0	1.2	12.6	6.3	—
17.0**	—	58.3	7.4	0.9	13.9	6.6	0.6

*Head separated by straight cut; **Head separated by semicircular cut.

The meat of the putassu is very pleasant, resembling the taste of navaga. It is a good table fish and can be used for preparing culinary products (in boiled and fried form).

Its liver is a good raw material for natural canned products. The liver oil of this fish (single analysis) contains 3,400 iu/g vitamin A and 68 γ/g vitamin E.

TABLE 97. CHEMICAL COMPOSITION, %

Body parts	Moisture	Protein	Fat	Ash
Whole fish	77.4	15.4	4.8	2.3
Meat	77.6–81.6 / 80.0	14.0–19.5 / 17.1	0.2–0.6 / 0.4	0.4–1.4 / 1.1
Head	75.1–79.6 / 78.2	10.0–13.9 / 12.2	0.6–2.4 / 1.4	5.7–7.3 / 6.7
Skin	69.4–73.8 / 71.7	20.5–26.3 / 22.6	1.2–6.9 / 3.8	0.6–1.4 / 1.0
Bones	70.6–73.4 / 72.1	15.5–16.8 / 16.0	1.1–4.3 / 2.4	8.3–9.9 / 9.2
Fins	65.1–75.2 / 70.1	11.3–14.5 / 12.9	0.9–1.3 / 1.1	— / —
Internal organs	75.3–80.0 / 80.6	4.1–14.0 / 10.2	3.7–8.5 / 6.2	0.5–1.6 / 1.2
Roe	80.5	14.2	4.0	1.2
Liver	22.4–40.2 / 30.5	4.0–8.6 / 5.5	48.2–73.1 / 60.7	0.2–0.5 / 0.3

Due to the small size of the putassu, the best method of processing it is as a whole fish.

Genus *Molva*

Blue ling—*Molva dypterygia**. Distributed along the coast of Norway, from Kattegat and Skagerrak to Finnmark. Often caught in the southwestern Barents Sea, on Bear Bank and near the south coast of Spitsbergen. Commonly found of the southern coast of Iceland; found southward near Ireland and Scotland. Length attained up to 1.5 m; however, individuals longer than 1 m caught rarely.

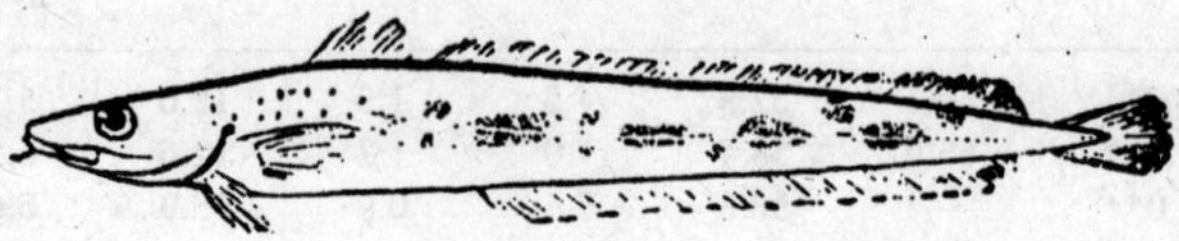

Fig. 94. Blue ling—*Molva dypterygia**.

Weight proportions of individual body parts (%) of blue ling 63.5 cm long and weighing 1,750 g (male, stage IV of maturation, caught off the southwest coast of Iceland), (%): head (separated along the jaw) 12.4, trunk 82.1, meat 65.2, skin with scales 5.9, fins 2.3, bones 10.7, viscera 2.6, including liver 0.6.

TABLE 98. CHEMICAL COMPOSITION, %

Body parts	Moisture	Protein	Fat	Ash
Head	73.7	16.8	0.3	8.0
Meat	78.8	19.0	0.1	1.1
Viscera	80.0	16.5	1.2	1.7
Liver	57.8	17.9	—	2.2
Bones	71.4	18.5	0.1	9.2

Genus *Melanogrammus*

Haddock—*Melanogrammus aeglefinus*. An important commercial fish. Widely distributed in the northern Atlantic Ocean: along the coast of America from Cape Hatteras to the Strait of Belle Isle. Found in small numbers off the coast of Labrador at Flemish Cape and at the southern tip of Greenland, but in large numbers off the coast of Iceland, in the North Sea, along the coast of Great Britain and Ireland, in the Norwegian and Barents seas. In the southern Barents Sea, haddock distributed toward the north, up to Spitsbergen Island and south to Goose Bank and Karsk Sea, in some years entering the White Sea. Commonly occurs in Kattegat and Skagerrak; it is

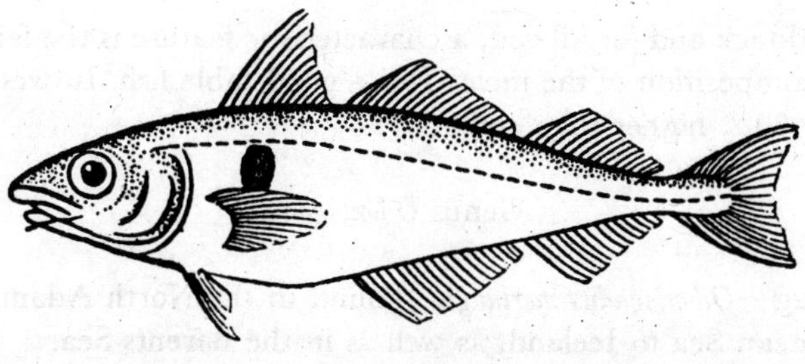

Fig. 95. Haddock—*Melanogrammus aeglefinus*

also found in the western Baltic Sea. Average weight of haddock 1,057 g, average length 45 cm.

Weight proportions of individual body parts (%): head 19.3, trunk 63.5, meat 50.6, skin 3.5, bones 7.4, fins 2.0, viscera 8.8, including liver 5.8.

The weight of haddock liver is slightly higher here since most of the analyses were carried out in 1965 when the fattiness of haddock was greater than usual. The weight of liver increases with the growth of the fish.

Length of fish, cm	21–30	31–40	41–50	51–60	61–70	71–80	81–90
Weight of liver, %	4.2	4.4	4.9	4.9	4.5	4.9	5.6

The potency of vitamin A in the liver is 250–720 iu/g, in the oil 900–1,610 iu/g, the quantity of aneurin (vitamin B_1) in the meat is 0.6–0.9 γ/g, riboflavin in the liver 1.2 γ/g, in the roe 5 γ/g, in testes 14 γ/g; nicotinic acid in the roe 15 γ/g.

TABLE 99. CHEMICAL COMPOSITION, %

Body parts	Moisture	Protein	Fat	Ash
Meat	78.9–81.7 / 80.8	15.8–18.5 / 17.2	0.1–1.0 / 0.2	1.1–1.4 / 1.3
Head	80.8–83.6 / 82.3	10.2–11.7 / 11.2	0.2–0.4 / 0.3	4.8–5.8 / 5.4
Liver	26.7–64.8 / 36.1	5.3	15.0–66.6 / 54.5	0.6
Bones	73.2–75.5 / 74.3	14.7–17.9 / 15.7	0.2–0.5 / 0.3	7.0–9.8 / 8.4
Roe in stage III of maturation	83.9	13.1	0.7	1.6
Testes in stage II of maturation	82.5	12.9	1.6	1.6
Viscera	80.3–81.0	11.8–12.3	0.7–1.1	3.0–4.5

For haddock and for all cod, a characteristic feature is the fairly constant chemical composition of the meat. It is a good table fish. In western Europe its value is 30% higher than cod.

Genus *Odontogadus*

Whiting—*Odontogadus merlangus*. Found in the North Atlantic, from the Mediterranean Sea to Iceland, as well as in the Barents Sea.

Reaches a length of 55 cm. Length of fish examined 18–25 cm, maximum

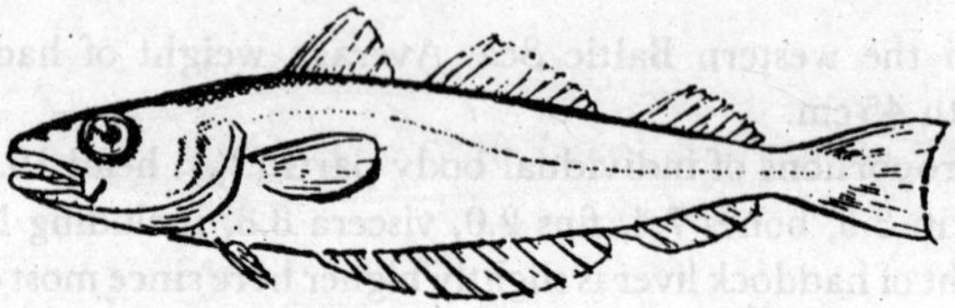

Fig. 96. Whiting—*Odontogadus merlangus*.

depth 3.5–5.5 cm, maximum thickness 2.0–3.0 cm, weight 56–161 g.

Weight proportions of individual body parts (%): head 23.9, with pieces of meat 2.6, trunk 58.0, meat with skin 51.9, bones 5.6, fins 2.0, viscera 14.5, including gonads 1.7 and liver 7.3.

Like haddock, the whiting is also a good table fish.

TABLE 100. CHEMICAL COMPOSITION, %

Body parts	Moisture	Protein	Fat	Ash
Meat	75.1–83.3 / 80.0	13.6–21.9 / 17.4	0.2–1.6 / 0.3	1.1–1.5 / 1.4
Liver	18.0–70.0 / 40.2	3.4–17.0 / 7.9	13.4–72.6 / 48.0	0.2–1.1 / 0.6
Head	78.2	14.7	0.3	5.3
Bones	72.2	19.2	0.6	9.1
Viscera	74.0	13.2	9.3	1.4

Genus *Pollachius*

This genus includes two species which live in the northern Atlantic Ocean.

Saithe or Pollock—*Pollachius virens*. A commercial fish, distributed in almost the same region of the North Atlantic as cod. Maximum standard length of pollock recorded 115 cm (at the age of 15 years).

Length of fish examined 44–88 cm, maximum depth 10 cm, maximum thickness 5.5–7.0 cm and weight 1,110–1,700 g.

Fig. 97. Saithe or Pollock—*Pollachius virens.*

Weight proportions of individual body parts (%): head 18.5, trunk 67.3, fins 3.5, meat 56.4, viscera 7.5.

TABLE 101. CHEMICAL COMPOSITION, %

Body parts	Moisture	Protein	Fat	Ash
Head	75.5	17.2	0.3	6.5
Meat	78.8–80.1 / 79.2	17.2–19.9 / 19.0	0.24–0.7 / 0.4	0.98–1.29 / 1.2
Liver	27.0	—	66.7	—

We have very little information about vitamin content in the body of pollock. The oil of the pyloric appendages of this fish has a potency of about 50,000 iu/g of vitamin A, the liver contains 7 γ/g and the roe 2 γ/g of riboflavin; the liver oil has a potency of 1,010 iu/g of vitamin A.

It is a valuable table fish. It is used for first and second courses as well as for cold snacks with various kinds of sauces. In western Europe stained with red coloring, it is used as imitation salmon.

Genus *Podonema*

Podonema—*Podonema longipes.* A non-commercial fish. Length of fish caught in September from Okhotsk Sea 51–63 cm, weight 630–810 g.

Weight proportions of individual body parts (%): head 21.8, bones and fins 12, liver 5.7, remaining viscera 3.3. Output of fillets 52.6%.

Chemical composition of meat (%): moisture 83.0, protein 11.0, fat 4.6, ash 1.3.

This fish resembles other cod, but the taste of its meat is inferior. Its liver is valuable as a raw material for oil extraction.

Genus *Theragra*

Walleye pollock—*Theragra chalcogramma.* A commercial fish, distributed in the Sea of Japan, the Okhotsk and Barents seas and the eastern Pacific

Fig. 98. Walleye pollock—*Theragra chalcogramma.*

Ocean. Length of the walleye pollock 31–55 cm, weight 200–1,400 g.

Weight proportions of individual body parts (%): head 15.8–25.7, trunk 52.7–58.2, fins 2.7–4.0, guts 5.0–11.0, liver 3.0–5.3, gonads 1.7–7.9.

Chemical composition of meat (%): moisture 81.1–83.2, protein 14.6–17.3, fat 0.4–0.9, ash 1.1–1.4.

The meat has high food value. However, in many fishing regions, for example, in the Sea of Japan, in the Tatar Strait, because of the larvae of *Liptostracom nabalia* it is not used as a food. Fish flour should be made from the pollock mince. The walleye pollock liver oil contains large quantities of vitamin A.

Genus *Urophycis*

This genus includes several species. They live mostly in the western Atlantic Ocean.

White hake—*Urophycis tenuis**. Found near Georges and the Newfoundland Banks. Length 120 cm, average length 70 cm, weight 18 kg, average weight about 4 kg.

Average length of specimens examined 41.6 cm (31–54), weight 840 g (220–1,650 g). Meat with skin 54.4%, viscera 12.0%, including liver 4.9%.

The meat of this hake is somewhat dry in consistency and in taste inferior to cod meat. It is used for first and second courses. The liver can be used for canning.

TABLE 102. CHEMICAL COMPOSITION, %

Body parts	Moisture	Protein	Fat	Ash
Meat	$\frac{81.0–83.0}{82.1}$	$\frac{15.2–17.6}{16.1}$	$\frac{0–0.4}{0.1}$	$\frac{0.8–1.3}{1.1}$
Liver	$\frac{31.7–61.7}{42.4}$	7.5	23.6–62.0	0.8
Roe in stage of III maturation	74.5	18.7	3.8	1.4
Testes in stage IV of maturation	85.3	12.3	0.7	1.1

Head	79.7–83.4 81.5	11.5–13.6 12.8	0.20–80.4*	4.2–6.0 5.9
Bones	75.9–80.5 78.2	13.7–16.0 14.6	0.3–0.5 0.4	5.1–7.8 6.5
Viscera	81.1–84.7 83.1	10.6–15.3 12.5	0.7–2.6 1.5	1.3–1.7 1.5

*Mistake in the original; should read 0.20 to 0.40.

Red hake—*Urophycis chuss*. Length of the red hake 22–75 cm, weight 140–6,400 g. Maximum length 1.2 m and maximum weight 18 kg. Found along the coast of North America.

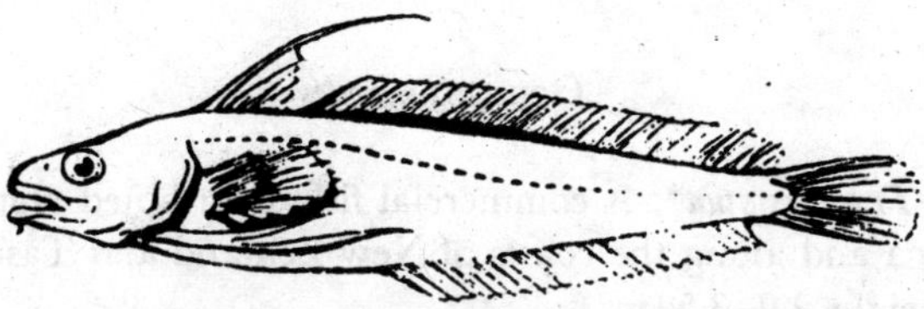

Fig. 99. Red hake—*Urophycis chuss.*

Weight proportions of individual body parts (%): head 18.6, trunk 67.6, meat 50.3, skin, bones and fins 21.5, viscera 8.5, including liver 2.0.

TABLE 103. CHEMICAL COMPOSITION, %

Body parts	Moisture	Protein	Fat	Ash
Meat	79.7–84.0 81.0	14.9–20.1 17.2	0.2–0.6 0.4	1.3–1.5 1.4
Liver	— 37.5	—	— 46.7	—
Viscera	— 80.2	— 15.6	— 2.0	— 1.5
Bones, fins and skin	— 78.8	— 14.8	— 0.9	— 5.5

The meat of the red hake is rough in consistency and in taste inferior to the meat of cod. First and second courses can be prepared from it. The liver can be used for canning. Oil may be extracted although its vitamin content is not very high; according to the result of analyses, it is 145 iu/g of oil.

Family GEMPYLIDAE—SNAKE MACKERELS

Genus *Rexea*

King barracuda—*Rexea solandri*. Distributed along the coast of southern Australia, Tasmania and New Zealand. Length of fish caught in December, 78 cm and weight 3,270 g.

Weight proportions of individual body parts (%): head 22.3, trunk 64.6, meat with skin 53.5, bones 10.1, fins 0.3, viscera 13.1, including liver 9.8.

Chemical composition of meat with skin (%): moisture 71.5, protein 19.5, fat 8.7, ash 1.35.

The fried meat of this fish is dense, somewhat dry with satisfactory taste qualities. It is best to freeze this fish with essential glazing for subsequent use in culinary products.

Genus *Thyrsites*

Snock—*Thyrsites atun**. A commercial fish, distributed along the Atlantic coast of Africa and along the coasts of New Zealand and Tasmania. Length 82–87 cm, weight 2.0–3.5 kg.

Weight proportions of individual body parts of fish 28–87 cm long weighing 200–3,460 g caught in December–May (%): head 14.8–22.8, trunk 65.3–73.1, fins 0.7–3.6, viscera 6.0–8.6, meat with skin 59.7–60.6, bones (backbone) 11.9–12.5. Output of fillets 55.0–57.4%. The skin of this fish is thin and extremely difficult to separate from the meat.

TABLE 104. CHEMICAL COMPOSITION OF MEAT, %

Place of catch	Moisture	Protein	Fat	Ash
Atlantic coast of Africa	64.2–75.5	21.1–22.3	0.8–13.0	1.0–1.8
Great Australian Bight	69.1–74.0	21.5–22.0	3.9–7.8	0.8–1.3

Data are available which suggest that the meat of large fish (length more than 60 cm) is infested with parasites. Because of the high protein content the meat is dense in consistency. After frying, it is dry. Smoked fish are high grade quality. The meat has long, compact intramuscular bones. Products canned in tomato sauce are mediocre in quality. From frozen fish shorter than 60 cm, hot smoked products should be prepared and fish longer than 60 cm in which the meat is infested with parasites should be used for fish meal.

Genus *Ruvettus*

Oilfish or Ruvetta—*Ruvettus pretiosus*. Caught with other fish in the southeastern Atlantic. Length of oilfish caught in December, 70 cm and weight 3 kg. Weight proportions of individual body parts (%): head 18, trunk 70, fins 2, viscera 10.

Chemical composition of meat (%): moisture 66.8, protein 14.9, fat 17.1, ash 1.2.

The taste qualities of this fish are high. It is an excellent table fish.

Family GOBIIDAE—GOBIES

These fish are distributed mainly in tropical and temperate waters. They are found in the Black and Caspian Seas and in the southern Soviet Far East waters. The important members are the sea raven and grubby.

Genus *Myoxocephalus*

Grubby—*Myoxocephalus aenaeus**. Length of fish caught in the northwestern Atlantic near Georges Bank 21 cm, maximum thickness 4 cm, maximum depth 6 cm and weight 175 g. Weight proportions of individual body parts (%): head 30, trunk with fins 57, viscera 13. Output of fillets 32%.

Chemical composition of meat (%): moisture 79.4, protein 19.1, fat 0.3, ash 1.2.

The taste of this fish is quite satisfactory. The meat should be used for culinary products.

Genus *Hemitripterus*

Sea raven—*Hemitripterus americanus*. Found in the northwestern Atlantic from Labrador to Chesapeake Bay. Length 35–40 cm, maximum length up

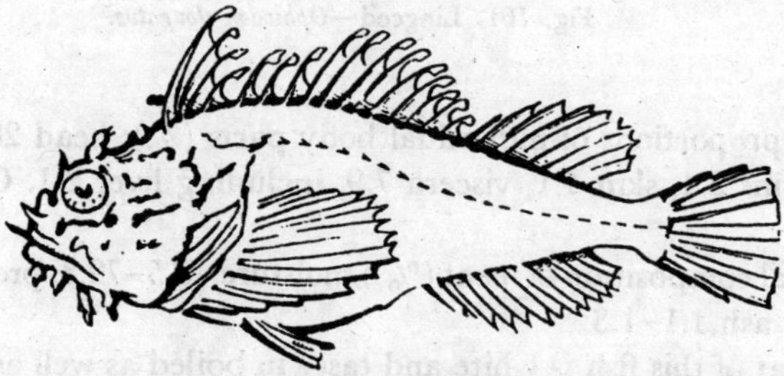

Fig. 100. Sea raven—*Hemitripterus americanus*.

TABLE 105. CHEMICAL COMPOSITION, %

Body parts	Moisture	Protein	Fat	Ash
Meat	81.0	16.1	1.7	1.2
Head, bones and fins	74.6	16.1	5.0	4.3
Viscera	82.6	10.7	5.6	1.1

to 70 cm. Length of goby caught in June, 36–41 cm, maximum depth 13–18 cm, maximum thickness 9–14, weight 1.4–2.1 kg.

Weight proportions of individual body parts (%): trunk with fins 50, head 28, intestines 22. Output of fillets 31%.

The taste of the meat of this goby is good. It should be used for culinary preparations.

Family HEXAGRAMMIDAE—GREENLINGS

Members of this family are widely distributed in the northern Pacific Ocean along the coasts of Asia and America as well as in the Bering and Okhotsk seas and the Sea of Japan.

Genus *Ophiodon*

Lingcod—*Ophiodon elongatus.* A commercial fish, widely distributed along the Pacific coast of North America, northwestern coast of Alaska to southern California. Length 32–62 cm, weight 1,100– 3,800 g.

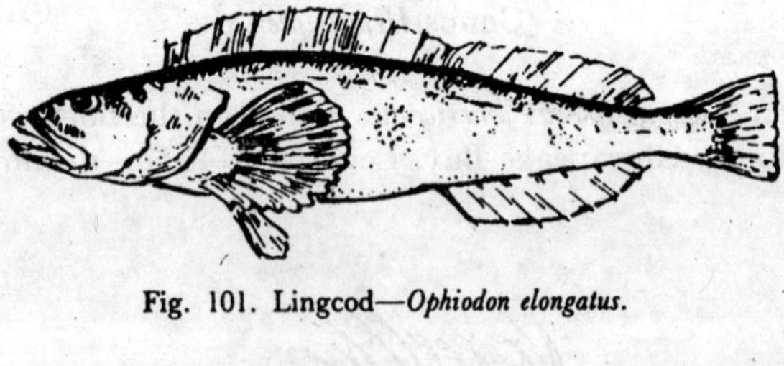

Fig. 101. Lingcod—*Ophiodon elongatus.*

Weight proportions of individual body parts (%): head 28.8, trunk 58.3, bones 6.7, fins 5.4, skin 4.1, viscera 7.9, including liver 2.1. Output of fillets 47.4%.

Chemical composition of meat (%): moisture 76.5–79.3, protein 16.4–18.7, fat 1.2–6.2, ash 1.1–1.3.

The meat of this fish is white and tasty in boiled as well as in fried forms. This is a good table fish. It should be used for culinary preparations.

Genus *Pleurogrammus*

Atka mackerel—*Pleurogrammus monopterygius*. A commercial fish, distributed in the Bering Sea, from the northern Kuril Islands to the Koryaksk coast and Chukotka and off the Aleutian Islands to the western tip of Alaska. Attains an average length of 38–44 cm and weighs 1 kg.

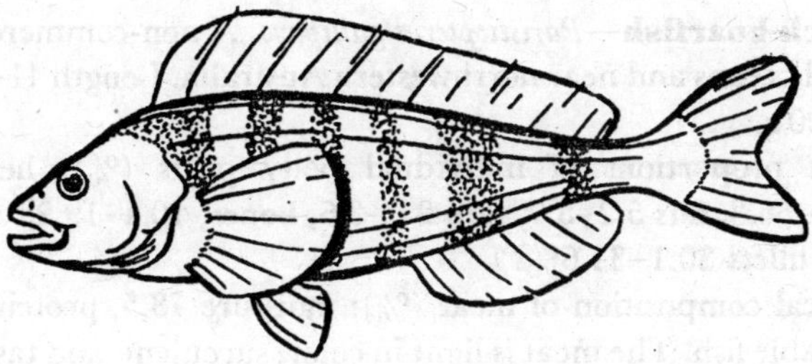

Fig. 102. Atka mackerel—*Pleurogrammus monopterygius*.

Length of atka mackerel examined 25.6–36.4 cm, weight 195–560 g.

Weight proportions of individual body parts (%): head 19.3–22.9, meat with skin 55.8–58.7, bones and fins 11.2–16.2, viscera 7.1 9.5, including liver 1.3.

Chemical composition of meat (%): moisture 74.1–81.9, protein 14.1–18.4, fat 2.2–5.5, ash 1.2–1.9.

The meat of atka mackerel is white. After boiling, it becomes mildly layered. It can be used for canned snacks and smoked products.

Genus *Hexagrammos*

Otaki's greenling—*Hexagrammos otakii**. A commercial fish, distributed along the coasts of Japan, Korea and northern China. Adult fish attain length of 30 cm and more. Length of greenling examined 32.1–40.1 cm, weight 385–705 g.

Weight proportions of individual body parts (%): head 19.5–21.0, meat with skin 57.5–57.9, fins and bones 9.5, viscera 7.7–10.6, including liver 1.6.

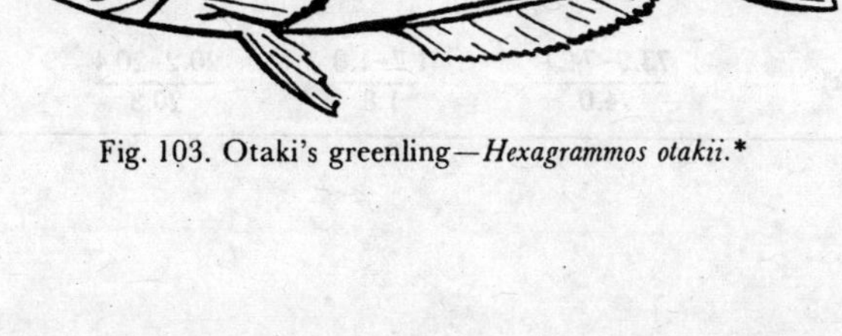

Fig. 103. Otaki's greenling—*Hexagrammos otakii*.*

Chemical composition of meat (%): moisture 76.6, protein 19.5–20.3, fat 1.7–3.8, ash 0.9–1.2.

The fish has white meat with a pleasant taste. It can be used as a table fish and for canned snacks.

Family HISTIOPTERIDAE—BOARFISHES

Peacock boarfish—*Paristiopterus gallipavo*. A non-commercial fish, distributed in all oceans and near northwestern Australia. Length 41–52 cm, weight 1,320–2,500 g.

Weight proportions of individual body parts (%): head 32.6–35.6, trunk 44.0–56.8, fins 5.2–8.7, skin 2.0–3.6, bones, 10.4–19.2, viscera 5.0–9.3. Output of fillets 30.1–34.6%.

Chemical composition of meat (%): moisture 78.5, protein 17.8, fat 1.2, ash. 1.4. Table fish. The meat is light in color, succulent, and tasty. Broth dark in color and not tasty.

Family HISTIOPHORIDAE—BILLFISHES

This family includes three genera (*Histiophorus, Makaira, Tetrapturus*) with eight species which are distributed mainly in the subtropical and tropical waters of the world's ocean. The members of this family are valuable commercial fish.

Genus *Makaira*

Blue marlin—*Makaira ampla*. Distributed in the tropical, subtropical and temperate waters of all oceans, along the coasts of northwestern Europe and near the northern equatorial currents, along the American coast from Montevideo to the Caribbean Sea and Bermuda Islands and near the eastern coast of America. In the equatorial zone found off the coast of Australia. A large pelagic fish attaining length of 4 m and weight of 300 kg.

Length of fish caught off northern Aden 200 cm and weight 50 kg.

TABLE 106. CHEMICAL COMPOSITION, %

Meat	Moisture	Protein	Fat	Ash
Abdominal	72.8–73.1 / 72.9	2.6–3.0 / 2.8	22.2–22.8 / 22.5	1.3–1.3 / 1.3
Dorsal (back)	73.9–74.1 / 74.0	1.7–1.8 / 1.8	20.2–20.4 / 20.3	1.2–1.2 / 1.2

The meat is light in color with good taste. It can be used for semi-finished products of culinary preparations and smoked products.

Striped marlin—*Makaira mitsukurii.* Distributed in the tropical and subtropical waters of the Pacific and Indian Oceans; in the Pacific Ocean, found in the west near southern Hokkaido Island, along the coast of Japan, in the South China and East China Seas, along the coast of Korea, near Taiwan as well as the Philippine and Hawaiian Islands; in the east found along the American coast fro... Chile to California, off the coast of New Zealand; in the Indian Ocean along the coast of Sri Lanka. Found rarely in the equatorial waters. Attains length of 3 m. Length of fish caught in the Pacific Ocean 177 cm and weight 23.1 kg.

Fig. 104. Striped marlin—*Makaira mitsukurii.*

Weight proportions of individual body parts (%): head 18.2, meat 59.5, skin 1.3, bones 3.9, fins 1.4, viscera 13.0, including liver 1.2.

This fish is best used to make culinary preparations.

TABLE 107. CHEMICAL COMPOSITION, %

Body parts	Moisture	Protein	Fat	Ash
Meat	79.5	19.8	0.1	1.0
Head	73.1	—	2.1	5.2
Bones	68.5	—	0.8	10.8

Black marlin—*Makaira mazara.* Found in the Indian Ocean, along the eastern coast of Africa and along the southern coast from Cape Town to East London, in the Pacific Ocean in the east from the Gulf of California to Peru, in the central part near the Hawaiian Islands, in the south around Australia and New Zealand and in the west from the Philippines to Korea.

Attains length of 4.8 cm, weight 500 kg (usually 130–180 kg). Length of fish caught in July near Madagascar Island 189 cm and weight 50 kg.

Weight proportions of individual body parts (%): head 17.2, trunk 78.2, light colored meat 61.4, dark colored meat 5.8, skin 5.4, bones 5.5, fins 4.1.

Chemical composition of light colored meat (%): moisture 77.4, protein 20.3, fat 0.3, ash 1.2.

Fig. 105. Black marlin—*Makaira mazara.*

The cooked meat of this fish is gray with a pinkish tinge, dense, rubber-like moderate in taste. The soup is sour. The meat may be infested with parasites. This fish can be used for culinary products and for fish meal.

Genus *Histiophorus*

Sailfish—*Histiophorus americanus*. Distributed in tropical and subtropical waters in all oceans. Found in large numbers in the Atlantic Ocean in the waters of the north and south equatorial currents as well as off the coast of west Africa near Cape Verde. Also found in the Gulf of Guinea. In the Indian Ocean commercially exploitable concentrations of sailfish spotted near Madagascar,

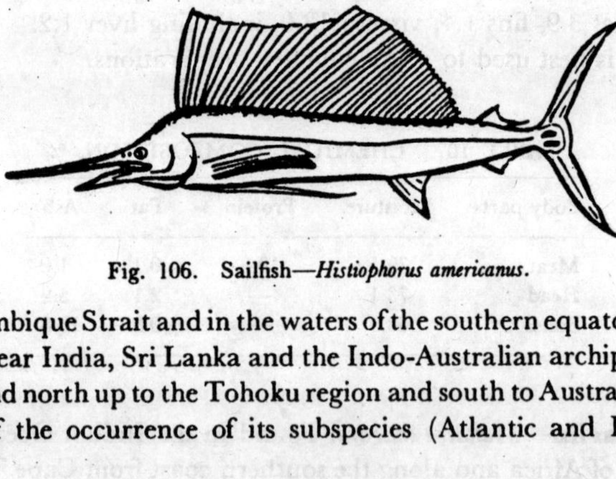

Fig. 106. Sailfish—*Histiophorus americanus.*

in the Mozambique Strait and in the waters of the southern equatorial currents. Also found near India, Sri Lanka and the Indo-Australian archipelago. In the west it is found north up to the Tohoku region and south to Australia. There is a possibility of the occurrence of its subspecies (Atlantic and Indo-Pacific).

TABLE 108. WEIGHT PROPORTIONS OF INDIVIDUAL PARTS OF SAILFISH WEIGHING 4–47 kg, %

Head	Trunk	Meat	Skin	Bones	Fins	Viscera		blood
						whole	including liver	
7.0–18.0 / 13.5	61.5–70.0 / 69.6	54.0–63.4 / 58.7	5.0	6.6	3.6–7.8 / 5.3	5.3–9.0 / 7.1	0.2	6.9–11.2 / 7.9

TABLE 109. CHEMICAL COMPOSITION, %

Body parts	Moisture	Protein	Fat	Ash
Meat				
total	74.0–78.2 / 76.2	19.6–22.8 / 21.8	0.1–1.5 / 0.8	0.3–1.4 / 1.2
abdominal part	69.0	14.2	15.8	1.9
middle of back	75.[illegible]	23.0	2.1	1.7
tail part	74.0	24.6	0.2	1.4
Offal (head, viscera, bones, fins)	72.4	18.5	1.4	7.7

Attains large size and weight. Length of body (with sword) usually 90–320 cm, weight 20–200 kg.

The cooked meat of the sailfish is pinkish. It has very good taste (without the typical fishy smell). The sailfish can serve as a raw material for conserves in oil and for cured balyks and culinary preparations.

Family HOPLEGNATHIDAE—KNIFEJAWS

Knifejaw—*Ostorhinchus conwaii*. Fish has little commercial value; distributed in the southern regions of Australia and Tasmania. Lives mainly at depths of 80–160 m, rarely at 330 m. Length 28–39 cm, weight 650–1,535 g.

Weight proportions of individual body parts of fish caught in January (%): head 24.2–32.0, trunk 49.8–62.7, meat with skin 41.7–44.1, bones 7.1–8.9, fins 4.2, viscera 12.8–15.3, including liver 0.9–1.4, gonads 3.5.

Chemical composition of meat with skin (%): moisture 76.0–79.3, protein 16.4–19.4, fat 1.2–6.2, ash 1.1–1.6.

This is a table fish. It has white, compact meat with good taste. This fish should be processed for freezing for subsequent culinary preparations.

Family LABRIDAE—WRASSES

These fish live mainly in tropical and temperate coastal waters. Some species may penetrate up to the coast of Norway along the European coast of the Atlantic Ocean and up to Cape Cod along the American coast. In our waters there are a few species of wrasses in the Black Sea and one species in the Far East.

Genus *Hemigymnus*

Tautog—*Hemigymnus melanopterus*. Inhabitant of tropical and temperate waters of the Atlantic Ocean. Weight proportions of individual parts of the

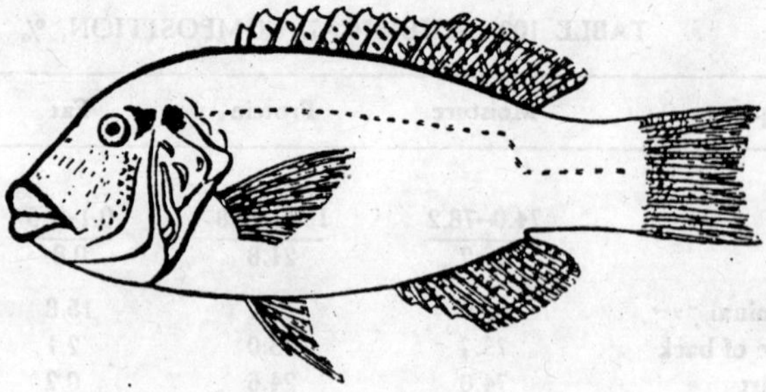

Fig. 107. Tautog—*Hemigymnus melanopterus.*

body of this fish caught in April along the coast of northwest Africa 30–49 cm long, with maximum depth 9–21 cm, maximum thickness 4–6 cm and weight 0.5–4.0 kg (%): head 18–19, trunk 66–68, viscera 13–16. Output of fillets 45–50%.

Chemical composition of meat (%): moisture 75–79, protein 18.8–20.6, fat 1.5–3.3, ash 0.7–1.4.

The quantity of macro- and micro-elements in the meat of this fish was (mg %): potassium 991, calcium 90, magnesium 19, copper 0.7, iron 3.4, phosphorus 400, iodine 0.21, manganese 0.07, cobalt 0.62.

Tautog is a good table fish. Fried or boiled, the meat is white, soft, juicy, tasty, without the specific fish smell. The soup is rich in fat and tasty.

Genus *Tautogolabrus**

Cunner—*Tautogolabrus adspersus**. Length of fish caught in northwest Atlantic, near Georges Bank, 24–27 cm, maximum depth 9–10 cm, maximum thickness 4–5 cm and weight 440–510 g. Weight proportions of individual body parts (%): head 22, trunk with fins 70, viscera 8. Output of fillets 41%.

The meat of fried fish is pink, soft, juicy and tasty, but it is bony and the bones are large in size. It can be used for preparing culinary products.

TABLE 110. CHEMICAL COMPOSITION, %

Body parts	Moisture	Protein	Fat	Ash
Head, bones and fins	69.6	18.8	3.6	8.0
Meat	79.8	18.6	0.3	1.3
Viscera	76.0	13.8	7.6	1.6

Genus *Lachnolaimus*

Hogfish—*Lachnolaimus maximus*. Found in the western Central Atlantic. Length of fish caught in the Gulf of Mexico 35–41 cm, maximum depth 12–16 cm, maximum thickness 4.0–4.5 cm, weight 0.8–1.95 kg. Weight proportions of individual body parts (%): head 19–26, trunk 67–76, viscera 5–6. Output of fillets 50–55%.

Chemical composition of meat (%): moisture 76.8–77.7, protein 19.4–21.0, fat 0.3–2.5, ash 1.0–1.3.

This is a good table fish. The meat is juicy and tasty when boiled or fried.

Family LATRIDAE—TRUMPETERS

Bastard trumpeter—*Latris forsteri*. Fish of low commercial value. Lives in the coastal waters of southern Australia, near Tasmania. Length 23–51 cm, weight 2–3 kg.

Weight proportions of individual body parts of fish caught in October–May, %: head 20.5–33.0, meat with skin 43.8–59.3, bones and fins 13.5–15.3.

Chemical composition of meat with skin (%): moisture 72.1–76.2, protein 20.3–21.0, fat 2.0–6.0, ash 1.4–1.5.

This is an excellent table fish. The meat is white with brown shading, sweet, excellent taste in fried and boiled forms. Excellent culinary preparations can be made from the meat. This fish should be processed for freezing with essential glazing.

Family LAMPRIDAE—OPAHS

Opah—*Lampris regius*. A very large fish (more than 1 m). Lives in the deep waters of the Indian Ocean. Constitutes an incidental catch in tuna fishing.

Length of opah caught in July in the Mozambique Strait about 100 cm, weight around 45 kg (44.7–46.2).

After cooking, the light colored opah meat is very tasty and has a soft consistency. The fat has a pinkish tinge. The dark colored meat is not tasty.

The opah is a good table fish. Cured balyks can be prepared from it. The dark meat and other waste obtained during dressing should be used for fish meal.

Family LEIOGNATHIDAE*—SLIMYS OR PONYFISHES

Ponyfish—*Leiognathus rivulata*. A non-commercial fish. Caught in the South China Sea and in the south-western Pacific Ocean. Weight of fish 150–180 g.

TABLE 111. WEIGHT PROPORTIONS OF INDIVIDUAL BODY PARTS OF OPAH, %

Head*	Trunk	Meat		Skin	Bones	Scales	Fins	Viscera	
		light	dark					whole	including liver
21.6–26.0 / 24.2	66.7–70.4 / 68.7	43.5–45.0 / 44.3	14.0–16.1 / 15.0	2.0–2.3 / 2.2	6.9–7.3 / 7.1	0.5	1.5–1.9 / 1.7	3.9–4.8 / 4.3	0.4–0.9 / 0.7

* Head separated by an oblique cut with the gills and shoulder bones.

TABLE 112. CHEMICAL COMPOSITION OF OPAH, %

Body parts	Moisture	Protein	Fat	Ash
Light meat from back	72.6–73.9 73.2	23.4–23.7 23.6	1.5–2.3 1.9	1.0
Dark meat	74.8	23.4	0.9	1.0
Light meat from abdomen	47.3–49.0 48.1	17.6–17.8 17.7	32.6–34.2 33.4	0.6–0.7 0.7

Fig. 108. Opah—*Lampris regius.*

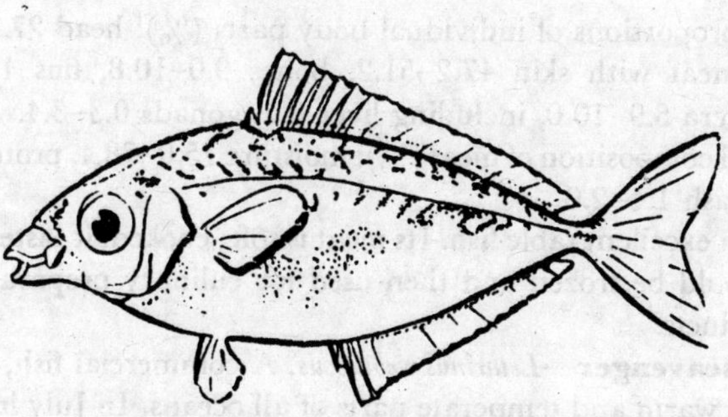

Fig. 109. Ponyfish—*Leiognathus* rivulata.*

Weight proportions of individual body parts (%): head 32.7, trunk 53.0, viscera 3.8, including liver 0.7, bones and fins 18.3–20.0. Output of fillets with skin 40.0%.

Chemical composition of meat (%): moisture 77.7–79.6, protein 19.0–21.2, fat 0.1–0.3, ash 1.0–1.2.

This is a small fish. It does not have much food value. It should be processed for fish meal.

Family LETHRINIDAE—SCAVENGERS

Genus *Lethrinus*

Longsnout scavenger—*Lethrinus miniatus*. A commercial fish, caught in the Gulf of Tonkin and eastern Indian Ocean.

Length of this scavenger caught throughout the year in the Gulf of Tonkin 18–56 cm, weight 145–2,900 g; for fish caught in September–December in the eastern Indian Ocean, 28.5–58.5 cm and 480–4,260 g.

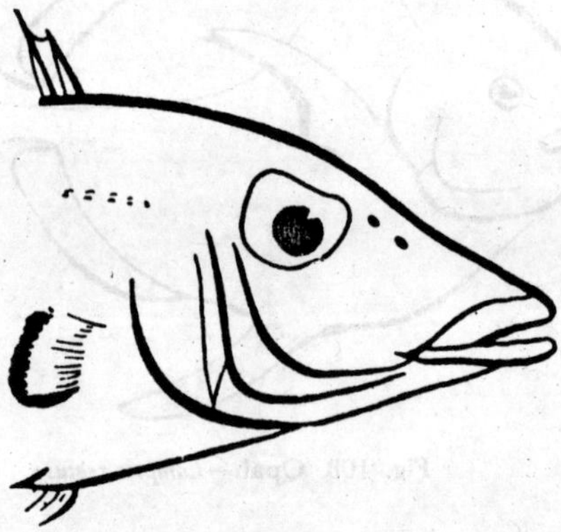

Fig. 110. Longsnout scavenger—*Lethrinus miniatus*.

Weight proportions of individual body parts (%): head 27.2–31.2, trunk 58.0–60.2, meat with skin 47.2–51.2, bones 9.0–10.8, fins 1.6–2.1, scales 1.0–2.0, viscera 6.9–10.0, including liver 0.7, gonads 0.3–3.4.

Chemical composition of meat (%): moisture 75.0–78.4, protein 19.0–20.8, fat 0.1–1.0, ash 1.4–2.0.

This is an excellent table fish. Its meat is soft; cooked, it tastes like chicken. The fish should be frozen and then used for culinary preparations and hot smoked products.

Dusky scavenger—*Lethrinus nebulosus*. A commercial fish, widely distributed in the warm and temperate parts of all oceans. In July in catches near Sri Lanka there was a predominance of fish about 28 cm long (25.4–32.3),

TABLE 113. WEIGHT PROPORTIONS OF INDIVIDUAL BODY PARTS, %

Head		Meat	Bones with skin	Scales	Fins	Viscera
whole	including meat of occiput and nape					
33.5–38.1 / 35.8	3.7	41.8–45.3 / 43.8	9.7–10.9 / 10.2	2.7	2.7	4.2

weight about 620 g (455–975), maximum depth 10.3 cm (9.5–12.5), maximum thickness 3.8 cm (3.0–4.8).

Chemical composition of meat (%): moisture 76.9, protein 20.5, fat 0.3, ash 1.4. After cooking, meat is white, soft, lean and mediocre in taste. The soup is tasteless. This fish is recommended for use as a table fish for second courses with spices and pungent sauces and for making hot smoked products.

Genus *Pentapodus*

Large eye bream—*Pentapodus dux*. Caught in the South China Sea. Length 18–56 cm, weight 145–2,900 g.

Weight proportions of individual body parts (%): head 18.0, trunk with fins 77.0, viscera 5.0, including liver 1.2.

Chemical composition of meat (%): moisture 75.0–78.0, protein 19.0–20.8, fat 0.1–1.0, ash 1.4–2.0.

This is a table fish. The meat is white and lean with mediocre taste.

Family LATILIDAE—TILEFISHES

Tilefish—*Lopholatilus chamaeleonticeps*. Distributed in the western Atlantic where it is caught in small numbers with other fish. Three specimens examined; two specimens 54–56 cm long, weighing 2.5–2.9 kg (May and December) and one specimen 73 cm long weighing 9.8 kg.

Chemical composition of this tilefish is characterized by fairly high protein and low fat content in the meat. The liver also contains relatively less fat (7.4%).

This is a table fish. Fish are fairly large with large scales; easy to descale; the absence of small bones makes it easy to dress at community kitchens and in homes. After boiling or frying the taste and consistency of the meat are good. This fish should be frozen as beheaded trunks and fillets.

TABLE 114. WEIGHT PROPORTIONS OF INDIVIDUAL BODY PARTS, %

Head	Trunk	Meat	Skin	Bones	Fins	Scales	Viscera			
							whole	including		
								gonads	liver	waste
20.6–23.0* / 21.5	67.3–69.7 / 68.7	51.9–57.4 / 55.4	2.9–3.2 / 3.0	7.5–11.5 / 9.5	1.0–1.6 / 1.3	1.0–1.3 / 1.1	6.1–7.8 / 6.7	0.1–1.0 / 0.3	1.3–1.9 / 1.6	1.5
30.0**	—	48.0	2.4	12.2	1.2	0.7	4.1	1.3	1.2	2.4

* Head separated by semicircular cut.
** Head separated by straight cut.

TABLE 115. CHEMICAL COMPOSITION, %

Body parts	Moisture	Protein	Fat	Ash
Head	71.4	15.2	6.2	6.5
Meat				
small individuals	78.1	18.2	1.8	1.2
large individuals	79.8	16.9	0.8	1.2
Bones	62.1	16.3	11.5	9.4
Viscera	80.4	14.1	2.2	1.3
Roe of stages				
II–VI	84.7	12.3	0.8	1.6

Family LUTIANIDAE—SNAPPERS

This family includes 20 genera and about 250 species. These are marine fish living in the tropical waters of the Indian, Atlantic and Pacific Oceans. They are usually caught with other commercial fish.

Genus *Aprion*

Members of this genus are distributed in the tropical and subtropical waters of the Indian, Pacific and Atlantic Oceans.

Snapper—*Aprion micropterus*. Distributed in the tropical and subtropical parts of the Indian and Pacific Oceans, from the southeastern coasts of Africa and Madagascar to the eastern coasts of Australia, the Philippine and the Hawaiian Islands. A large fish weighing more than 1 kg. In external appearance, resembles sea perch. Length of fish caught on the west Indian shelf 48.2 cm, weight 2,980 g.

Weight proportions of individual body parts (%): head 26.7, meat with skin 48.2, bones 12.0, scales 1.8, fins 3.4, viscera 5.8.

Chemical composition of meat (%): moisture 78.7, protein 18.3, fat 0.5.

After boiling, the meat has good taste, but a somewhat dry consistency. The meat can be sold in frozen form as well as after preparing canned products in tomato sauce, semi-finished products and culinary preparations.

Genus *Lutianus*

This is the most widely distributed genus. Its members live in the tropical zones of the Atlantic, Indian and Pacific Oceans.

Blue banded snapper—*Lutianus kasmira*. A widely distributed fish in the tropical Indo-Pacific zone. Reported from the Seychelles Islands and on the shelf of Somali and Oman. Caught in small numbers in trawl catches. Attains a length of 40 cm.

Weight proportions of individual body parts of fish 20 cm long, weighing 230 g caught in the Red Sea (%): head 32.6, trunk 54.0, scales 2.2, fins 2.2, viscera 8.6.

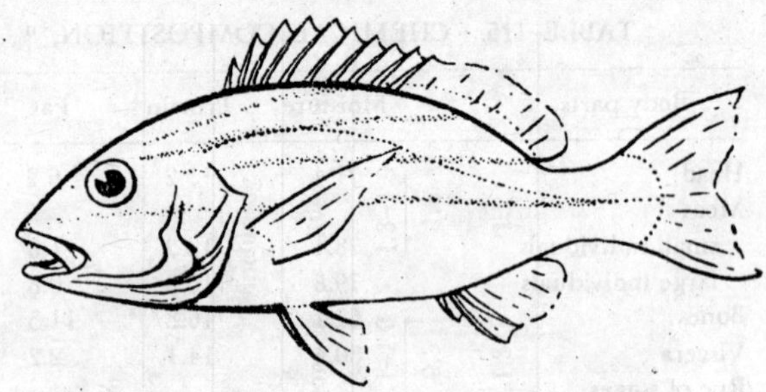

Fig. 111. Blue banded snapper—*Lutianus kasmira.*

Chemical composition of meat (%): moisture 77.6, protein 18.4, fat 0.7, ash 1.3.

The meat of this fish should be frozen and subsequently used for culinary products.

Chestnut snapper—*Lutianus agenes**. Length of fish caught in the eastern Central Atlantic, near Takoradi, 79 cm, maximum depth 26 cm, maximum thickness 12 cm and weight 10.2 kg.

Chemical composition of meat (%): moisture 78.8, protein 19.1, fat 0.7, ash 1.4.

The meat of this snapper is pinkish with dark streaks. The meat of fried snapper is somewhat dry but tasty. It is a good table fish which can be used for fillets.

Red emperor—*Lutianus sebae*. A widely distributed fish in the tropical Indo-Pacific zone. Reported from the Gulf of Aden and near Socotra Island. Attains a length of 90 cm.

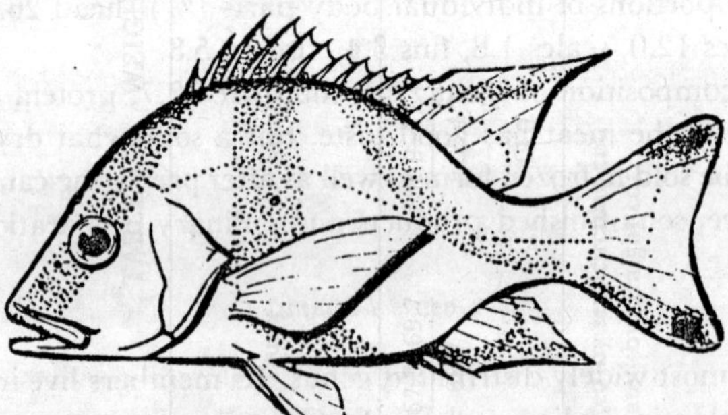

Fig. 112. Red emperor—*Lutianus sebae.*

Weight proportions of individual body parts of fish 37.8 cm long, weighing 1,253 g (%): head 38.0, trunk 48.0, scales 2.6, fins 3.0, viscera 6.5.

Chemical composition of meat (%): moisture 76.4, protein 20.7, fat 0.6, ash 1.3.

The meat of this fish should be frozen and then used for culinary products and semi-finished products.

Pensacola snapper—*Lutianus aya*. Distributed in the Gulf of Mexico especially along the Florida coast. In the western Atlantic Ocean this species is known from Massachusetts (USA) to Rio de Janeiro (Brazil). Average weight of this snapper 3 kg, maximum weight up to 25 kg.

TABLE 116. WEIGHT PROPORTIONS OF INDIVIDUAL BODY PARTS, %

Head	Meat	Fins	Bones	Viscera
27.5–31.9 / 29.5	46.6–51.4 / 49.0	5.7	9.1	6.2

TABLE 117. CHEMICAL COMPOSITION, %

Moisture	Protein	Fat	Ash
76.8–78.6 / 77.7	19.2–20.9 / 20.1	0.7–1.5 / 1.1	1.3–1.5 / 1.4

The meat has good taste qualities. It should be frozen and then used in culinary products.

Bright snapper or mutton snapper—*Lutianus analis*. Found in the Gulf of Mexico, in the western Atlantic Ocean, from Massachusetts to Rio de Janeiro (Brazil).

Weight varies between 0.2–9.0 kg, maximum weight 15 kg.

This is a good table fish. Its meat has good taste qualities. It should be frozen and then used for culinary products.

TABLE 118. WEIGHT PROPORTIONS OF INDIVIDUAL BODY PARTS, %

Head	Trunk	Meat	Skin	Scales	Bones	Fins	Viscera
28.5–34.2 / 31.3	56.5–63.5 / 60.0	41.8–50.0 / 44.1	3.0	2.3–2.9 / 2.6	8.3–12.6 / 10.2	2.2–4.5 / 3.4	5.0–5.7 / 5.4

TABLE 119. CHEMICAL COMPOSITION, %

Moisture	Protein	Fat	Ash
76.8–80.5 / 78.6	16.1–19.6 / 18.9	0.3–1.7 / 1.0	1.3–1.6 / 1.4

Lane snapper—*Lutianus synagris*. Distributed in the western Atlantic Ocean from the Bermuda Islands in the north to the Brazilian coast in the south. Also found in the Gulf of Mexico, especially in large quantities in the waters around Cuba. Length 18–20 cm.

TABLE 120. WEIGHT PROPORTIONS OF INDIVIDUAL BODY PARTS OF LANE SNAPPER WEIGHING 340–395 g, %

Head	Meat	Scales	Bones	Viscera	Fins
25.6–33.3 / 28.8	45.6–55.0 / 50.3	1.9	7.9	3.5–4.7 / 4.0	3.5–4.7 / 3.8

TABLE 121. CHEMICAL COMPOSITION OF MEAT, %

Months of catch	Moisture	Protein	Fat	Ash
Southwestern Central Atlantic				
	76.2	21.0	1.3	1.5
Gulf of Mexico				
April	76.7	20.0	1.9	1.4
March	75.5	21.8	1.3	1.4

This fish is recommended for culinary products.

Cubera snapper—*Lutianus cyanopterus*. This fish lives mostly in the Gulf of Mexico.

TABLE 122. RATIO OF INDIVIDUAL BODY PARTS OF LUTIANUS CYANOPTERUS WEIGHING 673–1,420 g, %

Head	Meat	Skin, scales, bones	Fins	Viscera
25.7–27.0 / 26.4	48.0–50.9 / 49.5	11.9	5.6	6.0–7.1 / 6.6

Chemical composition (%): moisture 76.6 (75.8–77.4), fat 1.6 (1.2–1.9), protein 20.3 (19.6–20.9), ash 1.3.

Gray snapper—*Lutianus griseus*. Found in the Gulf of Mexico; also occurs in large numbers off the coasts of Florida, the Dry-Tortugas and Cuba. Enters the less saline parts of the sea and the estuaries of rivers. In the western

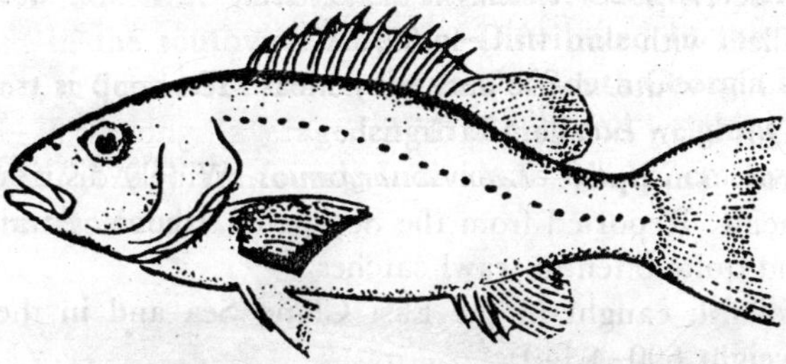

Fig. 113. Gray snapper—*Lutianus griseus.*

Atlantic Ocean, recorded from the Bermuda Islands and Massachusetts (USA) to Rio de Janeiro (Brazil). Average weight of about 1 kg, maximum 20 kg.

Weight proportions of individual body parts of fish weighing 820 g and 31 cm long (%): head 18.3, trunk 72.0, meat 50.0, viscera 4.3, fins 4.9, skin, scales and bones 20.3.

Chemical composition of meat (%): moisture 73.6, protein 20.6, fat 4.5, ash 1.3.

This fish should be used for culinary preparations and smoked products.

Olive striped snapper—*Lutianus vitta.* Lives near northwest Australia. Length 23.5–30.5 cm, weight 285–510 g.

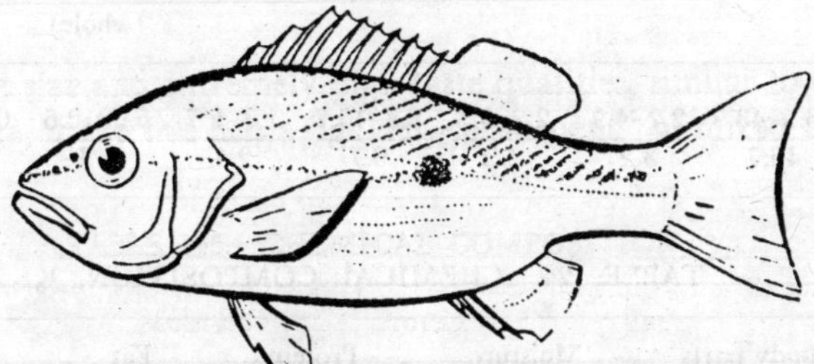

Fig. 114. Olive striped snapper—*Lutianus vitta.*

Weight proportions of individual body parts (%): head 24.7–37.6, trunk 52.0–64.4, fins 4.1–4.3, viscera 1.7–5.0, scales 1.7–3.9, bones 10.8–11.7. Output of fillets with skin 41.1–52.0%.

Chemical composition (%): moisture 76.0, protein 21.5, fat 1.0, ash 1.5.

The meat is white and slightly hard. The soup is rich and tasty. The skin is hard; it should be removed before processing.

Redfin snapper—*Lutianus erythropterus.* Found near northern Australia. Length 39–41 cm, weight 1,400–1,430 g.

Weight proportions of individual body parts (%): head 26.7–28.6, trunk

58.5–60.1, viscera 5.0–7.7, fins 3.9–5.5, scales 2.7–3.2, bones 12.1–13.2. Output of fillets with skin 46.1–46.5%.

This fish has white, dense and tasty meat. The soup is transparent, rich and tasty. This is an excellent table fish.

Blackbrow snapper—*Lutianus sanguineus*. Widely distributed species in the Indo-Pacific. Reported from the Seychelles Islands. Attains a length of 20 cm. Found quite often in trawl catches.

Length of fish caught in the East China Sea and in the Persian Gulf 33–67 cm, weight 600–4,140 g.

Of all the parts of the viscera, the liver is the most important. It has been established that the liver has the maximum fat content in July–August. In this period the liver oil is saturated with vitamin A (21,000–155,000 iu/g).

The meat of this fish is tasty, with soft consistency. The meat of the abdomen is fattier than the meat from the back side. The soup has mediocre taste. The meat is recommended for use in smoked curing and for semifinished and culinary preparations. In fish caught in the southwestern Pacific Ocean parasites have sometimes been observed in the abdominal cavity and at the attachment of the gillarch. Therefore, dressing should be done immediately after catching.

TABLE 123. WEIGHT PROPORTIONS OF INDIVIDUAL BODY PARTS OF BLACKBROW SNAPPER, %

Head	Meat	Skin	Scales	Bones	Fins	Viscera (whole)	Liver	Gonads
23.0–27.8 / 25.8	43.5–48.9 / 46.7	2.2–4.9 / 3.2	2.2–3.3 / 2.9	8.5–10.7 / 9.7	1.7–2.7 / 2.4	5.0–10.6 / 4.3	0.7–1.8 / 1.3	1.3

TABLE 124. CHEMICAL COMPOSITION, %

Body parts	Moisture	Protein	Fat	Ash
Meat	76.0–78.6 / 77.0	18.7–21.1 / 19.9	0.5–2.7 / 1.9	1.3–1.5 / 1.4
Viscera (whole)	73.0	13.2	8.7	2.5
Liver	79.8	23.6	5.2	1.4

Golden snapper—*Lutianus fulviflamma*. Lives near northwest Australia. Length of fish 24.0–29.5 cm, weight 350–570 g.

Weight proportions of individual body parts (%): head 29.7–33.0, viscera 3.1–3.3, fins 2.3–2.7, scales 1.8–2.6, bones 9.2–12.1. Output of fillets with skin 49.4–49.6%.

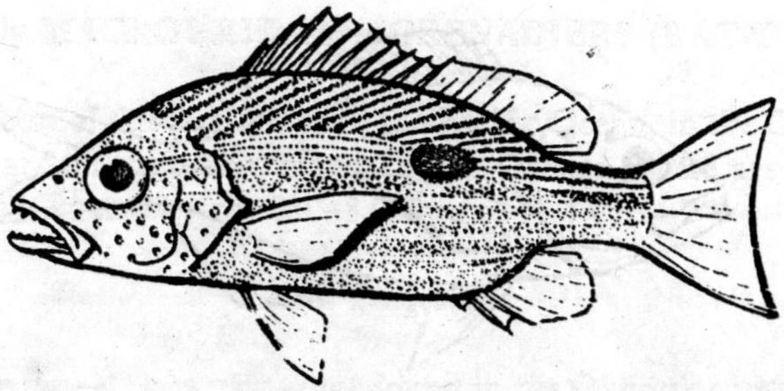

Fig. 115. Golden snapper—*Lutianus fulviflamma.*

Chemical composition of meat (%): moisture 77.8–78.0, protein 20.1, fat 0.1–0.2, ash 1.2.

It has denser and tastier meat. The soup is transparent, rich and tasty. Golden snapper is a good table fish.

Genus *Ocyurus*

Yellowtail snapper—*Ocyurus chrysurus.* Known to occur in the western Atlantic Ocean from the Bermuda Islands and Massachusetts (USA) to Rio de Janeiro (Brazil). Found in the Gulf of Mexico, especially in large numbers along the coasts of Cuba, Florida and the Yucatan Peninsula. In the eastern Atlantic Ocean found in the Bay of Biscay.

In catches individuals are usually caught 16–30 cm long (approximately

TABLE 125. WEIGHT PROPORTIONS OF INDIVIDUAL BODY PARTS, %

Head	Trunk	Meat	Scales	Bones	Fins	Viscera
19.5–22.6 / 21.0	61.0–72.1 / 65.4	48.4–59.2 / 52.5	2.3	9.6	5.4–7.3 / 6.4	6.8–9.8 / 8.2

three years old), rarely 27–30 cm (four and five years old).

The yellowtail snapper is a good table fish. This fish should be frozen and culinary preparations made later.

Genus *Pristipomoides*

Blue snapper—*Pristipomoides microdon.* Little commercial significance. Caught in the Gulfs of Siam and Tonkin. Length 16.2–51 cm, weight 470–2,300 g. Fish found rarely in catches.

Fig. 116. Blue snapper—*Pristipomoides microdon.*

Weight proportions of individual body parts (%): head 16.2–26.1, trunk 62.7–63.5, fins and bones 12.0, viscera 9.1–9.5, liver 0.8–1.2.

It has been reported that the liver of blue snapper contains 6.1% fat and 1 g of the liver oil contains 95,400 iu of vitamin A.

In the areas in which it is caught it is dressed into splitfish.

Genus *Rhomboplites*

Vermilion or large headed snapper—*Rhomboplites aurorubens*. Known in the western Atlantic Ocean from the Bermuda Islands and North Carolina (USA) in the north to Rio de Janeiro (Brazil) in the south. Also found in northern Mexico, often at the Campeche Bank. Caught at depths of 90–100 m.

TABLE 126. WEIGHT PROPORTIONS OF INDIVIDUAL BODY PARTS OF VERMILION SNAPPER CAUGHT IN THE CARIBBEAN SEA AND GULF OF MEXICO 18.5–24.4 CM LONG AND WEIGHING 175–286 g, %

Head	Trunk	Fins	Viscera	Skin, bones, scales
$\frac{23.0–25.4}{24.2}$	$\frac{54.3–57.8}{56.1}$	$\frac{5.4–7.1}{0.3}$	$\frac{6.7–12.8}{9.6}$	$\frac{8.6–11.0}{9.8}$

This is a good table fish. Its meat should be frozen for subsequent culinary preparations.

TABLE 127. CHEMICAL COMPOSITION OF MEAT, %

Moisture	Protein	Fat	Ash
80.1	17.3	$\frac{0.6–1.3}{1.0}$	1.7

Family MACROURIDAE*—GRENADIERS (RAT-TAILS)

The members of this family live in the Pacific Ocean, in the Okhotsk and Barents Seas and in the Atlantic and Indian Oceans. These are deep water fish. Length 53–91 cm, weight 650–3,250 g.

Genus *Licogramma*

Licogramma—*Licogramma zesta*. Found in the Okhotsk and Bering Seas. Weight proportions of individual body parts of licogramma caught in autumn in the Okhotsk Sea 52–61 cm long, weighing 750–950 g (%): head 24.8, bones and fins 10, viscera 4, including liver 0.8. Output of fillets 58.3%.

Chemical composition of meat (%): moisture 92.1, proteins 6.7, fat 0.2, ash 1.0.

The meat of this fish is watery and not tasty. The liver is rich in fat; however, its relative weight is very low so this fish is not important as a food nor as a raw material for preparation of vitamins. The fish should be used for fish meal.

Genus *Macrurus*

Roughhead grenadier—*Macrurus berglax*. A commercial fish, found in the Atlantic Ocean. Length 61–88 cm, weight 1,007–4,013 g.

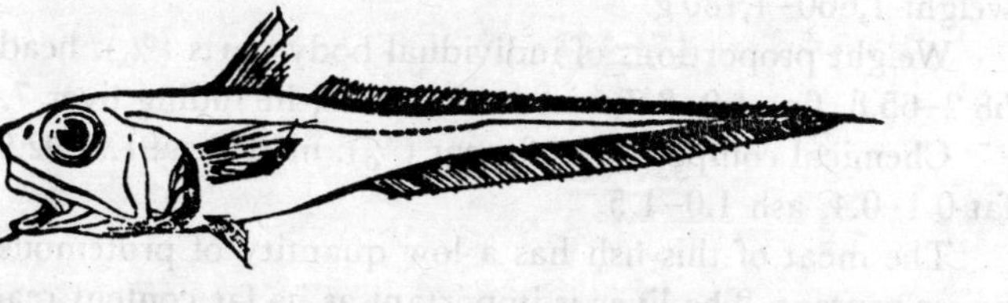

Fig. 117. Roughhead grenadier—*Macrurus berglax*.

Weight proportions of individual body parts of roughhead grenadier caught in April–December (%): head 20.2–31.7, meat 32.9–37.0, guts 8.9–14.3, gonads 0.5–12.0, liver 3.9–15.7, fins 0.6–0.9, scales 1.1–1.6, bones 6.9–7.3, skin 3.2–4.0.

TABLE 128. CHEMICAL COMPOSITION, %

Body parts	Moisture	Protein	Fat	Ash
Meat	81.3–83.1	15.4–16.8	0.1–0.3	0.9–1.2
Roe	74.7–87.6	8.8–21.1	0.5–2.5	1.1–1.6
Liver	28.4–67.8	4.7–12.7	17.3–64.2	1.1–1.4

Since the meat of the roughhead grenadier has high water content, it is not suitable for canning. The cooked meat has poor consistency; therefore, it is difficult to fry.

The most valuable part of this fish is its liver from which natural canned products can be prepared. Salted cavier of the roughhead grenadier can be used as a cheap product for snacks.

Rat-tail—*Macrurus bacvis*. A non-commercial fish in the Indian Ocean and around the northwestern coast of Africa. Length of rat-tail caught in July, 26–33 cm, weight 103–262 g.

Weight proportions of individual body parts (%): head 33.5, trunk without fins 52.9, fins 4.4, viscera 7.8. Output of fillets without skin 34.1%.

TABLE 129. CHEMICAL COMPOSITION, %

Body parts	Moisture	Protein	Fat	Ash
Fillets without skin	79.4	18	0.9	1.5
Bones, skin and fins	77.3	20.9	0.6	1.2
Viscera	76.5	10.1	11.9	1.5

Genus *Albatrossia**

Smalleye grenadier—*Albatrossia pectoralis*. Commercially not very important, found in the northwestern Pacific Ocean. Length 86.0–100 cm, weight 1,660–4,180 g.

Weight proportions of individual body parts (%): head 24.3–28.1, trunk 58.2–65.8, fins 1.9–2.7, viscera 8.4–11.8, including liver 7.0.

Chemical composition of meat (%): moisture 91.5–92.9, protein 5.8–8.8, fat 0.1–0.4, ash 1.0–1.5.

The meat of this fish has a low quantity of proteinous matter and high water content. The liver is important as its fat content reaches 60.3%. After salting (4–5% salt) the caviar resembles salmon caviar in taste. It is best to prepare natural canned products from the liver and salted products from the roe. The rest of the fish should be used for fish meal.

Family MEGALOPIDAE—TARPONS

Atlantic tarpon—*Megalops atlantica*. An important sport fish. In catches, it is taken with other fish from the eastern and western Central Atlantic. It is the largest member of the suborder Clupeoidei. Length in catches is 33–38 cm, weight 326–600 g. They may, however, attain a length of 2 m and weight of 50 kg.

The meat of this fish is characterized by high protein content and low fat

TABLE 130. SIZE CHARACTERISTICS

Size groups	Length, cm	Maximum depth, cm	Maximum thickness, cm	Length of head, cm	Weight of fish, kg
Small	33–38	6–7	3–3.5	8–9	0.4–0.6
Large	90	19	9	22	7.9
	170	—	—	—	6.3

TABLE 131. WEIGHT PROPORTIONS OF INDIVIDUAL BODY PARTS, %

Size groups	Trunk	Head	Viscera	Fins	Bones	Scales	Output of fillets
Small	79	14	7	3	—	—	58
Large	71	22	7	3	14	—	54

TABLE 132. CHEMICAL COMPOSITION, %

Size groups	Body parts	Moisture	Protein	Fat	Ash
Small	Meat	74.7	23.2	0.6	1.5
Large	Meat	77.4	21.3	0.9	1.2
	Meat	73.4	22.0	3.0	1.6
Small	Bones, fins, viscera	63.2	21.0	7.6	8.2
		70.0	17.7	10.3	2.0
Large	Bones, fins, viscera	61.4	22.7	5.7	10.2
		78.1	—	2.2	—

content. This fish is used as food in fresh, salted and smoke cured forms.

Family MORIDAE—MORID CODS

Australian ling—*Physiculus bachus**. A non-commercial fish, found near northern Australia.

Weight proportions of individual body parts of the few specimens caught in February–July weighing 1.1–2.2 kg, 44–51 cm long, with maximum depth 10–14 cm and maximum thickness 6.3–7.0 cm (%): head 21.0–25.0, bones and fins 10.8–13.3, viscera 8.8–11.5, including liver 6.1, gonads 2.4. Output of fillets 41.6–54.8%.

Chemical composition of meat (%): moisture 80.1–80.5, protein 17.5–18.9, fat 0.2–0.8, ash 0.6–1.6.

After boiling and frying the meat resembles the meat of the common cod but it is slightly more tasty. The heads and bones have no food value.

This is a good table fish. The large fish should be used for preparing dressed fish or fillets and the small fish should be frozen without gutting.

Family MENIDAE—MOONFISHES

Spotted moonfish—*Mene maculata*. Lives in the South China Sea. Length 13–23 cm, weight 62–360 g.

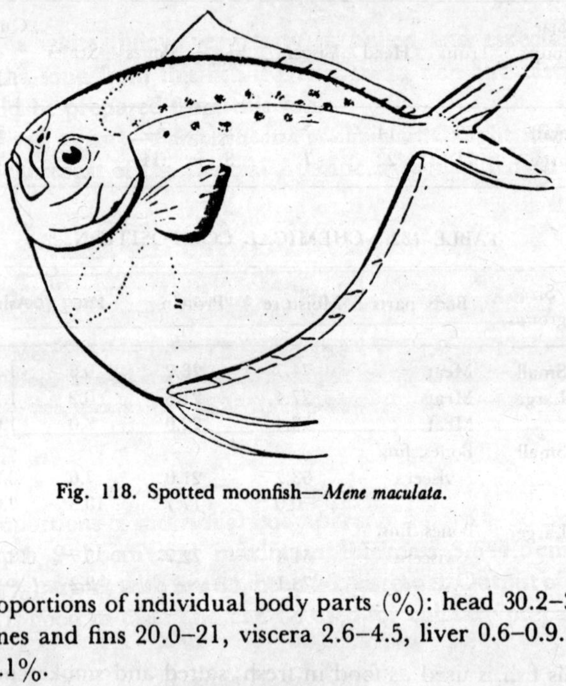

Fig. 118. Spotted moonfish—*Mene maculata*.

Weight proportions of individual body parts (%): head 30.2–33.0, trunk 54.1–57.0, bones and fins 20.0–21, viscera 2.6–4.5, liver 0.6–0.9. Output of fillets 34.3–37.1%.

Family MURAENESOCIDAE—PIKE CONGERS

These are distributed in the Atlantic, Indian and Pacific Oceans. Some species attain a length of 3 m. Only one genus is known.

Sharp toothed eel—*Muraenesox cinereus*. Lives in the Red Sea, in the seas and estuaries of India, the Malay archipelago and Australia, in the South China and East China Seas. Attains a length of 2 m, usual length 131–151 cm, weight 3.4–5.1 kg. Little commercial value as they are caught as isolated specimens.

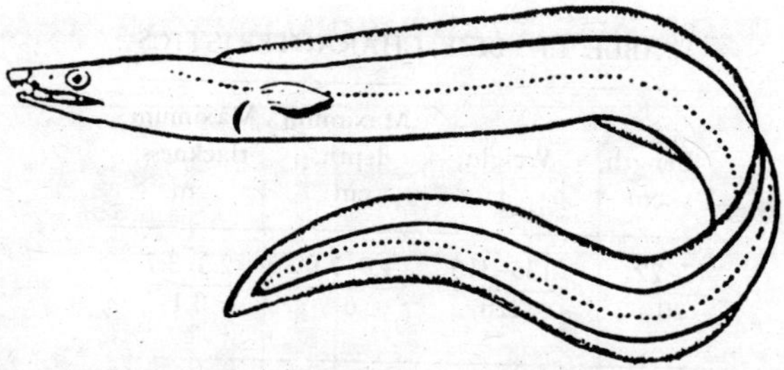

Fig. 119. Sharp toothed eel—*Muraenesox cinereus.*

Weight proportions of individual body parts (%): head 16.3, trunk 76.2, viscera 4.0, liver 0.9, gonads 2.6.

TABLE 133. CHEMICAL COMPOSITION OF MEAT, %

Source	Moisture	Protein	Fat	Ash
TINRO	79.6–81.5	15.8–18.3	0.2–0.3	1.7–1.8
AzcherNIRO	76.4	20.0	2.1	1.5

According to the data of AzcherNIRO, the meat of this fish after frying is white and very tasty. This fish is recommended for the preparation of hot smoked products. According to the data of TINRO, the meat is grayish, not tasty, rubber-like and inedible.

Family MULLIDAE—GOATFISHES

Six genera and more than 50 species of goatfish are known. Goatfish from the genus *Upeneus* belonging to about 20 species are found most frequently. Goatfish are small fish. They live in the tropical and subtropical regions of oceans and adjoining seas. In the temperate zones they are found in smaller numbers.

Goatfish are delicious in taste.

Genus *Pseudoupeneus*

Seychelles goatfish—*Pseudoupeneus seychellensis.* Caught with other fish near the Seychelles Islands (Indian Ocean).

Chemical composition of meat (%): moisture 77.5, protein 20.3, fat 0.3, ash 1.5.

After cooking, the meat is dark, hard, shrivelled with an unpleasant taste of medicine. The soup is colorless with a strong medicinal smell and taste. Because of the poor taste qualities, the medicinal flavor and unattractive

TABLE 134. SIZE CHARACTERISTICS

Length, cm	Weight, g	Maximum depth, cm	Maximum thickness, cm
17–22 / 20	135–315 / 220	4.8–7.0 / 6	2.5–3.7 / 3.1

TABLE 135. WEIGHT PROPORTIONS OF INDIVIDUAL BODY PARTS, %

Head		Trunk					
Total	including meat of occiput and nape	Total	including meat with skin	Bones	Scales	Fins	Viscera
28.6–37.5 / 32.9	3.2	46.0–62.5 / 54.6	37.5–53.3 / 46.0	8.6	2.7	2.1	6.1

appearance, Seychelles goatfish cannot be used for food purposes. It should be used for fish meal.

Genus *Upeneus*

Goatfish—*Upeneus gracilis*. Caught with other fish. In catches near Cape Fartak (Indian Ocean) most fish were about 14 cm (12.8–14.8) long and weighed about 65 g (48.2–80.0).

Chemical composition of meat (%): moisture 76.7, protein 19.8, fat 0.5, ash 1.6.

After cooking, the meat is white, hard and tasty. The soup is rich in fat but tasteless, with a mild iodine flavor.

TABLE 136. WEIGHT PROPORTIONS OF INDIVIDUAL BODY PARTS, %

Head separated by oblique cut	Trunk	Meat	Scales	Fins	Viscera
15.4–25.1 / 20.0	51.7–66.0 / 60.3	50.5	8.5	1.8	4.9–13.2 / 8.8

This fish can be used as a table fish for second course dishes and hot smoked products.

Spearnose goatfish—*Upeneus tragula*. Caught with other fish from the Red Sea and Gulf of Aden.

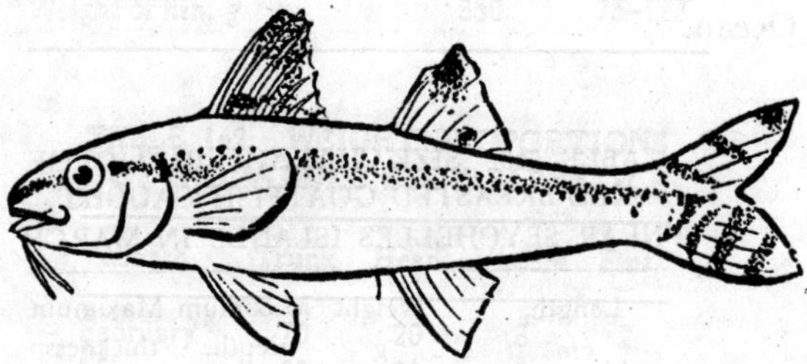

Fig. 120. Spearnose goatfish—*Upeneus tragula*.

TABLE 137. SIZE CHARACTERISTICS

Place of catch	Length, cm	Weight, g
Red Sea	21.6–28.0 / 25	250–540 / 400
Gulf of Aden	14–18 / 15	60–132 / 100

TABLE 138. WEIGHT PROPORTIONS OF INDIVIDUAL BODY PARTS, %

Head	Trunk	Scales	Fins	Viscera
23.5–32.9 / 28.3	52.3–60.5 / 57.4	1.5–7.7 / 4.8	0.7–2.1 / 1.5	2.8–12.6 / 7.2

TABLE 139. CHEMICAL COMPOSITION OF MEAT, %

Moisture	Protein	Fat	Ash
74.6–75.4 / 75.0	19.2–22.0 / 20.6	2.0–2.2 / 2.1	1.4–1.6 / 1.5

With increased length of the spearnose goatfish, there is an increase in the relative weight of the head, whereas the relative weights of the trunk, scales and viscera decrease.

After cooking, the meat is white, tasty, soft in consistency; however, there is an insignificant iodine flavor. The soup is flat and tasteless.

This fish should be used as a table fish for second course dishes and hot smoked products.

Red-breasted goatfish—*Upeneus bensasi**. Caught with other fish from the Indian Ocean.

TABLE 140. SIZE CHARACTERISTICS OF RED-BREASTED GOATFISH CAUGHT NEAR SEYCHELLES ISLANDS IN MARCH

Length, cm	Weight g	Maximum depth, cm	Maximum thickness, cm
11.8–15.0 / 13.0	30–70 / 45	2.4–3.8 / 3.0	1.5–2.0 / 1.7

Weight proportions of individual body parts (%): head 21.6, trunk 64.0, including meat with skin 51.9 and bones 12.1, scales 5.6, fins 2.2 and viscera 5.2.

Chemical composition of meat (%): moisture 77.0, protein 21.1, fat 0.3, ash 1.6.

After cooking, the meat becomes light in color with a pleasant taste, but there is a large number of fine bones. The soup is yellowish with a pleasant meat-like taste. After frying, the meat is light colored, pleasant in taste and the bones are not noticeable. Products canned in oil have good appearance and taste. This fish should be used for first and second course dishes (fried), smoked products and preparations canned in oil.

Yellow striped goatfish—*Upeneus vittatus*. Caught in small numbers with other fish. In catches from the Red Sea, most fish are about 13 cm long and weigh around 40 g.

Weight proportions of individual body parts (%): head separated by an oblique cut 25.7, trunk 60.9, scales 3.0, fins 1.7, viscera 5.5.

Chemical composition of meat (%): moisture 75.1, protein 19.2, fat 2.2, ash 1.6.

The taste qualities are similar to the spearnose goatfish. This goatfish is recommended as a table fish for second course dishes and hot smoked products.

African surmullet—*Upeneus prayensis*. Caught in small numbers with other fish near northwest Africa.

The raw meat of this fish is dense and orange. After frying, the meat is tasty but somewhat dry. It is recommended for canning.

TABLE 141. SIZE CHARACTERISTICS

Parameters	February	May
Length of fish, cm	24	15–18
Weight of fish, g	350	75–135

TABLE 142. WEIGHT PROPORTIONS OF INDIVIDUAL BODY PARTS, %

Month	Trunk	Head	Viscera	Fins
February	70	20	8	2
May	60	27	9	4

TABLE 143. CHEMICAL COMPOSITION OF MEAT, %

Month	Moisture	Fat	Protein	Ash
February	74.4	4.3	20.0	1.3
May	76.4	1.5	20.6	1.5

Genus *Parupeneus**

Golden yellow goatfish—*Parupeneus* luteus*. Caught in small numbers with other fish. In the February catches on the shelf of western India, most fish were about 20 cm (15.5–24.0) long and weighed about 150 g (70–266).

Weight proportions of individual body parts (%): head 31.9, trunk 56.0, scales 3.4, fins 2.8, viscera 5.3.

This goatfish is characterized by a large head (as a percentage of the weight of the whole fish), the head is larger than in most species of goatfish and it is poisonous.

Chemical composition of meat (%): moisture 78.3, protein 19.8, fat 0.3.

The meat is low quality. After cooking, it has a dry consistency and a poorly expressed fishy smell and taste. Because of the large and poisonous head and low taste properties, the meat of this goatfish should be prepared immediately after catching by removing the head; the trunk should be used for fish meal and the head should be discarded.

Family NEMIPTERIDAE—THREADFIN BREAMS

Besugo—*Nemipterus virgatus*. A non-commercial fish in the South China Sea and in the Great Australian Bight. Length 15–25 cm, weight 60–180 g.

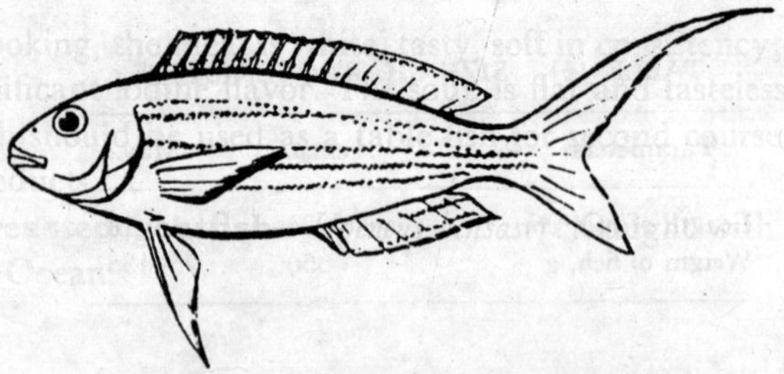

Fig. 121. Besugo—*Nemipterus virgatus.*

Weight proportions of individual body parts of fish caught in February–November (%): head 28.5–35.0, trunk 47.3–53.4, bones and fins 13.4–18.4, viscera 4.3–6.8, including liver 0.8. Output of fillets 32.4–40.8%.

Chemical composition of meat (%): moisture 76.8–78.7, protein 18.7–20.0, fat 1.0–1.5, ash 1.6–1.8.

The meat of this fish is loose, bony, lean, with mediocre taste properties. After washing to remove water soluble substances, mince can be made from this fish.

Japanese besugo—*Nemipterus japonicus.* A fish with little commercial value, found in the East China Sea, Indian Ocean, the Great Australian Bight. Length of fish caught in November–February 14–25 cm, weight 39–320 g.

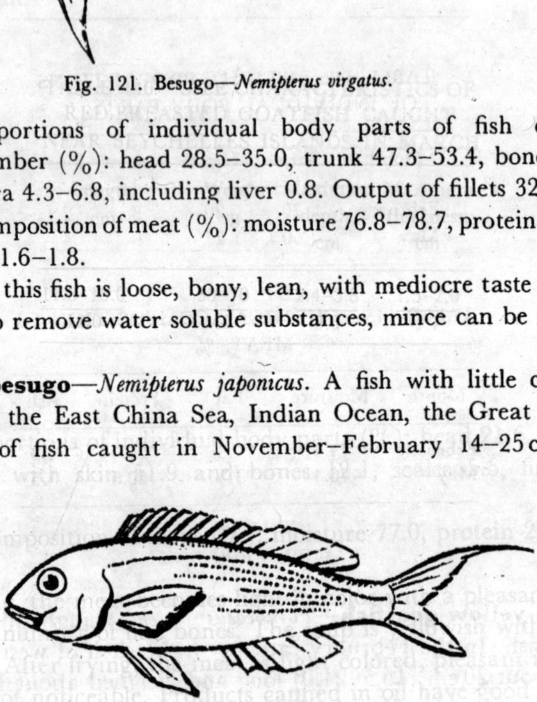

Fig. 122. Japanese besugo—*Nemipterus japonicus.*

Weight proportions of individual body parts (%): head 15.4–37.4, trunk 59.0–61.6, fins 2.0–3.6, scales 3.2–8.8, viscera 2.7–9.6. Output of fillets with skin 53.0%.

Chemical composition of meat (%): moisture 76.8–80.5, protein 15.4–20.0, fat 0.3–1.5, ash 1.6–1.8.

The fish have lean, loose and bony meat which in taste resembles fresh water carp. This fish can be frozen and used for hot smoked products.

Family NOTOTHENIIDAE—NOTOTHENIIDS OR COD ICE FISHES

This family consists of five genera. Outwardly they resemble Pacific greenlings. They are distributed only in the southern hemisphere. The members of the genus *Eleginops* are found outside Antarctica.

Patagonian toothfish—*Dissostichus eleginoides*. Caught in small numbers with other fish in the southwest Atlantic. Distributed in the Patagonian-Falkland region, in Scotia Sea and on the shelf of South Georgia Island. Often individuals 41–74 cm long (average length 63 cm) and weighing 720–3,750 g (average weight 2,170 g) are found. Individuals 81–127 cm long, weighing 7.3–20.5 kg are also caught.

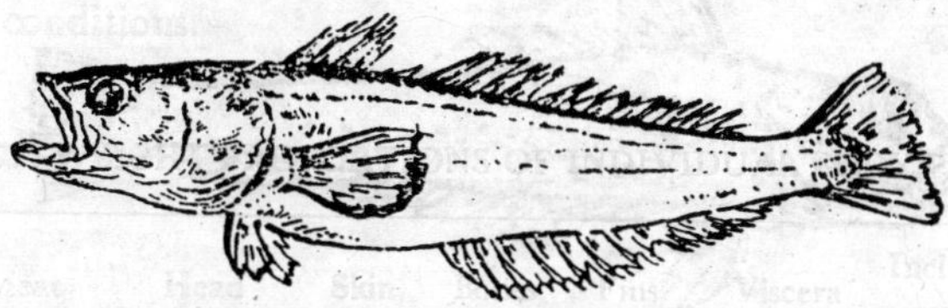

Fig. 123. Patagonian toothfish—*Dissostichus eleginoides*.

The meat is white, soft, juicy and good in taste after cooking. The soup is transparent, fatty with a pleasant taste and flavor. This is an excellent table fish. The large size and extremely high taste qualities, similar to those of the Siberian nelma, make this fish a "golden constituent" of mixed catches.

TABLE 144. WEIGHT PROPORTIONS OF INDIVIDUAL BODY PARTS, %

Head	Meat	Skin	Bones	Fins	Viscera
21.0–23.6 22.7	54.4–57.2 55.7	5.5–5.8 5.6	4.7–5.2 4.9	2.3–2.4 2.3	6.9–7.3 7.1

TABLE 145. CHEMICAL COMPOSITION, %

Body parts	Moisture	Protein	Fat	Ash
Head	59.7–61.3 60.5	9.3–11.7 10.5	24.2–25.8 25.0	2.4–2.6 2.5
Meat	61.4–74.2 65.0	12.4–14.4 13.6	12.4–23.9 20.4	0.9–1.2 1.0
Liver	57.7	17.5	23.2	1.1
Viscera (excluding liver)	77.8	9.8	10.8	1.4
Skin	53.7–58.3 56.0	20.0–21.8 20.9	20.1–22.6 21.3	1.8–2.6 2.2
Fins	57.1	13.3	22.2	6.7

The meat of this species is an excellent raw material for different culinary products: fried, jellied, with liquid sauces, roasted in dough, as cutlets, meat balls, rolls. This fish can be used for cured balyks and mild salt cured products.

Green notothenia—*Notothenia gibberifrons*. Distributed in the South Atlantic, near the southern Shetland, southern Sandwich, southern Orkney Islands and near South Georgia Island. Most widespread fish of Scotia Sea.

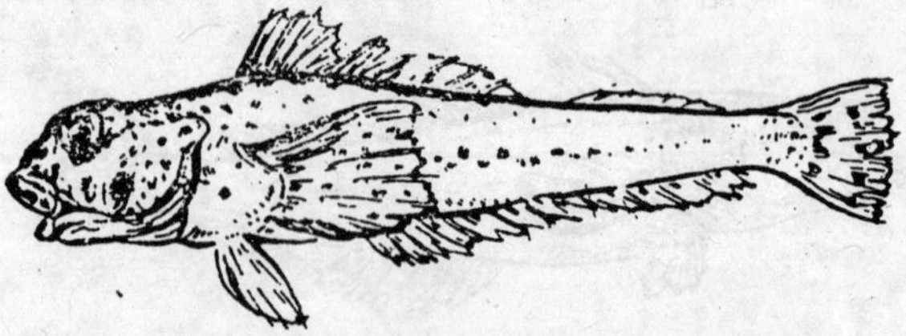

Fig. 124. Green notothenia—*Notothenia gibberifrons*.

Average length of green notothenia from the regions of South Georgia Island 29.3cm and in the case of fish from the southern Orkney Islands 28.7 cm.

Chemical composition of meat of green notothenia caught in February in the Scotia and Weddell Seas (%): moisture 78.5, protein 17.6, fat 1.0, ash 1.1.

TABLE 146. CHEMICAL COMPOSITION, %

Body parts	Moisture	Protein	Fat	Ash
Head	78.6	12.7	2.8	3.9
Meat	78.5–82.5	16.3–17.6	1.0–1.3	1.1
Skin	80.1	17.2	1.4	1.4
Fins	74.2	13.1	1.1	9.1

The meat of green notothenia is edible but it does not have very high taste properties. The meat of fish caught near the southern Orkney Islands is often infested with parasites, nematodes and cestodes; therefore, it should be used for fish meal.

Marbled notothenia—*Notothenia rossi marmorata*. An important commercial fish, found in the South Atlantic and in Scotia Sea. Average length

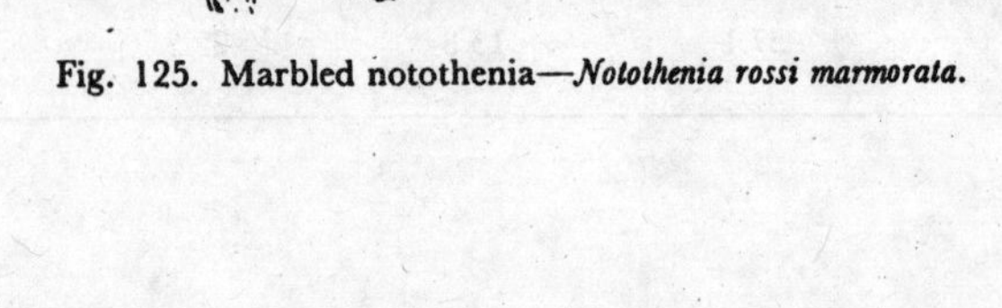

Fig. 125. Marbled notothenia—*Notothenia rossi marmorata*.

of the marbled notothenia 68 cm, average weight about 5 kg. Males slightly smaller in size than females. Average length of male specimens examined 67.9 cm, weight 4,120 g, while the average length of the females was 69.2 cm and weight 5,210 g.

Weight proportions of individual body parts of marbled notothenia fluctuate considerably depending upon the age of the fish, its sex and physiological conditions.

TABLE 147. WEIGHT PROPORTIONS OF INDIVIDUAL BODY PARTS, %

Trunk	Meat	Head	Skin	Bones	Fins	Viscera	Including gonads	Liver
38.0–54.1	30.1–39.2	22.0–30.0	4.2–5.5	4.5–6.0	3.0–4.0	10.7–40.0	9.6–29.0	2.2–5.2

TABLE 148. OUTPUT OF TRUNKS AS % OF THE WHOLE FISH WEIGHT

Size groups	March	April
Small (Less than 52 cm)	54.1	53.4
Medium (52–65 cm)	39.6	38.1
Large (more than 65 cm)	40.4	38.5

The low output of the trunks in large and medium sized marbled notothenia is explained by the relatively higher weight of the viscera. For example, if, in small fish, the viscera constitutes 10.7–14.5% in large and medium fish, it constitutes 40% of the weight of the whole ungutted fish. The increased weight of the viscera in marbled notothenia is mainly due to the gonads (testes and roe), the size of which also changes. In medium sized notothenia, the output of the testes in March–April was 25% and in large fish 29%. In March the roe constituted 10% of the weight of medium size fish and 16% of large fish; the values for April were 21.6 and 23.9%, respectively.

The weight of the liver in females is two to three times greater than in males; in small fish the output of the liver is less, 3.9%, in medium size fish it is 5.0% and in large fish 5.2%.

Chemical composition of meat (%): moisture 69.5–75.5, protein 12.8–16.6, ash 0.8–1.5. The fat content in the meat fluctuates markedly depending upon the size and time of catch. Between January and May, it is fairly constant and the average value is 10–12% (6.8–16.7).

In November, this fish is characterized by the minimum average fat content (8.9%); its maximum variation is 2.1–20.4%.

Fat content (%) in meat of marbled notothenia depending on size

Small	$\frac{6.4–11.5}{8.4}$
Medium	$\frac{10.0–12.3}{11.5}$
Large	$\frac{8.2–16.7}{12.6}$

TABLE 149. FAT CONTENT IN MEAT OF MARBLED NOTOTHENIA DEPENDING UPON TIME OF CATCH

Number of fish investigated	Month of catch	Fat content, %
10	November	$\frac{2.1–20.4}{8.9}$
1	January	11.2
5	February	$\frac{11.2–16.7}{13.0}$
6	March	$\frac{7.8–12.0}{9.9}$
4	April	$\frac{6.8–12.5}{10.7}$
2	May	$\frac{12.1–15.5}{13.8}$

After cooking, the meat is light colored, with large fibers, is juicy and tasty. The soup is rich in fat, transparent, with pleasant taste and smell. Fried notothenia has especially good taste qualities: the meat is dense, juicy, soft and tasty.

This is an excellent table fish. Second course dishes and fillets, and preparations from cured fillets should be made from this fish. Salt cured products and conserves can also be made. The heads should be used for ukha soups. The roe should be prepared by salting and testes, roe and liver should be canned in tomato sauce.

Gray notothenia—*Notothenia squamifrons*. A commercial fish, distributed in the Indian Ocean adjoining the Antarctic.

The gray notothenia has a relatively large liver (1.7% of the total weight of the fish).

TABLE 150. SIZE CHARACTERISTICS OF GRAY NOTOTHENIA

Place and time of catch	Stage of maturation	Length, cm	Maximum depth, cm	Maximum thickness	Weight, g
Near Crozet Island (November)	II–I[illegible]	[illegible] / 25.4	6.0–9.0 / 7.0	3.0–4.5 / 3.7	220–430 / 295
Near Kerguelen Island (January)	III	34.0–41.0 / 37.6	8.5–11.5 / 10.2	6.8–9.0 / 7.7	895–1,500 / 1,203

TABLE 151. WEIGHT PROPORTIONS OF INDIVIDUAL BODY PARTS, %

Body parts	Near Crozet Island (November)	Near Kerguelen Island (January)
Head	31.4	27.2–36.4 / 32.7
including meat of occiput and nape	6.1	5.1
Trunk	48.1	39.3–49.2 / 43.1
including meat with skin	38.0	32.5–40.8 / 36.4
Bones	10.1	6.7
Scales	3.4	1.4
Fins	2.4	2.2
Viscera, whole	13.2	11.1–28.4 / 20.2

After cooking, the meat of gray notothenia acquires a cream or light yellow tinge and rubber-like consistency (in small fish the meat is friable); the taste is typical of table fish. The soup is tasteless.

The meat of fried fish is white, dense and tastier.

The taste qualities of large fish caught near Kerguelen Island were better

TABLE 152. CHEMICAL COMPOSITION OF MEAT, %

Place and time of catch	Moisture	Protein	Fat	Ash
Near Crozet Island (Nov.)	83.3	14.7	0.5	1.1
Near Kerguelen Island (Jan.)	77.0	15.4	6.0	1.1

than those of small fish caught near Crozet Island. This fish should be used as a table fish for second course dishes, smoke cured products and products canned in oil from fried fish (natural canned products and canned preparations in tomato sauce do not have a marketable appearance).

Family OPHIDIIDAE—BROTULAS AND CUSK-EELS

Kingklip—*Genypterus capensis*. Caught in small numbers with other fish. Found in the southeast and southwest Atlantic. Along the African coast, distributed from Tiger Bay to the Agulhas Cape. Along the North American coast, found on the Patagonian shelf. Attains a length of 125 cm and weight of 11 kg.

Fig. 126. Kingklip—*Genypterus capensis*.

TABLE 153. SIZE CHARACTERISTICS

Parameters	Southeast Atlantic	Southwest Atlantic
Standard length, cm	87	32–73
Maximum depth, cm	15	4–11
Maximum thickness, cm	9	3–9
Weight, kg	8.6	0.2–1.9
Weight proportions of individual body parts, %		
trunk	56	57–64
head	27	27–28
viscera	17	8–15
Output of fillets	42	41

TABLE 154. CHEMICAL COMPOSITION OF MEAT, %

Place of catch	Moisture	Protein	Fat	Ash
Along the coast of southwest Africa	78.7	19	0.7	1.6
Along the coast of South America	80.0–82.2	17.2–17.7	0.1	1.2

The liver of this fish contains 55% moisture, 25% fat. One g of liver contains 1,000 iu of vitamin A.

After frying, the meat is white, semi-sweet, slightly watery and tasty. This is a good table fish. Fish weighing more than 0.5 kg should be used for fillets.

TABLE 155. CHEMICAL COMPOSITION OF WASTE, %

Body parts	Moisture	Protein	Fat	Ash
Bones and fins	75.5	17.5	0.5	6.5
Viscera	73.1	15.4	10.6	0.9

Black congrio—*Genypterus blacodes*. Caught in small numbers with other fish in the southwest Atlantic and near the Patagonian shelf.

TABLE 156. CHEMICAL COMPOSITION, %

Body parts	Moisture	Protein	Fat	Ash
Meat	79.5	19.0	0.3	1.2
Bones and skin	74.5	16.7	0.8	8.0
Viscera	74.1	18.8	5.8	1.2

Weight proportions of individual body parts of this fish 43–55 cm long with maximum depth 5–8 cm, maximum thickness 4–6 cm and weighing 330–880 g (%): trunk with fins 60, head 30, viscera 10. Output of fillets 40%.

After frying, the meat becomes white, juicy and slightly sweet. The taste qualities of the fish are good. This is a good table fish which can be used for fillets.

Family OSMERIDAE—SMELTS

Smelts are small fish found in the marine and fresh waters of the northern Atlantic and Pacific Oceans and in the Arctic Ocean. Among these fish are purely marine, migratory and fresh water species. There are four genera in the smelt family including *Mallotus,* with one species (capelin).

Capelin—*Mallotus villosus.* A commercial fish, distributed in the northern Atlantic Ocean and in the adjoining regions of the Arctic Ocean, in the Barents Sea up to Bear Island, in the White and Norwegian Seas, off the coast of Greenland up to 74°N and along the North American coast from Hudson Bay to the Gulf of Maine.

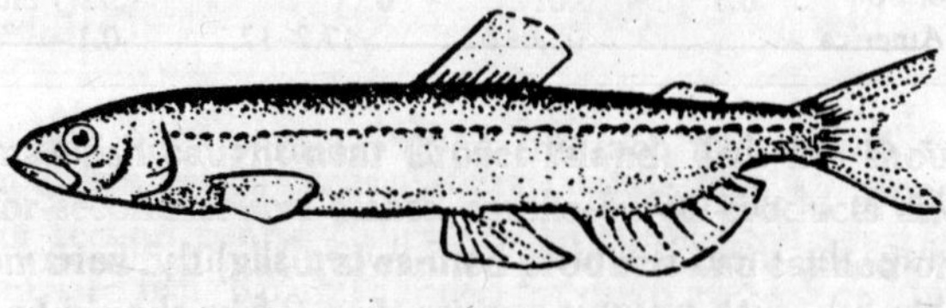

Fig. 127. Capelin—*Mallotus villosus.*

Weight proportions of individual body parts vary depending upon the maturation of the roe and fattiness of the fish. Thus, in mature females the proportion of the roe was 24.1% and the trunk 59.2%, while in the spent fish the trunk was 71.5–84%.

The meat of the capelin, especially in the spring, has a relatively low protein content with a large quantity of water. It also contains little fat so that the value of capelin at this time of year is low. However, after feeding (from July onward) the quality of capelin improves significantly and it becomes suitable for good quality canned products and for frying, marinades and smoke curing.

Family POMADASIDAE—GRUNTERS

These are relatively small marine fish which live in warm seas and are closely related to progies.

Genus *Anisotremus*

Pork fish—*Anisotremus virginicus.* Found along the Atlantic coast of America, from Brazil to Florida, in the Caribbean Sea and in the Gulf of Mexico. Attains weight of 0.9–1.2 kg.

Length of fish caught in July in the Gulf of Mexico 19–23 cm, maximum depth 9–11 cm, maximum thickness 3–4 cm, weight 200–375 g. Weight proportions of individual body parts (%): trunk with fins 63, head 29, viscera 8. Output of fillets 44%.

Data on taste qualities of this fish are not available.

TABLE 157. WEIGHT PROPORTIONS OF INDIVIDUAL BODY PARTS OF CAPELIN, %

Time of catch	Weight, g	Length, cm	Head	Trunk	Meat with skin	Bones and fins	Viscera		
							whole	roe	others
April									
mature females	17–19	—	13–17	62–60	—	—	21–28	16–24	6–1
spent females and males	16–27	—	11.5–20.0	60.0–81.6	—	—	4.4–7.7	—	—
August–February	35–48	16–18	12.2	72.3	64.6	7.7	12.9	—	—

TABLE 158. CHEMICAL COMPOSITION OF CAPELIN, %

Body parts	Season	Moisture	Protein	Fat	Ash
Meat with skin	April	71.2–82.6 / 78.8	12.0–14.8 / 13.1	1.4–11.7 / 5.4	1.1–2.2 / 1.9
	Autumn–winter	61.4–68.0 / 65.6	11.8–19.0 / 13.6	11.6–20.2 / 17.5	1.3–2.3 / 1.4
Head	April–winter	81.0–83.4 / 81.8	11.8–12.3 / 12.0	1.9–3.2 / 2.7	3.1–3.8 / 3.6
		72.2–79.1 / 76.1	10.0–13.6 / 11.6	6.0–8.8 / 7.7	2.9–3.8 / 3.4
Roe	April–winter	71.9–72.3	19.4–17.6	5.6–6.0	1.6–1.5
Bones		58.5–69.0 / 65.6	12.8–14.1 / 13.4	10.9–19.7 / 13.6	5.0–7.3 / 6.6
Viscera	April–winter	74.5–79.8 / 75.9	12.0–13.6 / 13.0	6.3–10.1 / 8.2	1.4–1.6 / 1.5
		41.0–73.2 / 55.8	6.2–12.2 / 10.1	11.0–51.3 / 31.9	0.8–2.0 / 1.4

TABLE 159. CHEMICAL COMPOSITION OF PORKFISH, %

Body parts	Moisture	Protein	Fat	Ash
Meat	77.9	20.3	0.5	1.4
Viscera	75.0	16.7	2.1	6.2

Genus *Pomadasys*

Silver grunter—*Pomadasys argyreus*. Found in the Red Sea. Length of fish caught in November–December 30.0–42.5 cm, weight 600–1,450 g. Weight proportions of individual body parts (%): trunk 48.1–52.5, head 32–40.6, scales 3.8–5.1, fins 1.6–2.6, viscera 3.3–5.2.

Chemical composition (%): moisture 77.8, protein 19.0, fat 1.3, ash 1.3.

This is a good table fish. Its meat has good taste qualities similar to the meat of fresh water fish. Semi-finished and culinary products should be prepared from this fish.

Striped grunter—*Pomadasys operculare*. Lives in the Indian Ocean and in the Persian Gulf. Length of fish caught in February 40–44 cm, weight 1,140–1,480 g.

Weight proportions of individual body parts (%): trunk 49.6–53.8, head

33.8–40.4, scales 7.5–8.6, viscera 4.4–7.5. Output of fillets with skin 41.7–45.1%.

Chemical composition of meat (%): moisture 77.6, protein 20.2, fat 0.6, ash 1.4.

This is a good table fish. It has white meat which is pleasant in taste but somewhat hard. The broth is lemon color and tasty.

These fish should be frozen onboard for subsequent preparations of semi-finished, culinary and smoked products.

Grunter bream—*Pomadasys hasta.* Found in the tropical regions of the eastern Indian Ocean and in the waters near Australia. Length 23–70 cm, weight 225–6,000 g.

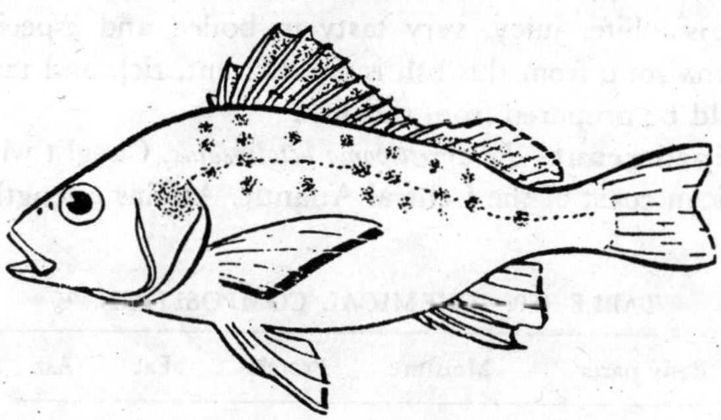

Fig. 128. Grunter bream—*Pomadasys hasta.*

Weight proportions of individual body parts (%): trunk 51.0–59.8, head 28.3–39.0, fins 2.0–2.7, scales 2.0–5.6, viscera 4.5–6.9, bones 10.0–15.5. Output of fillets with skin 41.0–44.3%.

Chemical composition of meat (%): according to the data of TINRO—moisture 75.1–76.8, protein 20.3–21.1, fat 1.5–2.7, ash 1.4–2.0; according to the data of VINRO—moisture 79–80, protein 18.8–19.3, fat 0.3–0.9.

The meat is juicy, white, tasty, especially in fried form. Semi-finished products and culinary preparations should be made from this fish.

Genus *Parapristipoma*

Three-striped grunter—*Parapristipoma trilineatum.* Comparatively less commercial importance. Found in the Korea Bay. Weight 250–435 g.

Weight proportions of individual body parts of fish caught in November (%): trunk with fins 75.1, head 18.3, bones 21.0, viscera 6.6. Output of fillets with skin 54.1%.

Chemical composition of meat (%): moisture 73.6, protein 20.4, fat 3.3, ash 2.7 (variable).

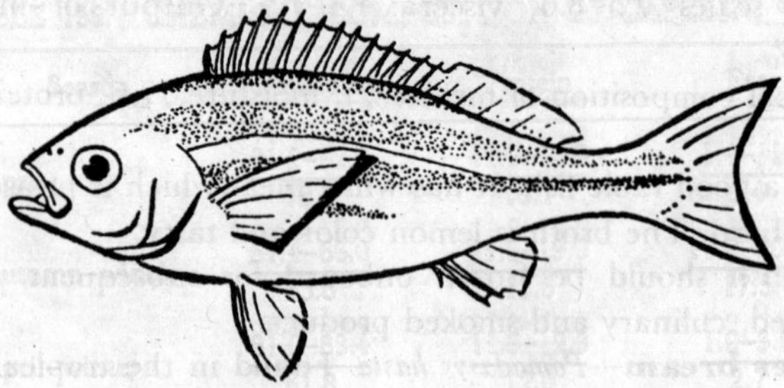

Fig. 129. Three-striped grunter—*Parapristipoma trilineatum.*

The meat is white, juicy, very tasty in boiled and especially in fried forms. The ukha soup from this fish is transparent, rich and tasty. Culinary products should be prepared from this fish.

Eight-striped grunt—*Parapristipoma octolineatus*. Caught with other fish, along the African coast of the Central Atlantic. Attains a length of 40 cm.

TABLE 160. CHEMICAL COMPOSITION, %

Body parts	Moisture	Protein	Fat	Ash
Meat	77.3	19.9	1.4	1.4
Head, bones, fins, viscera	66.2	20.6	4.6	8.6

Weight proportions of individual body parts of a fish 24–30 cm long, with maximum depth 9–11 cm and maximum thickness 3.5–4.5 cm, weighing 0.32–0.81 kg (%): trunk with fins 65, head 26, viscera 9. Output of fillets 41%.

The meat is good in taste and can be used for culinary preparations.

Genus *Hapalogenys*

Blackfin grunt—*Hapalogenys nigripinnis*. A fish of little commercial value, found in the East China Sea.

Length of fish caught in May 32 cm, weight 1,200 g.

Weight proportions of individual body parts (%): trunk 55.0, head 28.9, bones 10.0, fins 4.9, scales 3.3, viscera 8.0. The output of fillets (with skin) 45.0%.

This is a table fish. The meat is juicy, white, very tasty in boiled and especially in fried forms. The ukha soup from this fish is transparent, rich and tasty. The meat should be frozen for subsequent culinary preparations.

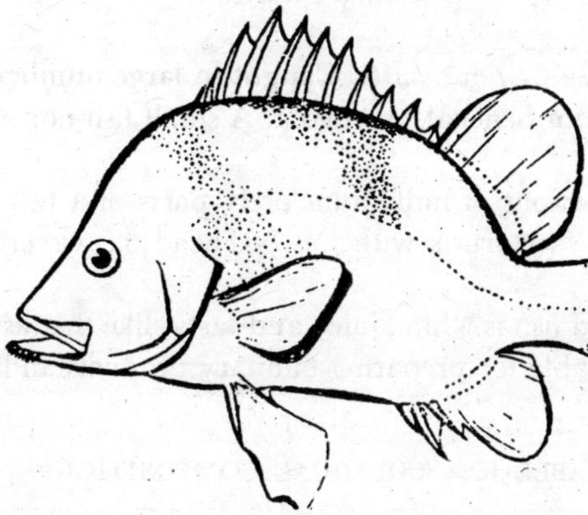

Fig. 130. Blackfin grunt—*Hapalogenys nigripinnis*.

Genus *Diagramma*

Hardlips—*Diagramma macrolepis*. Fish caught in Central Atlantic in April. Length 55 cm, weight 4.9 kg.

Weight proportions of individual body parts (%): trunk 56, head 34, viscera 7, fins 3.

TABLE 161. CHEMICAL COMPOSITION, %

Body parts	Moisture	Protein	Fat	Ash
Meat	76.4	20.7	1.8	1.1
Viscera	73.4	16.6	8.0	2.0
Skin, bones, fins	56.3	23.2	9.1	11.5

Mediterranean hardlips—*Diagramma mediterraneum*. Caught with other fish in the Central Atlantic, along the coast of northwest Africa. Length 30–60 cm.

Weight proportions of individual body parts of fish 40–54 cm long, with maximum depth 11–19 cm, thickness 6–6.5 cm, weighing 1.6–3.1 kg (%): trunk with fins 61, head 26–32, viscera 4–13. Output of fillets 36%.

Chemical composition of meat (%): moisture 76–77, protein 19.3–20.4, fat 1.2–2.4, ash 1.2.

Fried, the meat is white, soft, juicy and tasty. This is a good table fish.

Genus *Otoperca*

Green grunter—*Otoperca aurita.* Caught in large numbers in the eastern Central Atlantic, from Senegal to Angola. A small fish not more than 22 cm long.

Weighing proportions of individual body parts of a fish 12–17 cm long, weighing 50–125 g (%): trunk with fins 58, head 32, viscera 10. Output of fillets 38%.

The meat of fried fish is white, juicy and tasty, like the taste of fresh water fish. The fish is suitable for preparing culinary products and conserves.

TABLE 162. CHEMICAL COMPOSITION, %

Body parts	Moisture	Protein	Fat	Ash
Meat	79.8	17.8	1.2	1.2
Head, bones, viscera	69.1	17.5	6.8	6.6

Genus *Pseudopentaceros*

Pacific boarfish—*Pseudopentaceros richardsoni.* An important commercial fish, in large concentration near the Hawaiian Islands. Length of fish 22–31 cm, weight 210–680 g.

Weight proportions of individual body parts (%): trunk 54.0–58.2, head 28–33, skin with scales 5.4–9.3, bones 8.9–11.2, fins 3.0–5.4, viscera 7.0–11.0. Output of fillets (without skin) 35.0–42.0%.

Chemical composition of meat (%): moisture 57.4–73.6, protein 12.0–16.3, fat 9.4–27.0, ash 1.0–1.5. In most fish caught, the fat content was up to 18–20%.

This is a table fish. After cooking the meat is white, juicy in consistency, has a mildly sweetish taste similar to the meat of fresh water fish. The meat is very tasty after frying and smoke curing.

The fat from the meat of this fish is highly resistant to oxidation. Culinary products should be made from this fish.

Genus *Pristipoma*

Bennett's grunt—*Pristipoma bennetti.* Found with other fish in catches off the African coast. Length 15–30 cm.

Weight proportions of individual body parts of a fish caught near Wolffish Bay, 23–30 cm long, with maximum depth 9–12 cm, maximum thickness

3.2–4.5 cm, weighing 350–675 g (%): trunk 52–61, head 28–38, viscera 6–10, scales 2.9–3.8, fins 2.9.

Chemical composition of meat (%): moisture 76–77, protein 19–19.5, fat 0.3–2.0, ash 1.3–1.7.

This fish has light colored meat with good taste qualities. It is somewhat dry after frying. Semifinished products and culinary preparations should be made from this fish.

Genus *Bathystoma*

Golden lined grunt—*Bathystoma aurolineatum.* Widely distributed in the western Atlantic, from the Pensacola coast (USA) to Brazil. Length 9–23 cm. In catches, there is a predominance of fish 17–19 cm long.

TABLE 163. RATIO BETWEEN LENGTH AND WEIGHT

Parameters	Three-year old fish	Four-year old fish	Five-year old fish
Average length, cm	17.2	18.6	20.1
Average weight, g	105	125	152

Genus *Haemulon*

French grunt—*Haemulon flavolineatum.* Caught in the Gulf of Mexico. In catches it constitutes 15% of all species belonging to this genus.

White grunt—*Haemulon plumieri.* **Blue striped grunt**—*Haemulon sciurus.* In some cases the meat of these fish has a strongly expressed iodine taste. According to analyzed data such fish contained 235–362 mg% of iodine, that is, 100 times more than usually observed in the meat of fish.

TABLE 164. SIZE CHARACTERISTICS

Species	Standard length, cm	Maximum depth, cm	Maximum thickness, cm	Length of head, cm
Golden lined grunt	14–21	6.5–7.5	2.5–3	5.5–7
French grunt	15–21	6–8	2–4	—
White grunt	19–28	8–11	3–4	6–11
Blue striped grunt	18–19	6	2.5	—

TABLE 165. WEIGHT PROPORTIONS OF INDIVIDUAL BODY PARTS, %

Species	Weight, g	Trunk	Head	Internal organs	Output of fillets
Golden lined grunt	80–200	57–61	23–37	7–16	43
French grunt	110–250	64	25	11	44
White grunt	185–535	58	29–36	6–13	32
Blue striped grunt	135–145	57	28	15	34

After cooking, the meat of these fish is white, slightly sweet and somewhat bony. The taste properties are quite satisfactory. Often they are compared to small sparids. Culinary preparations can be made from these fish.

TABLE 166. CHEMICAL COMPOSITION, %

Species	Body parts	Moisture	Protein	Fat	Ash
Golden lined grunt	Meat	77.8–78.9	19.3–19.9	0.4–4.1	1.2–1.5
French grunt	Meat	72	21.1	5.8	1.1
White grunt	Meat	78.7–80.7	17–19.6	0.4–0.9	1.3–1.4
Blue striped grunt	Meat	76.3–76.6	20.6–20.8	1.4–1.6	1.3–1.4
Golden lined grunt	Whole fish	72.4	17.2	4.5	4.9
	Head	65.5–68.4	15.7–19	6–9.2	8.3–9.5
	Bones	66.6–84.0	18.7–18.8	3.5–8.9	8–11.1
	Fins	48.9	17.8	2.2	31.1
	Scales	52.8	23.8	0.3	23.1
	Viscera	75.8–80.8	12.4–16.9	1.2–4.0	3.1–5.6

Genus *Scolopsis**

Monocle-bream—*Scolopsis temiarali*. Fish of little commercial importance, caught in the Great Australian Bight. Length of fish caught in July–January 19–27.2 cm, weighing 150–625 g.

Weight proportions of individual body parts (%): trunk 54.2–55.8, head 33.4–33.5, scales 3.5, fins 2.2, viscera 7.1–8.0, bones 10.7–11.1. Output of fillets with skin 41.9–43.5%.

Chemical composition (%): moisture 77.8, protein 21.8–23.3, fat 1.2–1.5, ash 1.8–1.9.

The meat is white and juicy but it has an unpleasant taste. The soup is not very tasty. It has an unpleasant medicinal odor and flavor. It can be used for making washed mince.

Genus *Plectorhynchus*

Three banded sweetlip—*Plectorhynchus cinctus**. A commercial fish, caught in the tropical regions of the eastern Indian Ocean, in the East China Sea, along the coasts of Australia and New Zealand. Length of fish caught in July–January 50.5–67.5 cm, weight 2.4–6.0 kg.

Weight proportions of individual body parts (%): trunk 52.0–66.9, head 20.6–33.6, bones 6.5–15.3, scales 0.3–4.3, fins 0.8–3.4, viscera 8.0–16.4. Output of fillets with skin 41.0–60.4%.

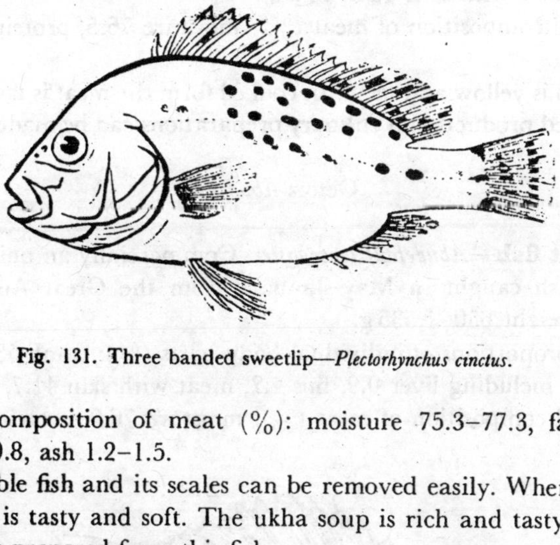

Fig. 131. Three banded sweetlip—*Plectorhynchus cinctus*.

Chemical composition of meat (%): moisture 75.3–77.3, fat 0.3–0.7, protein 20.5–20.8, ash 1.2–1.5.

This is a table fish and its scales can be removed easily. When boiled or fried, the meat is tasty and soft. The ukha soup is rich and tasty. Culinary products can be prepared from this fish.

Genus *Plectropomus**

Spotted coral trout—*Plectropomus maculatus*. Commercially of little importance, caught near northwest Australia. Length of fish caught in July–February 42–61 cm, weight 910–2,900 g.

Weight proportions of individual body parts (%): trunk 54.9–59.4, head 28.5–33.7, viscera 4.1–12.0, fins 3.5–4.4, skin 4.4–4.8, bones 6.2–10.4. Output of fillets 40.1–48.4%.

Chemical composition of meat (%): moisture 77.5, protein 19.8, fat 0.5, ash (variable) 2.3.

The meat of this fish is tasty, white, somewhat dense. The soup is transparent, rich and tasty. This is a good raw material for culinary processing.

The thick and strong skin of this fish must be removed before boiling or frying. Culinary products can be prepared from this fish.

Family POMACENTRIDAE—DAMSELFISHES

Genus *Rhonciscus*

Grunt—*Rhonciscus strideus*. Species of little commercial value, caught in the Indian Ocean near Madagascar Island.

Length of fish caught in May–November 13.5–18.5 cm, weighing 72–148 g.

Weight proportions of individual body parts (%): trunk 47.3–51.0, head 30.4–32.0, scales 4.7–5.0, bones 10.5–10.6, fins 3.0–3.4, viscera 7.6–11.6. Output of fillets with skin 36.8–40.8%.

Chemical composition of meat (%): moisture 76.5, protein 19.9, fat 1.9, ash 1.5.

The soup is yellow and tasty. In cooked form the meat is tasty and gray in color. Smoked products and culinary preparations can be made from this fish.

Genus *Abudefduf*

Sergeant fish—*Abudefduf leucogaster*. Commercially an unimportant fish. Length of fish caught in May–January from the Great Australian Bight 28–39 cm, weight 650–1,535 g.

Weight proportions of individual body parts (%): trunk 53.0, head 32.0, viscera 15.0, including liver 0.9, fins 4.2, meat with skin 41.7, bones 7.1.

Chemical composition of meat (%): moisture 76.0, protein 19.4, fat 3.0, ash 1.6.

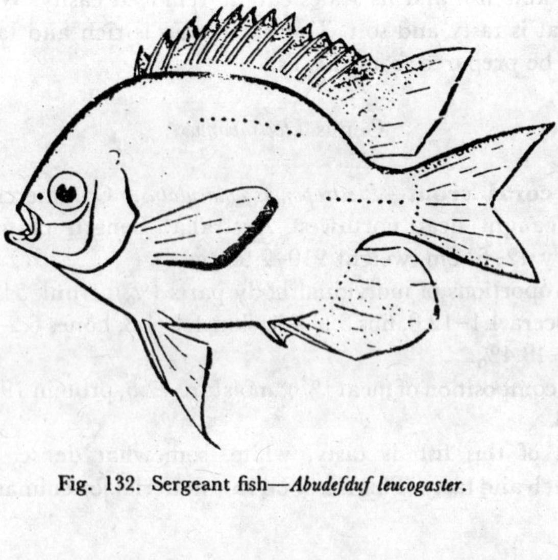

Fig. 132. Sergeant fish—*Abudefduf leucogaster*.

The meat of this fish is white and dense with good taste properties. Good culinary preparations are made from this fish.

Spotted sergeant fish—*Abudefduf anabatoides*. A non-commercial fish, caught in the East China Sea. Length of the fish caught in May 8.5–10.5 cm, weight 15–30 g.

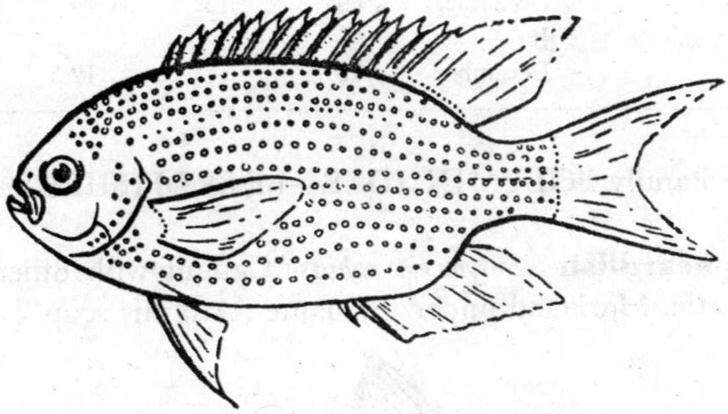

Fig. 133. Spotted sergeant fish—*Abudefduf anabatoides*.

Weight proportions of individual body parts (%): trunk with fins 57.6–62.4, head 25.6–28.7, viscera 8.9–14.1.

Chemical composition of meat (%): moisture 73.7, protein 23.9, fat 0.9, ash 1.5.

The meat of this fish is not tasty and it is bony; it should be used only for fish meal.

Genus *Pomacentrus*

Demoiselle—*Pomacentrus soltatrix*. A non-commercial fish, caught in the South China Sea. Length 8.5–10.5 cm, weight 14–32 g.

Weight proportions of individual body parts (%): trunk 52.0–54.3, head 22.8–30.3, fins 3.1–4.2, viscera 3.8–8.6.

Chemical composition of meat (%): moisture 73.7, protein 22.1–24.0, fat 0.9, ash 1.0–1.4.

The meat is bony with a strange taste; it should be used for fish meal.

Family PTEROTHRISSIDAE—GISSUS

African gissu—*Pterothrissus belloci*. Caught in small numbers with other fish in the eastern tropical zone of the Atlantic Ocean at depths of 150–300 m. Weight proportions of individual parts of fish 25–35 cm long, weighing 0.1–0.3 kg (%): trunk 71, head 16, viscera 12. Output of fillets 50%.

When fried, the meat is white, soft and tasty but has a large quantity of minute bones; it should be used for canning.

TABLE 167. CHEMICAL COMPOSITION, %

Body parts	Month	Moisture	Protein	Fat	Ash
Meat	June–December	76	19.4	3.4	1.5
Head, bones, fins	June–December	69	13	14	4
Viscera	June–December	72	14	12.5	1.4

Family POLYMIXIIDAE—BEARDFISHES

Common beardfish—*Polymixia nobilis.* Caught with other fish in the southeast Atlantic. Marketed under the name "Atlantic scup".

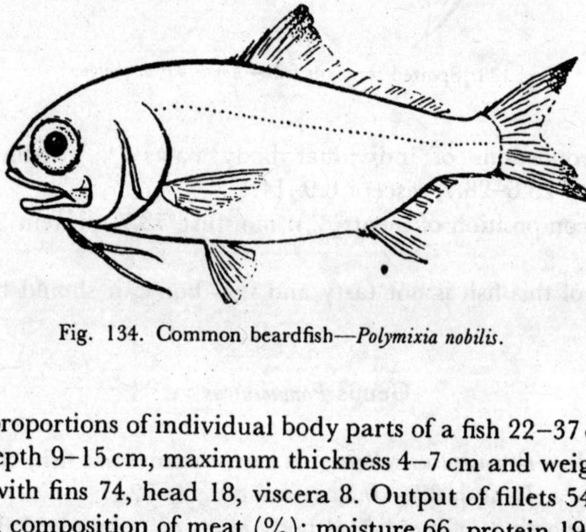

Fig. 134. Common beardfish—*Polymixia nobilis.*

Weight proportions of individual body parts of a fish 22–37 cm long, with maximum depth 9–15 cm, maximum thickness 4–7 cm and weight 0.3–1.8 kg (%): trunk with fins 74, head 18, viscera 8. Output of fillets 54%.

Chemical composition of meat (%): moisture 66, protein 19.9, fat 13, ash 1. The meat is tasty and fatty. It should be used for culinary preparations.

Family POLYNEMIDAE—THREADFINS

These fish are distributed in the tropical waters of the Indian and Pacific Oceans. The members of this family are not caught in large numbers, but specimens of large size may constitute a significant part of a catch.

Genus *Galeoides*

Longnose threadfin—*Galeoides polydactylus.* A fish which should have significant commercial value in the future. Length of fish caught in May near the Gulf of Guinea 16–24 cm, weight 125–345 g.

Weight proportions of individual body parts (%): trunk 60, head 25, viscera 10, fins 5. Output of fillets 46%.

Chemical composition of meat (%): moisture 76.7–77.0, protein 20.6–20.7, fat 1.2–1.4, ash 1.2–1.4.

When fried the meat is tasty, somewhat sweet, resembling in taste the meat of sander. This is a good table fish. It can be used for culinary preparations.

Genus *Polynemus*

Dara threadfin—*Polynemus indicus*. Length of fish caught on the western shelf of India 86.2–87.5 cm, average length 86.7 cm, weight 9,260–10,250 g, average weight 9,740 g.

Chemical composition of meat (%): moisture 75.6–76.6, protein 21.9–23.7, fat 0.7–2.0, ash 1.3.

The meat is white. After frying, the taste qualities are good. The consistency of the meat is dense. This is a good table fish from which semifinished products and culinary preparations can be made.

Silver six-thread tassel fish—*Polynemus xanthonemus*. Length of fish caught on the western shelf of India 78–108 cm, weight up to 16 kg.

Weight proportions of individual body parts (%): meat with skin 53.2, head 17.6, viscera 5.1, bones 11.7, scales 4.6, fins 5.5.

Chemical composition of meat (%): moisture 80.3, protein 18.3, fat 0.4.

Genus *Eleutheronema*

Four fingered threadfin—*Eleutheronema tetradactylum**. Found in the Indian Ocean.

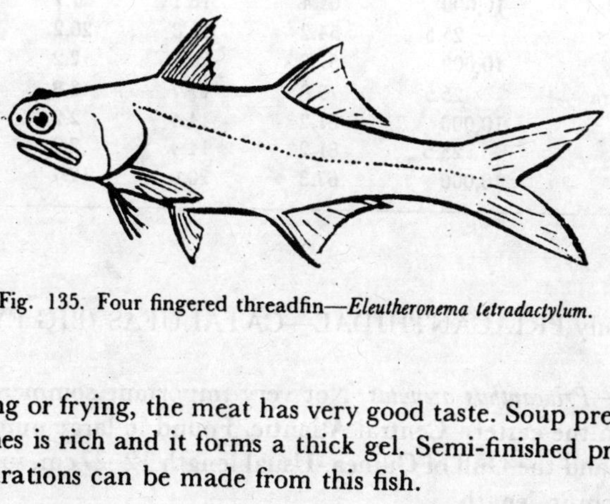

Fig. 135. Four fingered threadfin—*Eleutheronema tetradactylum*.

After boiling or frying, the meat has very good taste. Soup prepared from heads and bones is rich and it forms a thick gel. Semi-finished products and culinary preparations can be made from this fish.

TABLE 168. SIZE CHARACTERISTICS OF FOUR FINGERED THREADFIN CAUGHT IN THE INDIAN OCEAN NEAR AUSTRALIA

Parameters	Small fish	Large fish
Length, cm	24.0	92.5
Weight, g	25.5	10,000

TABLE 169. WEIGHT PROPORTIONS OF INDIVIDUAL BODY PARTS, %

Size groups	Trunk	Head	Meat	Scales	Bones	Fins	Viscera	Including liver
Small	67.3	21.5	59.3	—	8.0	2.3	8.9	—
Large	64.8	19.6	47.8	3.5	14.0	4.1	7.8	0.9

TABLE 170. CHEMICAL COMPOSITION, %

Body parts	Weight of fish, g	Moisture	Protein	Fat	Ash
Meat	25.5	74.7	21.2	2.5	1.5
	10,000	79.0	19.4	0.4	1.2
Head	25.5	67.6	18.5	7.3	5.3
Bones	25.5	62.9	8.1	21.0	7.8
	10,000	64.4	16.1	7.1	11.8
Fins	25.5	54.2	11.8	20.2	14.0
	10,000	57.9	20.1	2.2	19.3
Viscera	25.5	75.8	17.7	3.8	2.3
	10,000	81.2	14.4	2.2	1.6
Liver	25.5	81.2	14.4	2.2	1.6
	10,000	67.3	20.0	12.7	—

Family PRIACANTHIDAE—CATALUFAS (BIG EYES)

Big eye—*Priacanthus arenatus*. Not very important commercially, widely distributed in the eastern Central Atlantic. Found in large numbers south of Cape Verde and the Gulf of Guinea. Usual length 12–27 cm, some specimens reaching 35 cm in length.

TABLE 171. CHEMICAL COMPOSITION, %

Body parts	Moisture	Protein	Fat	Ash
Meat	76.7	20.2	1.5	1.6
Head, bones, viscera	74.1	14.5	4.6	6.8

Weight proportions o: individual body parts of fish caught in June along the coast of northwest Africa and 12–15 cm long, weighing 50–85 g (%): trunk with fins 57, head 35, viscera 8. Output of fillets 36%.

The meat of this fish is tasty. It can be used for culinary preparations and smoked products.

Bulleye—*Priacanthus boops*. Not very important commercially, caught in the East China Sea. Length of fish caught in March–April 19.5–30.0 cm, weight 150–440 g.

Fig. 136. Bulleye—*Priacanthus boops*.

Weight proportions of individual body parts (%): trunk 56.5–57.3, 28.4–29.0, viscera 10.3–11.3, bones 8.5–11.9, fins 3.1–3.5. Output of fillets 45.4–48.0%.

Chemical composition of meat (%): moisture 74.3, protein 21.0, fat 3.2, ash 2.0.

The meat in boiled form is soft and juicy; it becomes slightly compact after frying which diminishes its taste. Hot smoked products can be made from the meat; however, in smoke curing, because of the compactly located scales, good color is not obtained although the meat is fairly tasty. This fish should be frozen for subsequent preparation of culinary products.

Japanese big eye—*Priacanthus macracanthus*. Not very important commercially, caught in the East China Sea and in the western Pacific Ocean. Length of fish caught in March–September 17–24 cm, weight 108–240 g.

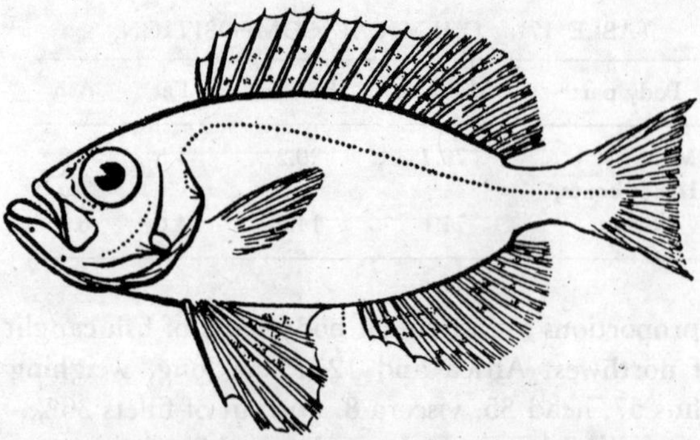

Fig. 137. Japanese big eye—*Priacanthus macracanthus*.

Weight proportions of individual body parts (%): head 21.0; output of fillets with skin 61.9%.

Chemical composition of meat (%): moisture 75.5, protein 21.6, fat 1.7, ash 1.3.

This fish has dense, hard and not very tasty meat and the fat is quickly oxidized. It can be used for culinary preparations.

Spot finned bullseye—*Priacanthus tayenus*. Fish of little commercial value, caught in the western Pacific Ocean and South China Sea, Sometimes significant in the composition of the catch.

Fig. 138. Spot finned bullseye—*Priacanthus tayenus*.

Length of fish caught in March–September 17–25 cm, weight 100–260 g.

Weight proportions of individual body parts (%): trunk with fins 62.0, head 29.4, viscera 8.6.

Chemical composition of meat (%): moisture 76.5–78.9, protein 19.0–20.4, fat 0.2–0.4, ash 1.5–1.6.

The meat is not very tasty. It is processed in chilled and frozen forms with subsequent use in culinary products.

Family PLEURONECTIDAE—RIGHTEYE FLOUNDERS

This family includes 44 genera distributed in almost all high seas (some fish enter inlets, gulfs and rivers). Members of this family are found in large numbers, forming concentrations, predominantly in the temperate latitudes.

Genus *Acanthopsetta*

Nadezhny's flounder—*Acanthopsetta nadeshnyi.* Caught in small numbers in the Sea of Japan and Okhotsk Sea. Output of trunk after gutting 71% and output of fillets 39%. Chemical composition of meat (%): moisture 80.5, protein 14.0, fat 2.0–4.6.

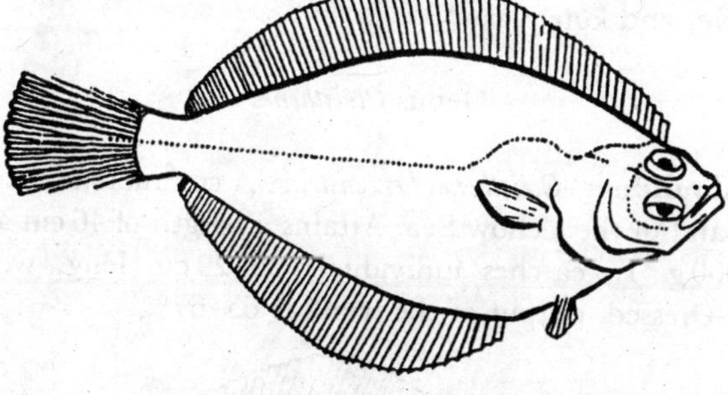

Fig. 139. Nadezhny's flounder—*Acanthopsetta nadeshnyi*

This fish should be used for culinary preparations or for canning such as "fried flounder in oil" and "smoked flounder in oil".

Genus *Atheresthes*

Kamchatka flounder—*Atheresthes evermanni.* Distributed in the Okhotsk and Bering Seas, extending to the Navarin Cape and in the east to St. Matthew Island.

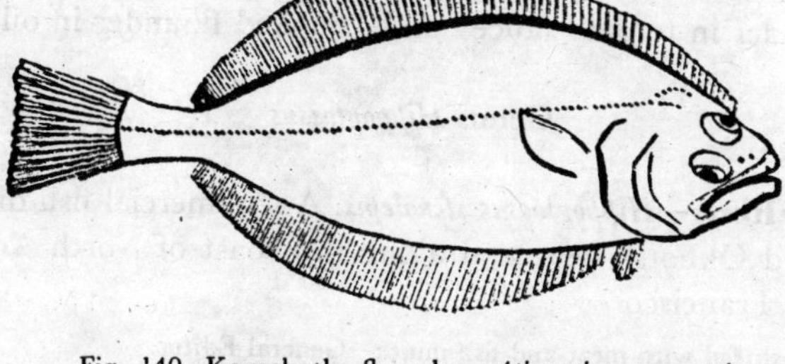

Fig. 140. Kamchatka flounder—*Atheresthes evermanni.*

American arrowtooth flounder—*Atheresthes stomias.* Commercially important along the coast of the USA, south of San Francisco. Usual length 30–75 cm, weight 1–3 kg.

Weight proportions of individual body parts (%): meat with skin 52–67, head 18–22, bones and fins 6–13, viscera 6–19

The meat of this flounder like the Greenland halibut contains 10–17% protein and 4–22% fat. Unlike other flatfish the arrowtooth flounder is known for low mineral content in the bones (4–5% ash).

Unlike the Greenland halibut, the meat of the arrowtooth flounder is easily separated during salting and heat-processing and its condensation is poor; the meat becomes friable and its structure is destroyed. Therefore, the meat of arrowtooth flounder is best frozen and thereafter fried in the form of breaded pieces at a temperature of 160° C; it can also be prepared as mince or stuffing material for pies and kulebyaka*.

Genus *Cleisthenes*

Sohachi flounder—*Cleisthenes herzensteini.* A commercial fish found in the Sea of Japan and in the Yellow Sea. Attains a length of 46 cm and weight of more than 500 g. In catches individuals 25–29 cm long, weighing 280 g predominate. Dressed, output of the trunk is 65–67%.

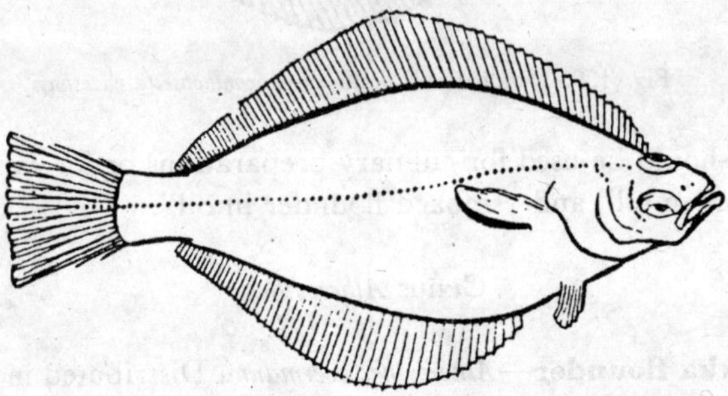

Fig. 141. Sohachi flounder—*Cleisthenes herzensteini.*

Chemical composition of meat (%): moisture 79, protein 15.2, fat 0.8–9.9.

This fish should be used for culinary preparations or canned snacks such as "fried flounder in tomato sauce" and "smoked flounder in oil"

Genus *Hippoglossus*

Pacific halibut—*Hippoglossus stenolepis.* A commercial fish distributed in the Bering and Okhotsk Seas and along the coast of North America from Alaska to San Francisco.

* A special pie stuffed with meat and fish mince—General Editor.

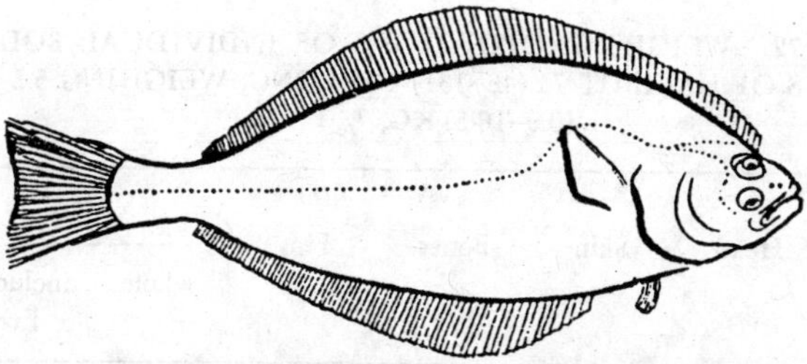

Fig. 142. Pacific halibut—*Hippoglossus stenolepis*.

Attains a length of 230 cm and weight of 239 kg. In catches fish 40–85 cm long weighing up to 8 kg predominate.

Weight proportions of individual body parts of Pacific halibut 70–125 cm long, weighing 6–25 kg (%): meat with skin 61.8, head 11.5–18.0, backbone and fins 7.5–13.0, viscera 8.5–13.5, including liver 0.9–2.6.

The meat of the Pacific halibut, unlike other flatfish, contains a relatively high quantity of protein, 17.3–20.8%, fat 0.9–9.8%. The fat is mainly concentrated around the fins and in the bones.

The Pacific halibut is a highly valuable fish.

It can be used for first and second courses.

Atlantic halibut—*Hippoglossus hippoglossus*. Found with other fish in catches. Widely distributed in the North Atlantic, off Greenland and in the Norwegian and Barents Seas. The largest fish among flatfish attains a length of 2 m and weight of 300 kg.

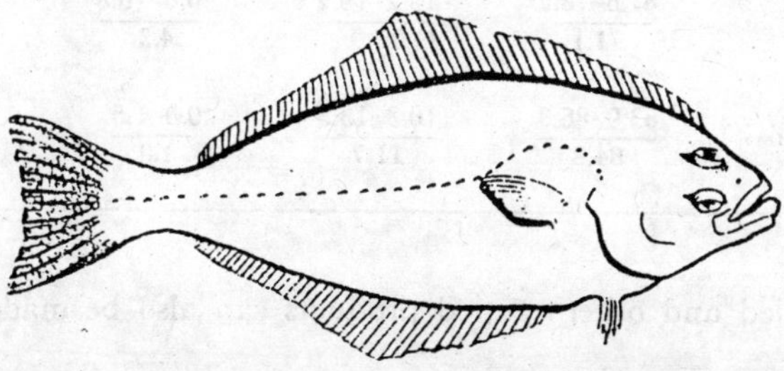

Fig. 143. Atlantic halibut—*Hippoglossus hippoglossus*.

The vitamin A potency of its liver oil is 440–135,000 iu/g and that of the oil of other internal organs 600,000–1,000,000 iu/g.

This is a valuable table fish. Its meat can be used for snacks and tasty first and second course dishes. Because of the significantly high protein content in the meat and its water content, it has a dense consistency after boiling and frying and does not disintegrate. The meat gives ukha soup a good flavor.

TABLE 172. WEIGHT PROPORTIONS OF INDIVIDUAL BODY PARTS OF HALIBUT 75 (48–131) CM LONG, WEIGHING 5.5 (0.9–16.5) KG, %

Meat	Head	Skin	Bones	Fins	Viscera whole	Viscera including liver
54.1	14.4	2.0	7.1	2.0	5.3	0.1
60.0	19.6	6.5	12.1	4.7	16.9	2.9
58.0	16.9	4.3	9.6	3.1	7.4	1.2

TABLE 173. CHEMICAL COMPOSITION, %

Body parts	Moisture	Protein	Fat	Ash
Meat	73.8–80.8 / 78.1	15.4–20.3 / 18.0	0.0–6.2 / 2.0	1.2–1.6 / 1.3
Liver	66.3–75.7 / 69.3	12.4–17.4 / 14.4	2.8–17.8 / 13.6	1.2–1.7 / 1.5
Head	65.4–79.7 / 73.0	13.7–15.9 / 14.5	0.1–15.0 / 7.7	3.2–5.5 / 4.2
Bones	55.6–74.4 / 66.1	13.8–17.5 / 15.7	traces–23.9 / 9.5	6.7–9.8 / 7.2
Fins	61.6–78.2 / 71.1	15.2–19.2 / 17.0	0.0–10.8 / 4.3	5.7–7.7 / 6.7
Viscera without liver and gonads	83.0–86.3 / 84.8	10.3–13.3 / 11.7	0.0–1.5 / 1.0	0.9–2.3 / 1.5

Very tasty jellied and other canned products can also be made from this halibut.

Genus *Hippoglossoides*

American plaice—*Hippoglossoides platessoides*. Caught with other fish in the North Atlantic. Distribution in the northern Atlantic Ocean, in the western and central Barents Sea and in the Kara Sea.

Maximum length of this fish in the Barents Sea 51.5 cm and weight not more than 1.5 kg. In catches, a predominance of individuals 20–45 cm long.

Weight proportions of individual body parts of fish 30–43 cm long, weighing 311–1,322 g (%): head (separated by semicircular cut) 15.3, trunk

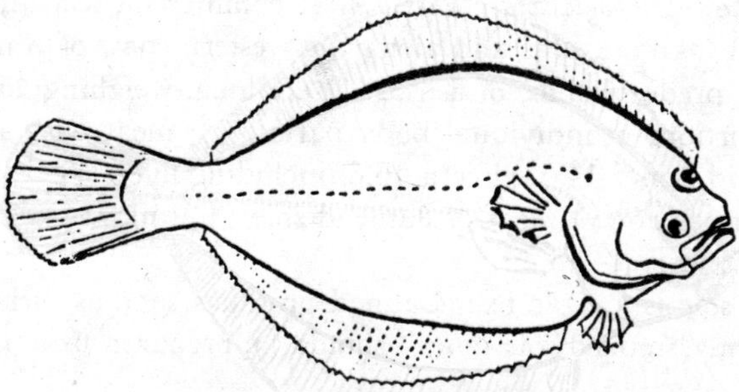

Fig. 144. American plaice—*Hippoglossoides platessoides*.

TABLE 174. CHEMICAL COMPOSITION, %

Body parts	Moisture	Protein	Fat	Ash
Meat	75.4–83.7 / 80.4	14.1–16.8 / 15.5	0.1–5.9 / 2.7	1.2–1.8 / 1.4
Jelly-like meat	87.0	10.9	0.8	—
Roe	70.4–82.4	14.2–24.7	0.1–2.0	0.7–2.7
Liver	62.3–79.7 / 73.4	12.9–15.7 / 14.7	1.3–20.4 / 7.3	1.1–1.8 / 1.5
Head	75.2–82.1 / 78.4	11.3–13.6 / 12.5	0.4–7.6 / 2.4	5.2–7.0 / 6.0
Bones	54.8–73.2 / 67.9	10.0–17.0 / 13.8	0.5–17.7 / 6.8	8.9–13.9 / 10.4

65.3, meat 49.4, skin 3.9, bones 11.6, fins 2.4, scales 1.3, viscera 13.2, including liver 1.7.

In the northwest Atlantic individuals are often caught with jelly-like watery meat. Culinary preparations should be made from this plaice.

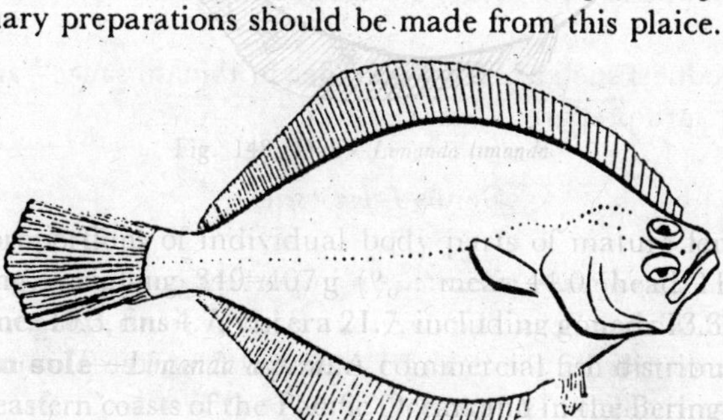

Fig. 145. Flathead sole—*Hippoglossoides elassodon*.

Flathead sole—*Hippoglossoides elassodon.* A commercial fish, distributed in the Bering and Okhotsk Seas and along the western coast of America. In catches there is a predominance of fish 32–38 cm long, weighing 450–650 g.

Weight proportions of individual body parts (%): meat with skin 52.7, head 16.7, fins and bones 14.0, viscera 16.6, including liver 1.6.

Chemical composition of meat of flathead sole (%): moisture 79.3–81.5, protein 14.6–16.7, fat 1.4–3.3.

The flathead sole is a table fish. Canned products such as "fried sole in tomato sauce" and "smoked sole in oil" should be prepared from this fish.

Genus *Glyptocephalus*

Steller's small mouth flounder—*Glyptocephalus stelleri.* Distributed in the Sea of Japan, in the southern Okhotsk Sea; also found off the southern coast of Kamchatka. Maximum body length 60 cm. In catches there is a predominance of individuals 38–44 cm long, weighing 800 g. Output of trunk after gutting 68–75% and fillets 48%.

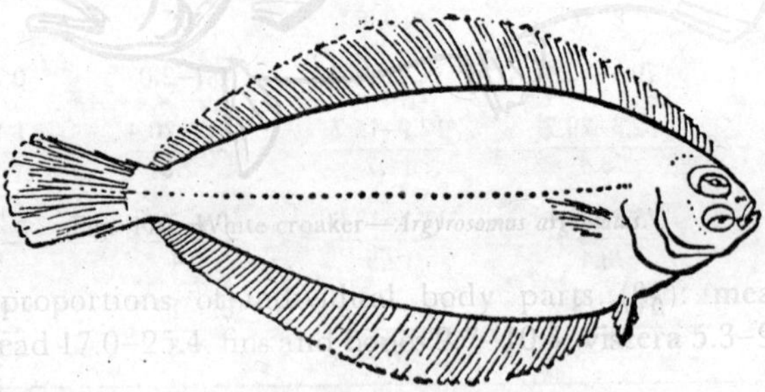

Fig. 146. Steller's small mouth flounder—*Glyptocephalus stelleri.*

Chemical composition of meat (%): moisture 73.2–80.0, protein 17.3–18.0, fat 0.8–7.0.

Canned products such as "fried flounder in tomato sauce" and "smoked flounder in oil" are prepared from this fish.

Genus *Lepidopsetta*

Rock sole—*Lepidopsetta bilineata.* An important commercial fish found in the Bering and Okhotsk Seas; also distributed in the south along the American coasts and is known from the Sea of Japan and from the Tatar Strait up to Korea and near the coast of Japan.

Weight proportions of individual body parts (%): meat with skin 50.3, head 14.8, fins and bones 15.0, viscera 19.9, including gonads 15.0, liver 1.7.

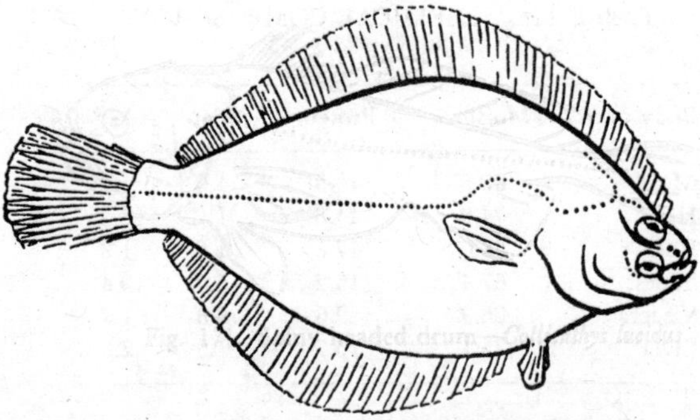

Fig. 147. Rock sole—*Lepidopsetta bilineata.*

Chemical composition of meat (%): moisture 76.6–82.6, protein 13.2–17.6. This fish should be used for canned products such as "fried sole in tomato sauce" and "smoked sole in oil".

Genus *Limanda*

Dab—*Limanda limanda.* Distributed in the northern Atlantic Ocean from Iceland and Norway in the north to the Bay of Biscay in the south, in the Barents and White Seas and also found in the Baltic Sea. Attains a length of 40 cm and weight of 720 g.

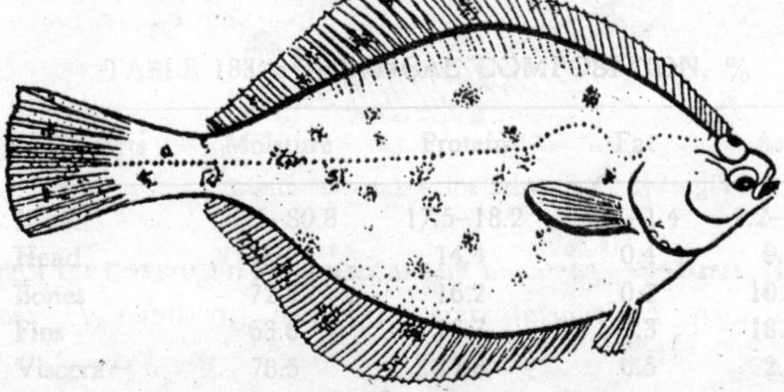

Fig. 148. Dab—*Limanda limanda.*

Weight proportions of individual body parts of mature females in June 32–33 cm long, weighing 349–407 g (%): meat 44.0, head 11.6, skin with scales 5.7, bones 10.3, fins 4.7, viscera 21.7, including gonads 13.3 and liver 3.0.

Yellowfin sole—*Limanda aspera.* A commercial fish distributed along the western and eastern coasts of the Pacific Ocean and in the Bering and Okhotsk Seas; also found in the Sea of Japan. A demersal fish. Attains a length of 46–48 cm. Average length of yellowfin sole in commercial catches 33–34 cm,

TABLE 175. CHEMICAL COMPOSITION OF DAB, %

Body parts	Moisture	Protein	Fat	Ash
Meat	80.6	16.5	1.6	1.2
Head	74.9	14.8	1.7	6.3
Roe	68.0	17.8	1.5	1.3
Bones	67.4	15.2	5.0	10.4
Viscera	86.9	9.6	0.8	1.7
Liver	75.1	16.7	1.6	2.3

weight 400–450 g. After dressing the output of trunk is 65–69% and fillets 35–49%.

Chemical composition of meat (%): moisture 78.2–82.6, protein 13.7–16.3, fat 0.7–1.5.

These fish are used for preparing canned products such as "fried sole in tomato sauce" and "smoked sole in oil" and for culinary preparations.

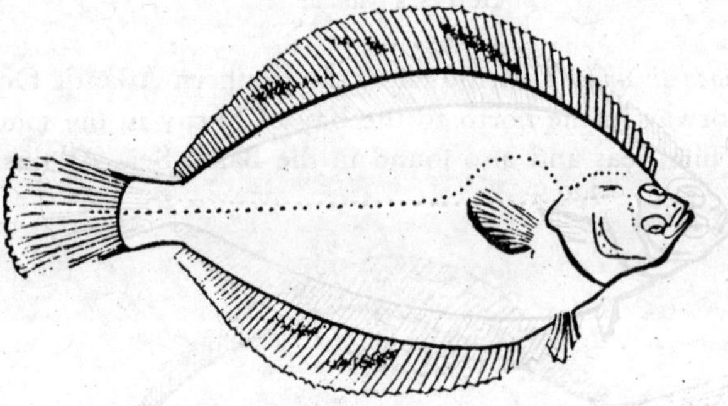

Fig. 149. Yellowfin sole—*Limanda aspera.*

Longsnout flounder—*Limanda punctatissima.* There are two forms in this species: the southern *L. punctatissima* and the northern *L. punctatissima*

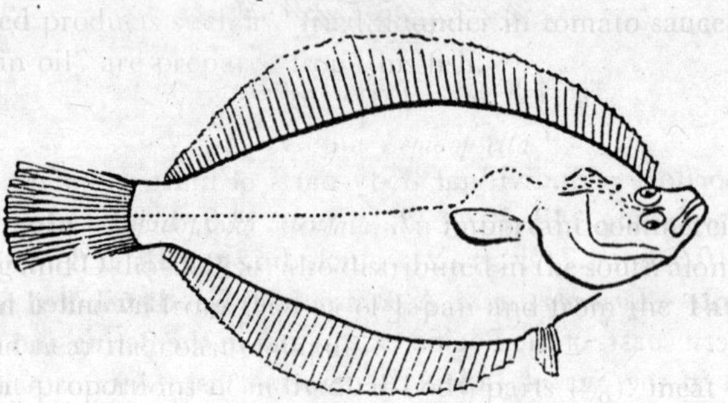

Fig. 150. Longsnout flounder—*Limanda punctatissima.*

proboseidea. The southern form lives in the northern Sea of Japan (from the estuaries of Tumanula River) and southern Okhotsk Sea and the northern form (with proboscis) lives in the Bering and Okhotsk Seas.

A commercial fish, attains a length of 39 cm. In catches the average weight is 260–279 g with a body length of 20–32 cm. Output of trunks after gutting 68–73% and fillets 48.5%. The fat content of meat 0.7–2.3%.

These fish are used for canning products such as "fried flounder in tomato sauce" and "smoked flounder in oil" and for culinary products.

Yellow striped flounder—*Limanda herzensteini*. A commercial fish distributed in the Sea of Japan from Cape Lazarev to the coast of Korea, passing through the La Perouse Strait (Soya Strait) into the Okhotsk Sea. Attains a length of 43 cm. In commercial catches the average length of fish in the summer is 25- 30 cm, in winter 28–32 cm, average weight 270–330 g.

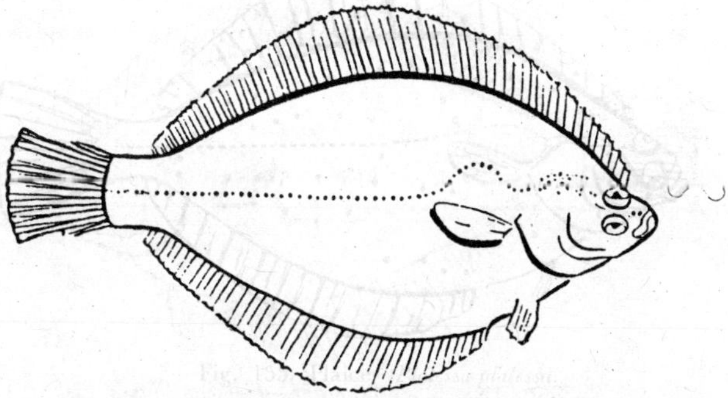

Fig. 151. Yellow striped flounder—*Limanda herzensteini*.

Output of trunks after gutting 69–76% and fillets 47.5%. Fat content of meat 0.6–2.6%.

These fish are used for canned products such as "fried flounder in tomato sauce" and "smoked flounder in oil" and for culinary preparations.

Yokohama flounder—*Limanda yokohamae*. Species found along the coast

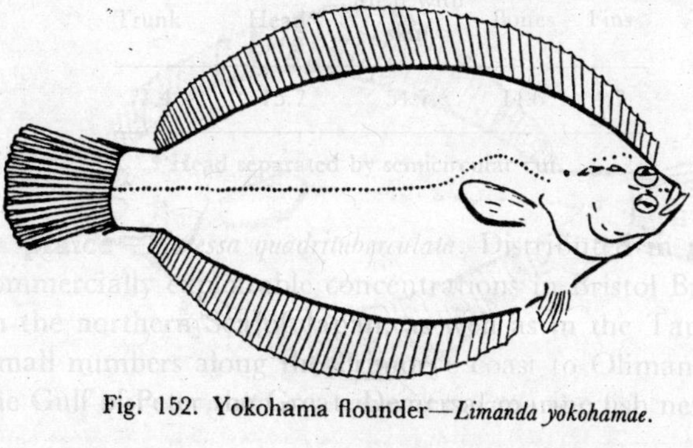

Fig. 152. Yokohama flounder—*Limanda yokohamae*.

of Japan, in the Sea of Japan, along the coast of the USSR from Cape Lazarev to the Korean coast. Attains a length of 49 cm. The output of trunks after dressing 68–69% and fillets 48%.

Chemical composition of meat (%): moisture 78, protein 19, fat 0.3–4.2.

The yokohama flounder is used for culinary preparations and canned snacks such as "fried flounder in tomato sauce" and "smoked flounder in oil".

Genus *Liopsetta*

Black flounder—*Liopsetta obscura*. Distributed in the Yellow Sea, along the coast of Korea, in the Sea of Japan and in the southern Okhotsk Sea. Attains a length of 40 cm.

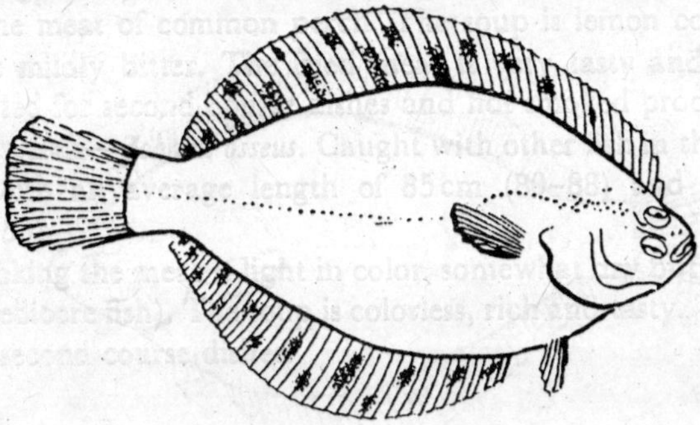

Fig. 153. Black flounder—*Liopsetta obscura*.

Chemical composition of meat (%): moisture 82, protein 14, fat 2.5.

The fish are first frozen and then used for culinary products and canned preparations such as "fried flounder in tomato sauce" and "smoked flounder in oil".

Banded flounder—*Liopsetta pinnifasciata*. Distributed in the Sea of Japan. Catches are not very large. Output of trunks after dressing is 68–75% and

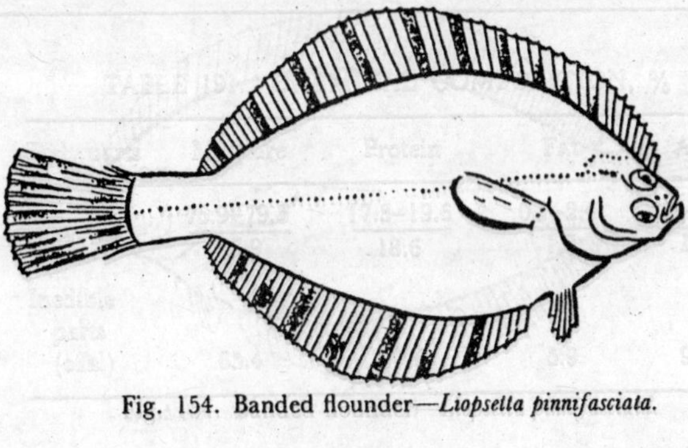

Fig. 154. Banded flounder—*Liopsetta pinnifasciata*.

fillets 45%. Chemical composition of meat (%): moisture 80, protein 16, fat 2.1–3.4.

Canned products such as "fried flounder in tomato sauce" and "smoked flounder in oil", and culinary preparations are made from this fish.

Genus *Platessa*

Plaice—*Platessa platessa.* Caught with other fish in the northern Atlantic Ocean from Portugal to Iceland, in the central and western Baltic Sea (rarely in the Gulf of Finland), in the Barents and White Seas. Attains a length of 1 m, in the Barents Sea up to 86 cm. In catches the fish is generally 30–40 cm long and weighs 0.5–1.0 kg.

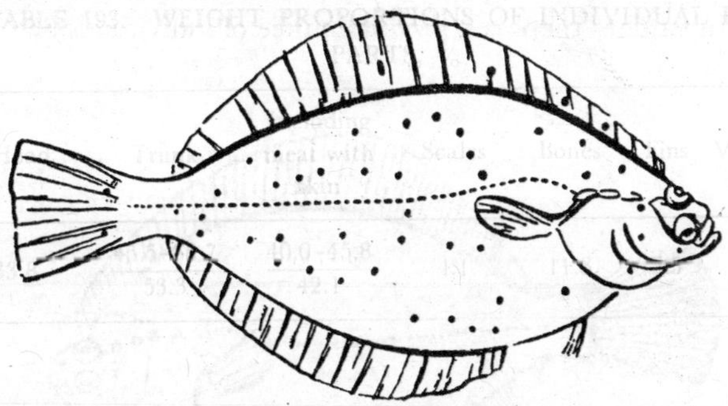

Fig. 155. Plaice—*Platessa platessa.*

The plaice should be used for canned products and salt cured and smoked products.

TABLE 176. WEIGHT PROPORTIONS OF INDIVIDUAL BODY PARTS OF PLAICE 30–44 CM LONG, WEIGHING 480–1,800 g, %

Trunk	Head*	Meat with skin	Bones	Fins
71.4	15.7	54.7	14.6	4.7

* Head separated by semicircular cut.

Alaska plaice—*Platessa quadrituberculata.* Distributed in the Bering Sea forming commercially exploitable concentrations in Bristol Bay, in Okhotsk Sea and in the northern Sea of Japan, as well as in the Tatar Strait. Also found in small numbers along the Primorye coast to Olimanda Cape, then entering the Gulf of Peter the Great. Demersal marine fish never found near

TABLE 177. CHEMICAL COMPOSITION OF PLAICE, %

Body parts	Moisture	Protein	Fat	Ash
Meat	77.9–81.7 / 80.1	13.7–16.3 / 15.5	1.6–4.6 / 2.8	1.1–1.3 / 1.2
Roe	70.0–85.4	11.3–24.8	0.3–1.5	1.1–1.4
Milt roe	84.3–86.2	10.3–11.1	0.2–2.3	1.5
Viscera	73.0–89.5	7.8–12.2	1.2–5.5	—
Liver	72.1–78.1 / 76.5	13.1–15.1 / 14.1	0.6–12.0 / 6.2	1.2–1.9 / 1.6

fresh water. In catches there is a predominance of individuals 36–39 cm long (up to 50 cm).

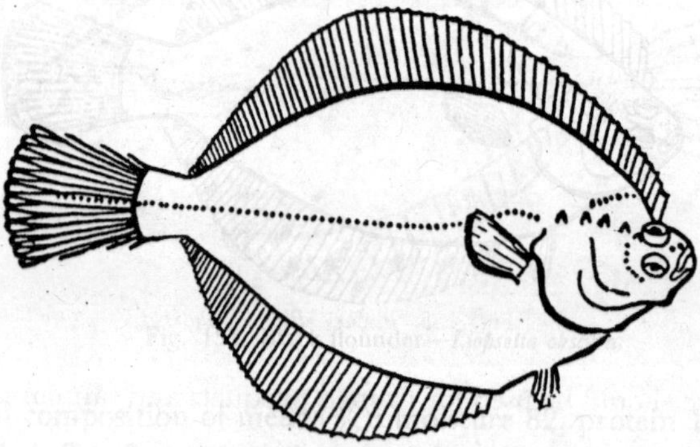

Fig. 156. Alaska plaice—*Platessa quadrituberculata.*

Chemical composition of meat (%): moisture 79.3–81.6, protein 13.5–17.1, fat 2.4–3.8.

This plaice should first be frozen and then used for canned products such as "fried plaice in tomato sauce" and "smoked plaice in oil" and culinary preparations.

Genus *Pleuronectes*

Starry flounder—*Pleuronectes stellatus.* Widely distributed in the northern Pacific Ocean; found in the Bering Sea and penetrating the Chukotsk Sea through the Bering Strait. Along the Pacific coasts of North America, south to San Francisco. Found throughout the Okhotsk Sea. In the Sea of Japan known up to Tokyo. Attains a length of 54 cm, usually the length is 23–24 cm

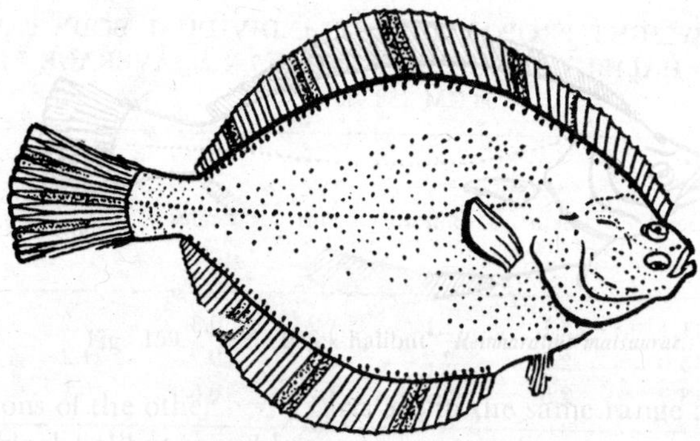

Fig. 157. Starry flounder—*Pleuronectes stellatus.*

and weight 2–4 kg. After dressing, the output of trunks is 59–64% and fillets 45%.

Chemical composition of meat (%): moisture 79.1–79.6, protein 14.3–16.2, fat 3.0–3.7.

The starry flounder should first be frozen and then used for culinary preparations and canned products such as "fried flounder in tomato sauce" and "smoked flounder in oil".

Genus *Reinhardtius*

Greenland halibut—*Reinhardtius hippoglossoides.* Distributed in the northern Atlantic Ocean. Found in large numbers west of Greenland, an important commercial fish. Attains a length of 120 cm, weight 44.5 kg. In the Barents Sea there is a predominance of individuals in catches weighing 2–4 kg. With the development of deep sea trawling, the catch of this halibut has increased in recent years.

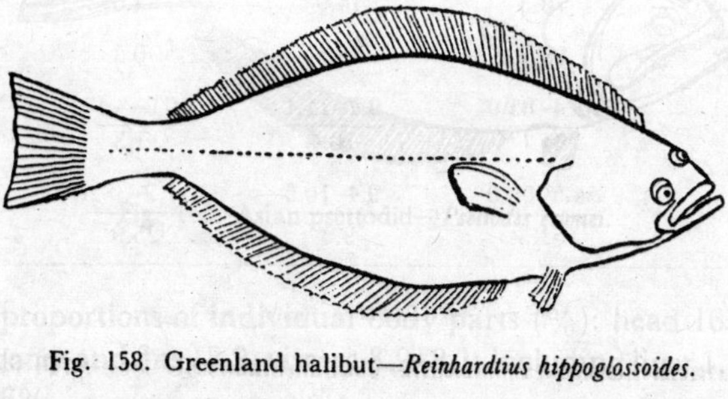

Fig. 158. Greenland halibut—*Reinhardtius hippoglossoides.*

The vitamin A content in the liver oil is 690–1,370 iu/g or 2,800–64,200 iu/g. The meat fat contains vitamin F (1.8 iu/g) which is also present in the liver oil (4.8 iu/g).

TABLE 178. WEIGHT PROPORTIONS OF INDIVIDUAL BODY PARTS OF GREENLAND HALIBUT WEIGHING 2.3 KG (0.7–4.2), AVERAGE LENGTH 54 CM (38–66), %

Meat	Head	Skin	Bones	Fins	Scales	Viscera	
						whole	including liver
56.1	11.4	2.1	4.1	1.2	0.3	4.7	0.9
70.9	15.0	2.9	7.2	2.3	0.7	21.2	3.3
64.5	13.7	2.5	5.6	1.7	0.6	10.1	1.9

The Greenland halibut is used for cured balyks after cold smoking. In fresh (frozen) form the fish is less valuable because of high water content and the poor consistency of the meat. The salted roe has pleasant taste.

TABLE 179. CHEMICAL COMPOSITION, %

Body parts	Moisture	Protein	Fat	Ash
Meat	64.6–71.8 / 69.4	12.3–13.2 / 12.8	13.1–21.0 / 16.1	0.8–1.0 / 0.9
Head	60.3–67.9 / 64.2	10.1–11.3 / 10.5	16.6–25.4 / 20.1	2.3–2.7 / 2.5
Bones	53.4–59.1 / 55.5	10.4–11.3 / 10.8	24.0–31.5 / 27.5	4.0–4.8 / 4.4
Fins	38.0–46.4 / 42.9	9.2–9.6 / 9.4	36.0–46.0 / 40.6	4.2–4.4 / 4.3
Roe in stage III of maturation	78.3	14.0	2.6	1.0
Testes	83.7	12.5	0.8	1.3
Viscera	83.4–85.9 / 84.1	9.7–11.1 / 10.4	0.0–3.1 / 1.3	1.3–2.4 / 1.7
Liver	54.5–68.0 / 61.5	9.4–10.5 / 9.8	17.2–33.4 / 25.9	0.8–1.1 / 1.0

Pacific black halibut—*Reinhardtius matsuurae.* Fish form shoals in those regions where conditions are congenial for transporting eggs and larvae to shallow waters. Attain weight of 7–10 kg, larger fish occur rarely.

The head of this fish is relatively larger than in halibut (18–21% of the weight); the fillets together with the skin constitute about 60%, the weight

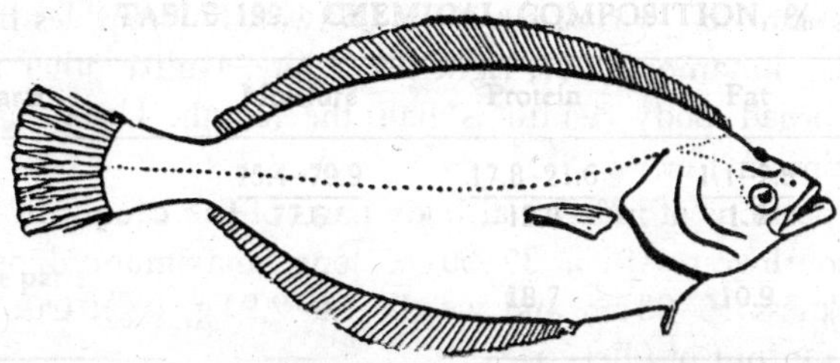

Fig. 159. Pacific black halibut—*Reinhardtius matsuurae*.

proportions of the other body parts are in the same range as in halibut. The meat of black halibut has a high water content (it contains up to 84% moisture and 5.4–13% protein), but a lot of fat is deposited in the tissues: in the meat up to 19%, in the bones and fins up to 30%. The bones and fins contain a total of 3–4% mineral matter. The meat has good taste. The fish are suitable for salting and subsequent cold smoking for first and second course dishes. During boiling or salting the meat becomes condensed, moisture is released and the fat is retained fairly well in the tissues.

Family PSETTODIDAE—SPINY-RAYED FLOUNDERS OR PSETTODIDS

Asian psettodid—*Psettodes erumei*. Not important commercially. Caught with other fish in the western Pacific Ocean and in the South China Sea. Length 26.5–43.0 cm, weight 245–1,110 g.

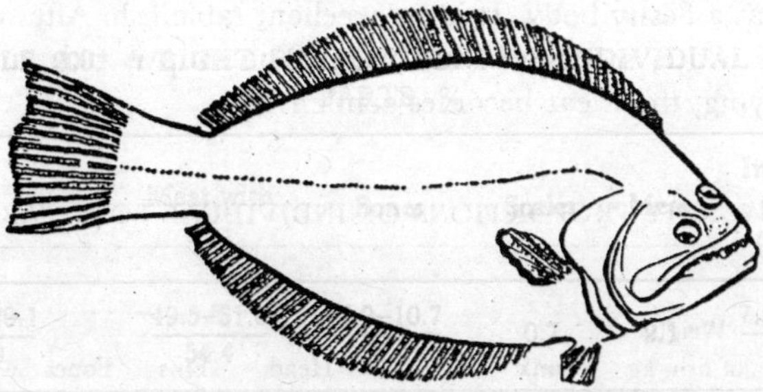

Fig. 160. Asian psettodid—*Psettodes erumei*.

Weight proportions of individual body parts (%): head 16.8–28.7, trunk 62.5–74.8, bones and fins 18.9, viscera 3.9–8.1, including liver 1.1–1.2. Output of fillets 47.6%.

Chemical composition of meat (%): moisture 74.4, protein 19.3, fat 0.3, ash 1.0.

This fish resembles common flatfish. The meat has medium taste properties.

African psettodid—*Psettodes belcheri*. Outwardly resembles halibut; left side of the body cinnamon color, right side white. Unlike other flatfish this species has a broad body: width is half the length. Usual body length 40–50 cm, maximum 1 m.

Weight proportions of individual body parts of fish caught in June along the coast of southwest Africa 39–50 cm long, maximum depth 4–6 cm, maximum thickness 16–23 cm and weight 1.2–2.9 kg, (%): trunk 72, head 15, viscera 13. Output of fillets 54%.

After frying, the meat is soft and tasty. This is a good table fish.

Family RACHYCENTRIDAE—COBIAS (CRABEATERS)

Cobia (Crabeater)—*Rachycentron canadum*. Distributed in the western Pacific Ocean, eastern Indian Ocean and in the Atlantic Ocean. Forms concentrations in the southeast Atlantic.

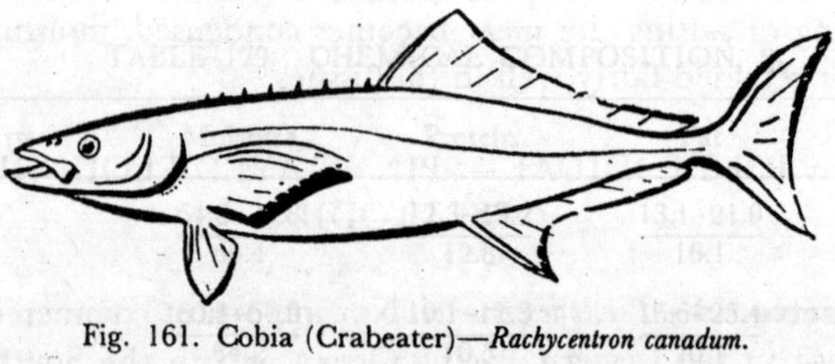

Fig. 161. Cobia (Crabeater)—*Rachycentron canadum*.

Usual length of fish 0.6–1.0 m, sometimes up to 2 m.

The fish has a fleshy body. It is an excellent table fish. After cooking, the meat becomes white, soft, juicy and tasty. The soup is rich, aromatic and tasty. After frying, the meat becomes condensed.

TABLE 180. WEIGHT PROPORTIONS OF INDIVIDUAL BODY PARTS, %

Place of catch	Weight of fish, kg	Trunk	Meat	Head	Fins	Bones	Viscera whole	Viscera including roe	Viscera including testes
Arabian Sea	8.0	—	52.5	20.3	4.2	9.0	8.8	—	0.9
Northwest Atlantic	3.2	—	56.0	20	—	—	—	—	—
Great Australian Bight	1.6	63.4	55.0	25.1	4.5	8.4	8.0	—	—
Southeast Atlantic	16.0	63.0	—	23.0	4.0	—	10.0	—	—

Cobia should be used for smoke cured balyks and culinary preparations.

TABLE 181. CHEMICAL COMPOSITION, %

Place of catch	Moisture	Protein	Fat	Ash
Arabian Sea	78.1	19.9	0.8	—
Northwest Atlantic	77.0	19.1	2.9	1.1
Great Australian Bight	77.4–79.3	20.6–22.1	0.3–0.4	1.7–1.9
Southeast Atlantic				
April	82.0	16.5	0.2	
January	77.6	18.7	2.5	1.2

Family SALMONIDAE—SALMONIDS

Members of this family are distributed in the seas and rivers of Europe, north and east Asia and North America. The family includes nine genera. Recently, some species of this family have been acclimatized in the southern hemisphere and the pink salmon has adapted to the Barents Sea.

Genus *Oncorhynchus*

Pink salmon—*Oncorhynchus gorbuscha*. Widely distributed in the northern Pacific Ocean. Caught in the Arctic Ocean south to Lena and along the Asian coast south to Korea and Hokkaido Island and along the North America coast.

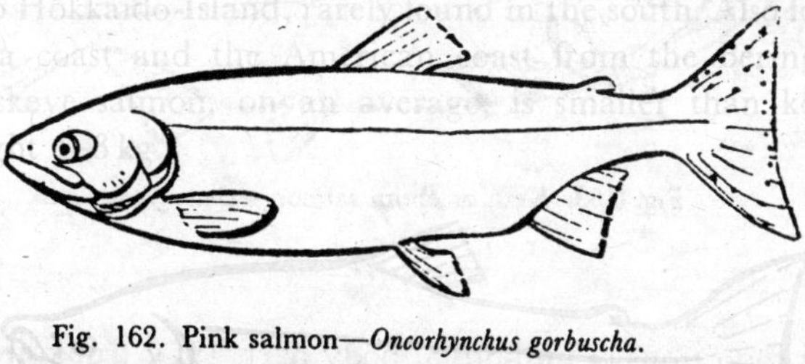

Fig. 162. Pink salmon—*Oncorhynchus gorbuscha*.

Average length of the Amur pink salmon 44 cm, Kamchatka pink salmon 49 cm and average weight 1.0–1.5 kg and in some years less than 1 kg.

Weight proportions of individual body parts of pink salmon 62 cm long (%): trunk 57–60, head 14–15, liver 2, roe 3.5–4.5, testes 2.5–3.0, remaining viscera 6, fins 3. Output of fillets 50%.

In the prespawning period the pink salmon contains 4–9% and 19–24% protein. Depending upon the maturation phase, the food values of the fish decrease.

The meat of the pink salmon is pinkish and softer than that of keta. Mildly salted or smoked products and canned products and culinary preparations can be made from this salmon. A part of the head and nape are used for canned products such as "ragout of salmon". Fish paste can be made from the soft roe and liver with the addition of meat.

TABLE 182. CHEMICAL COMPOSITION OF MEAT, %

Stage of sexual maturation	Moisture	Protein	Fat	Ash
Without changes (smelt)	65.3	24.1	9.7	1.7
I	68.5	22.9	7.1	1.5
II	72.2	21.2	2.6	1.5
IV (spawning colors)	71.4	21.7	3.6	1.5
V (spent fish)	87.0	12.6	0.6	0.8

Keta or chum salmon—*Oncorhynchus keta.* Distributed along the Asian coasts south to Korea and northern Japan and in the north from the estuary of Mna River along the American coast south to San Francisco. Length of keta usually caught 45–80 cm and weight 1–10 kg.

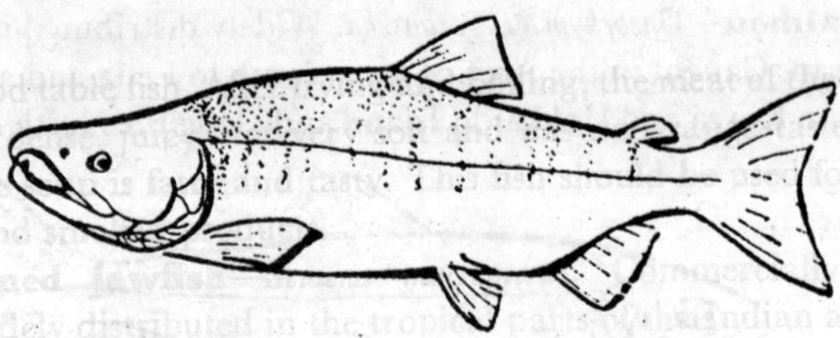

Fig. 163. Keta or chum salmon—*Oncorhynchus keta.*

Weight proportions of individual body parts (%): trunk around 73 (with machine dressing for canning not more than 63), head 10–11, including gills 2.5, liver 1.5–2.0, roe 3.5–4.0, testes 2.0, remaining viscera 5.5–6.5, fins with pieces of meat 3.0, skin with scales 3.0, backbone 4.0.

The chemical composition of keta meat in other parts of the body depends upon the area of fishing and the stage of maturation of the fish. Meat of marketable keta contains 19–23% protein and 4–9% fat, whereas the Amur keta may have up to 20% fat in the meat. In keta entering the upper parts of a river, in spawning and spent fish, the protein content is reduced to 10–12% and the fat to 0.5%. The fat content is higher in the head, bones and fins than in the meat. The roe in stage IV of maturation contains 10–12%

fat and 30% protein. The testes and other organs of the viscera contain low quantities of fat, 1–4%.

Keta is used to prepare mildly salted or smoked products as well as canned and culinary products. A part of the head and nape are used for canned products such as "ragout of salmon". Fish paste can be made from the testes and liver with the addition of meat.

Silver or Coho salmon—*Oncorhynchus kisutch.* Distributed mainly along the coast of the Pacific Ocean of North America (from Alaska to California) as well as along the Asian coast from Anadyr to Hokkaido Island. Found very rarely in the northern Okhotsk Sea.

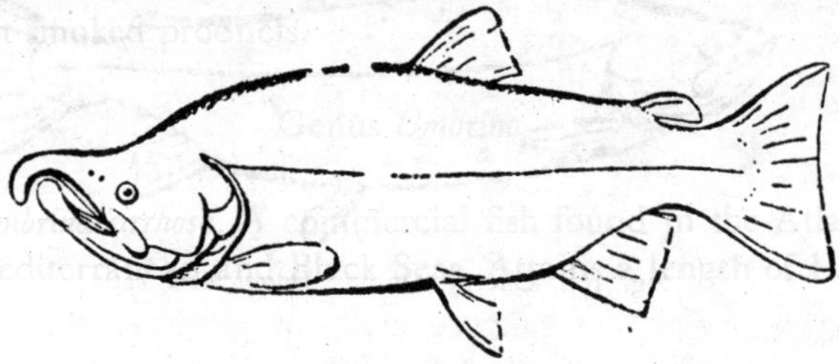

Fig. 164. Silver salmon—*Oncorhynchus kisutch.*

The protein content in the meat reaches 20–23% and fat 6–9%.

Silver salmon is best used in preparing canned products and mildly salted preparations. The roe of this salmon has a mildly bitter flavor.

Sockeye salmon—*Oncorhynchus nerka.* Distributed along the Asian coast from Anadyr to Hokkaido Island, rarely found in the south. Also found along the Kamchatka coast and the American coast from the Bering Strait to California. Sockeye salmon, on an average, is smaller than keta: length 40–60 cm, weight 1–3 kg.

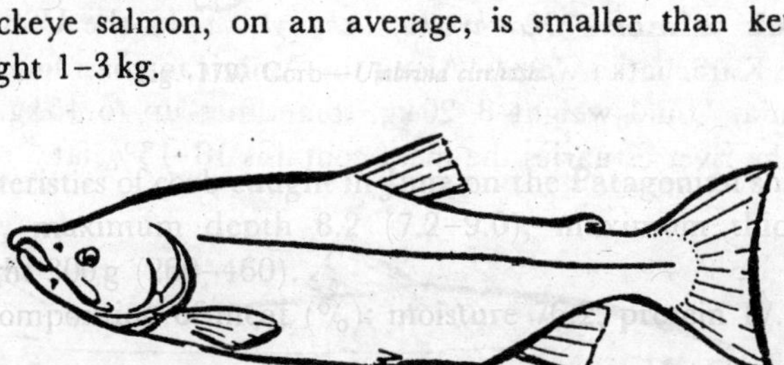

Fig. 165. Sockeye salmon—*Oncorhynchus nerka.*

The quantity of offal (%) on an average, in dressing is: head 10, liver 2, testes 2.0, remaining viscera 7.5, fins 3.0. Output of trunks after dressing for canning 61% of the weight of the intact fish.

In the prespawning period, when the salmon approach the rivers the fat content in the meat may reach 9% and the protein 19–22%.

Canned products are made from this salmon. The meat is tastier than the meat of keta and pink salmon. The cavier of sockeye is brighter than that of keta but is smaller and mildly bitter.

Cherry salmon—*Oncorhynchus masu*. A valuable fish but commercial catches from the Far East are small. In the waters of the USSR, distributed from the estuary of Amur River to the boundary with Korea. Also found in Sakhalin and rarely found south of Kamchatka.

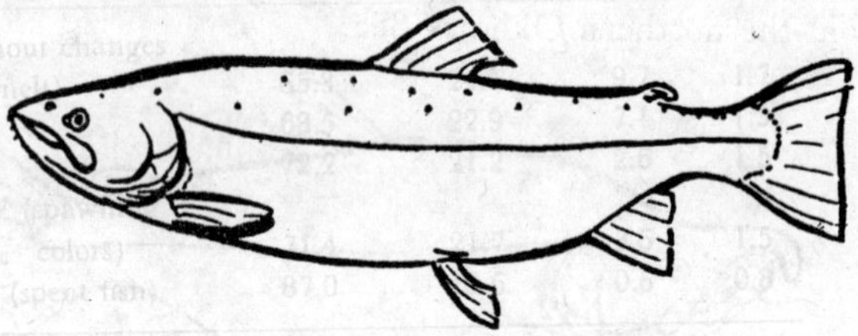

Fig. 166. Cherry salmon—*Oncorhynchus masu.*

Length attained up to 71 cm and weight 9 kg. Usual weight 1.6–9.0 kg.

Weight proportions of individual body parts of cherry salmon weighing 3,140 g (%): trunk 75.2, head 7.2, roe 3.9, testes 2.1, liver 1.0, other waste and losses 10.2.

Chemical composition of meat (%): moisture 66.7, protein 18.2, fat 13.6, ash 0.9.

The cherry salmon should be used for making salt cured products and conserves.

Chinook salmon—*Oncorhynchus tschawytscha*. In the USSR, distributed only in the Kamchatka waters. Along the American coast found from Alaska to California. Usual weight 8–20 kg, sometimes up to 45 kg. During their ascent in the river estuaries the meat contains 10–13% fat.

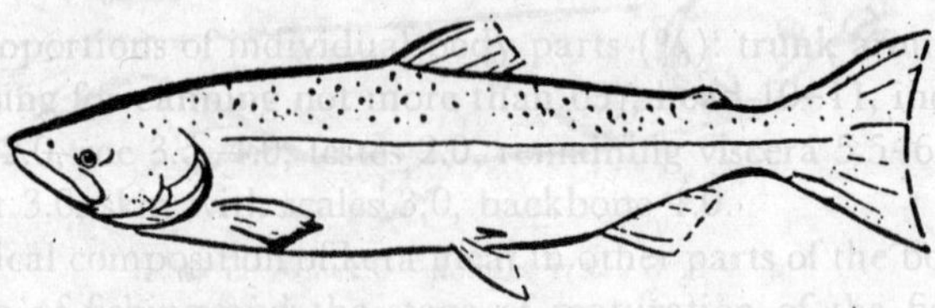

Fig. 167. Chinook salmon—*Oncorhynchus tschawytscha.*

During dressing of this salmon the waste comprises (%): head 10, liver 2, roe 5, testes 2, remaining viscera 5, fins 2.

Cold, mildly salted, delicious products are prepared from this salmon.

Genus *Salvelinus*

Sakhalin char—*Salvelinus leucomaenis.* **Dolly warden or malma**—*Salvelinus malma.* These fish live in the Bering and Okhotsk Seas and in the Sea of Japan; they are also found near northern Japan.

The Sakhalin char reaches a length of 66 cm and weight of 4–6 kg. The usual length of the fish is 30–47 cm, weight 1–4 kg (northern Primorye).

The malma is of greater commercial importance than the Sakhalin char. Length of the malma reaches 50 cm and weight 1.5 kg.

The output of trunks during dressing is 51%. The protein content reaches 18–19%, the fat 4%, that is, less than any other Pacific salmon.

These fish should be used in fresh form and for natural canned products. If stored in frozen form without careful glazing, rapid oxidation of the fat occurs.

Genus *Salmo*

Kamchatka semga—*Salmo penshinensis.* **Kamchatka steel head**—*Salmo mykiss.* The commercial significance of these fish is not very great. They are found along the Kamchatka and Primorye coasts. The Kamchatka semga is larger than the Kamchatka steel head. The average lengths of the semga and steel head are 70 and 51 cm and weight 5.6 and 1.6 kg, respectively. Their meat contains 21.6–24.2% protein and 9–13% fat. Both species constitute good raw material for delicious, mildly salt cured products.

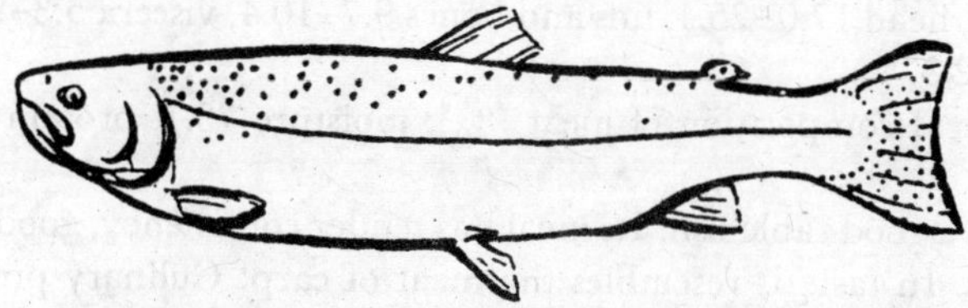

Fig. 168. Kamchatka semga—*Salmo penshinensis.*

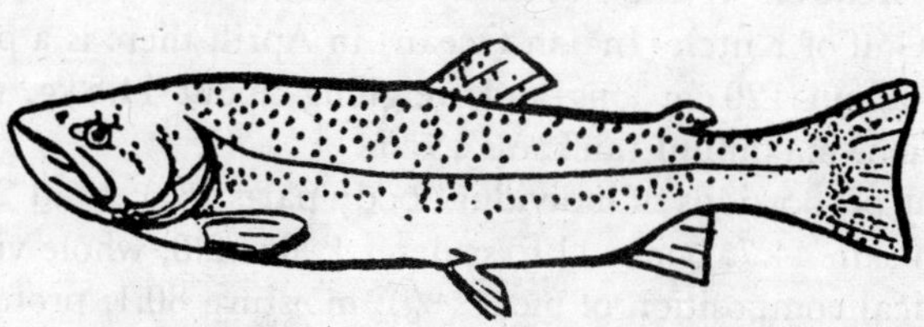

Fig. 169. Kamchatka steel head—*Salmo mykiss.*

Family SCIAENIDAE—CROAKERS OR DRUMS

This family includes valuable fish. The members live in the coastal regions of warm seas, mainly in the tropics, Atlantic, Indian and Pacific Oceans, near the coasts of southern Europe, Africa, South and Southeast Asia, Australia and America. There are about 150 species, some of which inhabit fresh water.

Genus *Argyrosomus*

White croaker—*Argyrosomus argentatus*. Commercially not very important. Found along the southern coast of Japan, southern Korea and possibly India. Length of white croaker caught in January 23–24 cm, weight 270–295 g.

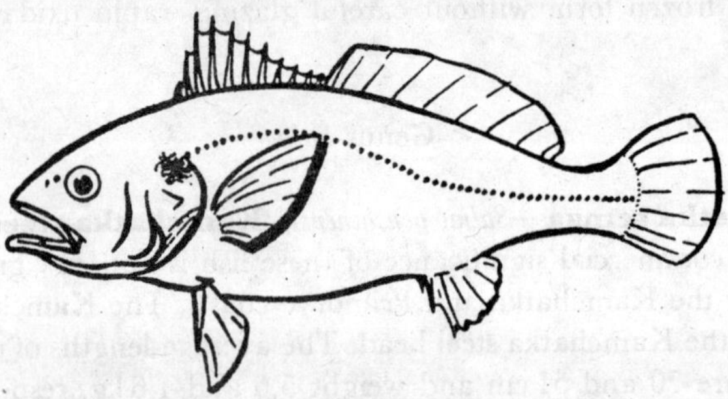

Fig. 170. White croaker—*Argyrosomus argentatus*.

Weight proportions of individual body parts (%): meat with skin 58.4–62.0, head 17.0–25.4, fins and bones 9.7–10.4, viscera 5.3–9.1, including liver 1.5–2.3.

Chemical composition of meat (%): moisture 73.1, protein 20.5, fat 5.1, ash 1.3.

This is a good table fish. Its meat has tender consistency, good organoleptic properties. In taste it resembles the meat of carp. Culinary products should be prepared from this fish.

Genus *Collichthys*

Spiny headed drum—*Collichthys lucidus**. A commercial fish. In catches from the Gulf of Kutch (Indian Ocean) in April, there is a predominance of large fish about 120 cm long and weighing about 14.5 kg, maximum body height 24 cm, maximum thickness 15 cm.

Weight proportions of individual body parts (%): head 29.6, trunk 56.4, meat with skin 44.7, bones 11.6, scales 1.1, fins 2.8, whole viscera 9.6.

Chemical composition of meat (%): moisture 80.1, protein 17.8, fat 0.7, ash 1.1.

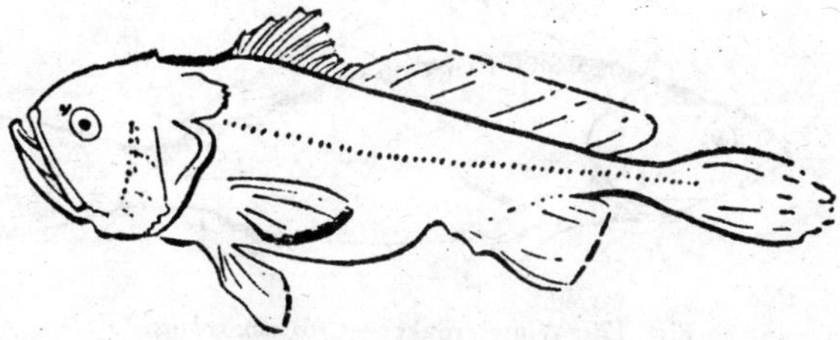

Fig. 171. Spiny headed drum—*Collichthys lucidus.*

The meat was highly infested with cysts (parasites of bone and cartilage); therefore, tests for boiling and frying were not conducted.

Because of infestation with cysts the meat of this fish is not recommended for food purposes. Fish meal can be prepared from the meat.

Genus *Cynoscion*

Striped drum—*Cynoscion striatus.* Caught with other fish from the southwest Atlantic, near the Patagonian shelf. Found in the Atlantic Ocean near the coasts of Argentina and Uruguay.

Size characteristics: length 20–38 cm, length of head 7–8 cm, maximum depth 4.2–10.0 cm, maximum thickness 2.3–5.5 cm, weight 110–780 g.

Weight proportions of individual body parts (%): trunk 57–66, head 24–33, viscera 6–9, fins 2–4, meat with skin 49–52, bones 4.7, scales 3.

TABLE 183. CHEMICAL COMPOSITION, %

Body parts	Moisture	Protein	Fat	Ash
Meat	79.5–80.8	17.5–18.2	0.1–1.4	1.2–1.6
Head	76.7	14.4	0.4	9.0
Bones	72.4	16.2	0.7	10.3
Fins	63.6	16.7	0.3	18.7
Viscera	78.5	17.8	0.5	2.1

This is a good table fish. After boiling, the meat is white, soft and juicy. In taste, the striped drum resembles fresh water fish. It can be used for smoked products and culinary preparations.

White croaker or white weakfish—*Cynoscion nobilis.* Commercially not very important, caught along the Pacific coast of America from Victoria to the Gulf of California. Usual weight about 10 kg, sometimes up to 35 kg.

Length of white croaker caught in December–January 34–62 cm and weight 385–2,360 g.

Weight proportions of individual body parts (%): trunk 68.1–70.4, head

Fig. 172. White croaker—*Cynoscion nobilis.*

20.4–24.1, fins 2.2–2.7, scales 0.3, viscera 4.7–6.8, meat with skin 51.0–52.1.

Chemical composition (%): moisture 77.1–72.4, protein 18.4–20.4, fat 0.3–0.6, ash 1.6–2.3.

This is a table fish. After cooking, the meat is white, dense, pleasant in taste with a mildly acid taste.

Culinary preparations are made from the white croaker.

Genus *Johnius*

Black spot jawfish—*Johnius axillaris.* A commercial fish, caught in August on the Sofala shelf.

TABLE 184. SIZE CHARACTERISTICS

Length, cm	Weight, g	Maximum depth, cm	Maximum thickness, cm
33.5–36.5 / 35.0	680–820 / 750	8.0–9.5 / 8.9	4.0–5.0 / 4.7

TABLE 185. WEIGHT PROPORTIONS OF INDIVIDUAL BODY PARTS, %

Trunk	Including meat with skin	Head	Including meat of occiput and nape	Scales	Bones	Fins	Viscera
47.8–56.0 / 50.7	41.1–47.6 / 44.0	37.8	9.0	0.7	6.7	2.9	6.8

Chemical composition of meat (%): moisture 78.0, protein 19.0, fat 1.7, ash 1.2.

This is a good table fish. After cooking, the meat is white, soft and tastier. The soup is greenish and excellent in taste. The fried meat is soft and tasty. Products canned in oil (prepared from steam blanched fish) have a good appearance and taste. The meat in these products is cream colored and soft. Meat in products canned in tomato sauce is soft and unpleasant in taste. This fish should be used as a table fish for first and second course dishes and canned in oil.

Captain fish—*Johnius heterolepis*. A commercial fish, caught off the Travancore coasts (Indian Ocean) in May.

Chemical composition (%): moisture 78.7, protein 18.0, fat 1.2, ash 1.3.

This is a good table fish. After boiling, the meat is white, soft and very

TABLE 186. SIZE CHARACTERISTICS OF FISH

Length, cm	Weight, g	Maximum depth, cm	Maximum thickness, cm
27.8–39.0 / 30.0	320–765 / 500	7.0–9.8 / 7.9	3.8–5.4 / 4.5

TABLE 187. WEIGHT PROPORTIONS OF INDIVIDUAL BODY PARTS, %

Trunk	Including meat with skin	Head	Including meat of occiput and nape	Bones	Fins	Viscera	
						whole	including roe
57.9–60.3 / 59.0	51.5	30.5	5.8	7.5	2.4	7.3	1.5

tasty. The broth is colorless, not rich, mediocre in taste. This fish should be used for second course dishes.

Drab jawfish—*Johnius sina*. A commercial fish, widely distributed in the tropical parts of the Indian and Pacific Oceans.

TABLE 188. SIZE CHARACTERISTICS OF DRAB JAWFISH CAUGHT IN NOVEMBER OFF THE COAST OF OMAN

Length, cm	Weight, g	Maximum depth, cm	Maximum thickness, cm
24–29 / 25	190–300 / 230	5.0–7.0 / 5.8	3.0–3.5 / 3.1

TABLE 189. CHEMICAL COMPOSITION, %

Body parts	Moisture	Protein	Fat	Ash
Meat	78.3	19.8	0.5	1.3
Inedible parts (total quantity of offal)	73.1	16.9	2.2	5.9

Weight proportions of individual body parts (%): head 26.5, trunk 62.3 (55.8–66.7), including meat with skin 51.5, scales 3.2, bones 10.8, fins 2.3, total viscera 4.9.

This is a good table fish. After cooking the meat is white, soft, tasty and resembles the meat of common perch. The soup is lemon colored, pleasant in taste but mildly bitter. The fried meat is very tasty and soft. This fish should be used for second course dishes and hot smoked products.

Bony jawfish—*Johnius osseus*. Caught with other fish in the Persian Gulf, in March with an average length of 85 cm (80–88) and weight 9,500 g (6,810–11,750).

After cooking the meat is light in color, somewhat dry but tasty (like that of typical mediocre fish). The soup is colorless, rich and tasty. This fish should be used for second course dishes.

TABLE 190. WEIGHT PROPORTIONS OF INDIVIDUAL BODY PARTS, %

Head	Trunk	Including meat	Skin	Scales	Bones	Fins	Internal organs		
							whole	including roe and testes	liver
17.8–23.4 20.1	62.8–65.6 64.7	47.7–53.2 51.5	2.4–3.0 2.7	2.2–2.9 2.5	9.7–12.1 10.5	1.3–1.6 1.4	10.2–10.6 10.3	0.3–0.4 0.4	0.7–2.1 1.7

TABLE 191. CHEMICAL COMPOSITION, %

Body parts	Moisture	Protein	Fat	Ash
Meat	76.9–79.5 77.9	17.3–19.6 18.6	0.1–2.4 1.2	1.3–1.5 1.4
Inedible parts (offal)	65.4	17.6	5.9	9.3

Dussumier's jawfish—*Johnius dussumieri.* A commercial fish inhabiting the shelves of all warm seas.

TABLE 192. SIZE CHARACTERISTICS OF FISH EXAMINED

Length, cm	Weight, g	Maximum depth, cm	Maximum thickness, cm
16–21 / 18	90–180 / 120	4.7–5.8 / 5.1	2.6–3.2 / 2.8

TABLE 193. WEIGHT PROPORTIONS OF INDIVIDUAL BODY PARTS, %

Head	Trunk	Including meat with skin	Scales	Bones	Fins	Viscera
33.8	45.5–66.7 / 53.3	40.0–45.8 / 42.1	4.1	11.3	2.3	5.1

TABLE 194. CHEMICAL COMPOSITION, %

Body parts	Moisture	Protein	Fat	Ash
Meat	78.0	19.5	0.7	1.3
Inedible parts (offal)	67.1	17.8	5.0	8.2

After cooking the meat is gray with soft consistency and pleasant taste. The soup is rich and tasty with a large quantity of jelly-forming substances. In fried form the meat is also very tasty.

The fish is recommended for use as a table fish for second course dishes (especially jellied) and hot smoked products.

Captain fish—*Johnius belengerii.* A commercial fish distributed from the eastern coast of Africa through the coasts of India to the Philippines. Lives in all warm seas. In catches from the western shelf of India, there is a predominance of individuals 80–90 cm long, with maximum depth 25.1 cm, maximum thickness 13 cm, weight about 9,000 g.

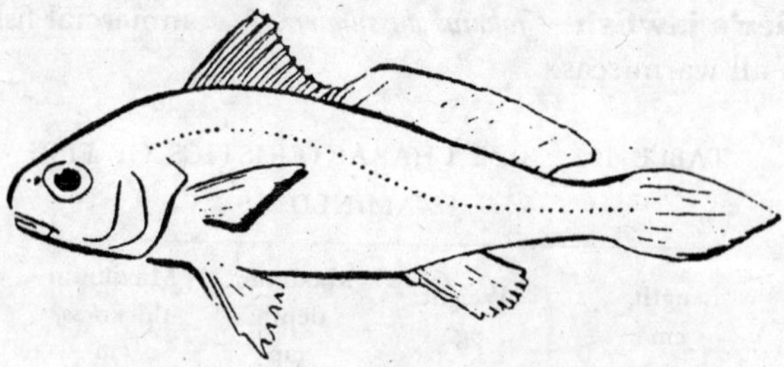

Fig. 173. Captain fish—*Johnius belengerii.*

Weight proportions of individual body parts (%): head 25.8, trunk 57.4, including meat with skin 44.9, scales 4.7, fins 3.7, whole viscera 6.3.

Chemical composition of meat (%): moisture 78.0, protein 20.8, fat 0.5, ash 1.5.

After boiling and frying the meat is white, dense and very tasty. This is a suitable table fish for first and second course dishes.

Genus *Micropogon*

American croaker—*Micropogon opercularis*. A commercial fish from the southwest Atlantic, Patagonian shelf. Found along the coasts of Brazil, Uruguay and Argentina. Size characteristics: length 27–43 cm, head length 8–9 cm, maximum depth 8.0–12.7 cm, maximum thickness 4–7 cm, weight 330–1,580 g.

Weight proportions of individual body parts (%): trunk 50–65, including meat with skin 36–52, head 23–36, fins 1.8–2.6, bones 6–11, scales 0.6–3, viscera 6–12, including roe 7, testes 0.4, liver 1.2.

TABLE 195 CHEMICAL COMPOSITION, %

Body parts	Moisture	Protein	Fat	Ash
Meat	75.4–78.6	19.1–20.5	0.5–3.4	0.8–1.3
Head	65.2	14.9	9.2	8.7
Bones	60.1	18.0	8.9	12.7
Skin	61.8	30.4	6.0	1.8
Fins	55.0	19.5	2.2	22
Scales	38.8	30.2	0.6	29.8
Roe	83.1	12.6	2.1	1.0
Viscera	70.9	15.8	10.2	1.9

Mineral content of meat (mg%): potassium 199, calcium 48, magnesium 59, phosphorus 157, iron 3.1.

This is a good table fish. The meat fried and boiled is white, juicy and somewhat hard. The soup is rich and tasty. This fish can be used for first and second course dishes.

Genus *Otolithoides*

Croaker—*Otolithoides miiuy.* Not very important commercially, found in the East China Sea and near Australia. Length of fish caught in January–February 28–50 cm and weight 210–1,500 g.

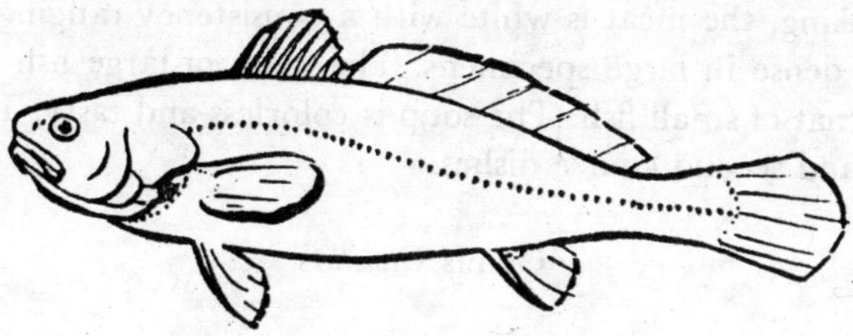

Fig. 174. Croaker—*Otolithoides miiuy.*

Weight proportions of individual body parts (%): head 17.0–25.4, meat with skin 58.4–62.0, fins and bones 9.7–10.4, viscera 5.3–9.1, including liver 1.5–2.0.

Chemical composition of meat (%): moisture 75.4–78.0, protein 18.9–21.6, fat 0.3–1.8, ash 1.3–1.6.

This is a good table fish. The meat is soft, tasty and resembles the meat of carp. The fish is used for fried preparations.

Giant drum—*Otolithoides biauritus.* A commercial fish, distributed in the same regions as other members of this genus.

TABLE 196. WEIGHT PROPORTIONS OF INDIVIDUAL BODY PARTS, %

Head	Trunk	Including meat with skin	Bones	Scales	Fins	Viscera
22.0–28.1 23.8	57.2–64.7 62.8	46.5–55.1 48.7	5.3–8.0 7.2	1.8	2.2	5.6–10.2 8.6

In catches off the coast of Pakistan (October) there is a predominance of individuals 45 cm (38–52) long, weighing 940 g (570–1,440); in catches on the shelf of western India (January) there is a predominance of fish 100 cm (81–115) long, weighing 9,500 g (8,325–10,700).

TABLE 197. CHEMICAL COMPOSITION, %

Body parts	Moisture	Protein	Fat	Ash
Meat	75.9–77.5 / 76.7	19.9–20.4 / 20.2	1.0–1.9 / 1.5	1.2–1.3 / 1.3
Inedible parts (offal)	75.1	15.2	4.0	5.0

After cooking, the meat is white with a consistency ranging from soft in small fish to dense in large specimens. The meat of large fish is fattier and tastier than that of small fish. The soup is colorless and tasty. This is a table fish for first and second course dishes.

Genus *Otolithes*

Silvery croaker—*Otolithes argenteus*. A commercial fish, inhabiting the coastal waters of tropical Indian Ocean. Distributed from the eastern coast of Africa, Madagascar Island to Australia and the Philippines. Attains a length of 75 cm. In catches off the coast of Pakistan (in October) and off the Australian coast (in February), there is a predominance of fish 34 cm (28–38) long, weighing 350 g (290–800).

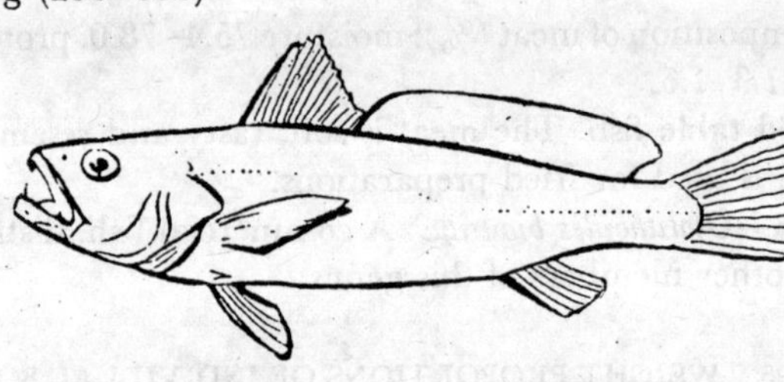

Fig. 175. Silvery croaker—*Otolithes argenteus.*

After cooking, the meat is gray and soft; in smaller fish it is tastier than in larger ones. The soup is lemon colored and tasty. In taste, the meat of this fish is akin to the meat of fresh water fish. This fish should be used for first and second course dishes and for hot smoked products.

TABLE 198. WEIGHT PROPORTIONS OF INDIVIDUAL BODY PARTS, %

Head	Meat with skin	Bones	Scales	Fins	Viscera
17.0–23.8 / 22.0	50.0–62.0 / 53.5	6.4–7.0 / 6.7	2.7	1.6	8.6–13.1 / 12.4

TABLE 199. CHEMICAL COMPOSITION, %

Body parts	Moisture	Protein	Fat	Ash
Meat	75.4–79.9 / 77.6	17.8–21.6 / 19.7	1.1–2.3 / 1.7	1.2–1.6 / 1.4
Inedible parts (offal)	64.4	18.7	10.9	4.7

Bigmouth croaker—*Otolithes macrophthalmus*. A commercial fish, caught with other fish from the eastern Central Atlantic, especially from the Canary Islands to Angola. A predominance of large individuals 50–120 cm long is found in catches.

Size characteristics: length 27–60 cm, maximum depth 7.0–10.4 cm, maximum thickness 3.6–6.1 cm, weight 0.3–10.0 kg.

Weight proportions of individual body parts (%): trunk 60–69, head 20–27, viscera 4–11, fins 1.5–2.5, scales 1.6–3.4.

Chemical composition of meat (%): moisture 75.8–78.4, protein 17.7–20.4, fat 0.7–3.2, ash 1.3–1.4.

This is a good table fish. The meat is light in color. It has high taste qualities after cooking and smoke curing. It can be used in smoked form and for culinary products.

Red croaker or rosy jawfish—*Otolithes ruber*. A commercial fish, widely distributed in the tropical Indian and Pacific Oceans. Attains a length of 90 cm.

TABLE 200. WEIGHT PROPORTIONS OF INDIVIDUAL BODY PARTS, %

Head	Meat with skin	Bones	Scales	Meat	Inedible parts (offal)
21.3–29.1 / 24.9	49.5–61.6 / 54.4	6.9–10.7 / 8.1	0.7	2.1	7.5–10.7 / 9.3

TABLE 201. CHEMICAL COMPOSITION, %

Body parts	Moisture	Protein	Fat	Ash
Meat	76.2	19.8	2.7	1.2
Inedible parts (offal)	71.4	13.8	5.6	7.9

In catches taken in January from the Persian Gulf, there is a predominance of fish 30 cm (27.1–36.5) long weighing 400 g (320–790).

This is a good table fish. After cooking, the meat is white and very tasty. The soup has a pleasant pungent taste. This fish should be used for first and second course dishes and hot smoked products.

Croaker fish—*Otolithes lateoides*. Found in the South China Sea and in the Gulf of Siam. In catches taken in January from the Bay of Bengal (Andaman Sea), there is a predominance of fish 100 cm long, weighing 7,000–9,700 g.

Weight proportions of individual body parts (%): head 25.2, meat with skin 46.1, bones 11.9, viscera 6.1, including liver 1.3, scales 3.1, fins 3.9.

Chemical composition of meat (%): moisture 77.4, protein 18.6, fat 0.4.

After cooking, the meat is dry but good in taste, typical of table fish. This fish should be used for first and second course dishes.

Genus *Pogonias*

Black drum—*Pogonias chromis*. Caught with other fish in the southwest Atlantic near the Uruguay shelf. Length of black croaker caught in March, 110 cm and weight 20.2 kg.

Weight proportions of individual body parts (%): trunk 52, head 32, viscera 13, fins 3. Output of fillets 32%.

Chemical composition of meat (%): moisture 75.5, protein 20.7, fat 2.6, ash 1.2.

This is a good table fish. After frying, the meat is cream colored with layers of black color; it is tasty and soft. This fish can be used for first and second course dishes.

Genus *Pseudosciaena*

Big yellow croaker—*Pseudosciaena crocea*. Commercially not very important, caught in the East China Sea. Length of fish caught in March–November 38–55 cm.

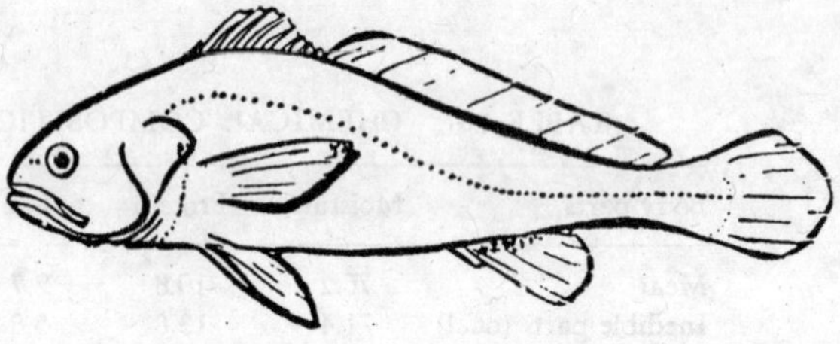

Fig. 176. Big yellow croaker—*Pseudosciaena crocea*.

Weight proportions of individual body parts (%): head 17–25.4, meat with skin 58.4–62.0, bones and fins 9.7–10.4, viscera 5.3–9.1, including liver 1.5–2.0.

Chemical composition of meat (%): moisture 81.2–83.3, protein 15.8–16.2, fat 0.4–0.7, ash 1.0–1.4.

This is an excellent table fish. It has soft and tasty meat which resembles the meat of carp. Culinary products can be prepared from this fish.

Little yellow croaker—*Pseudosciaena polyactis*. Not very important commercially, distributed in the East China Sea and near Australia. Length of fish caught in May 31 cm, weight 260 g.

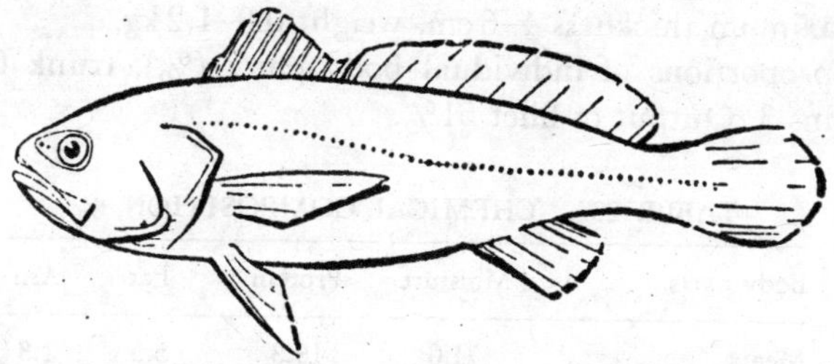

Fig. 177. Little yellow croaker—*Pseudosciaena polyactis*.

Chemical composition of meat (%): moisture 76.4–81.0, protein 14.9–18.8, fat 1.8–3.5, ash 0.9–1.6.

This is an excellent table fish. The meat is soft, tasty and resembles the meat of carp. This fish should be used for culinary items.

Genus *Sciaena*

African drum or Meagre—*Sciaena aquila*. Caught with other fish from the eastern Central Atlantic. Weight of fish caught in May off the North African coast 1.8–4.1 kg. Weight proportions of individual body parts (%): trunk 61–64, head 18–24, viscera 9–17, fins 1.2–2.0, scales 0.5–1.3. Output of fillets 46.5%.

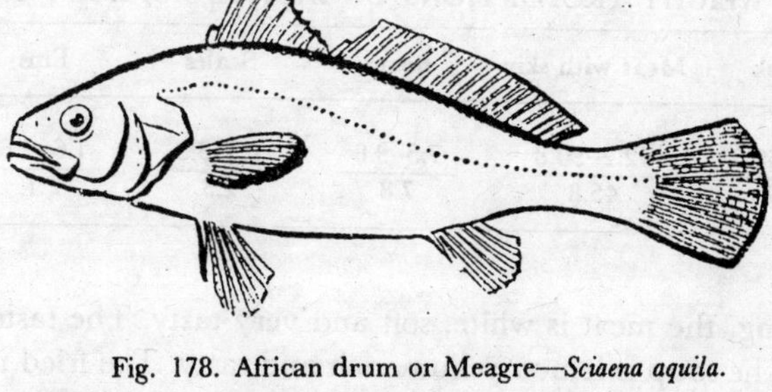

Fig. 178. African drum or Meagre—*Sciaena aquila*.

Chemical composition of meat (%): moisture 75, protein 19.5, fat 2.3, ash 1.4.

Mineral composition of meat (mg%): potassium 370, calcium 37, magnesium 48, phosphorus 241 and iron 3.8.

This is a good table fish. The meat is light colored, the taste qualities are high after frying. It resembles sander in taste. Semi-finished products, culinary preparations and smoke cured products can be prepared from this fish.

American drum—*Sciaena hololepidota.* Caught with other fish in the Southwest Atlantic. The size characteristics of fish caught in January in the Patagonian shelf: length 35–42 cm, length of head 10–13 cm, maximum depth 8–11 cm, maximum thickness 4–6 cm, weight 0.8–1.2 kg.

Weight proportions of individual body parts (%): trunk 60, head 27, viscera 10, fins 3. Output of fillet 51%.

TABLE 202. CHEMICAL COMPOSITION, %

Body parts	Moisture	Protein	Fat	Ash
Meat	74.0	19.3	5.5	1.3
Bones and skin	69.7	15.9	5.4	9.1
Viscera	78.6	15.7	4.4	1.3

This is a good table fish. After frying and boiling, the meat of the American drum is white, dense, juicy, not very soft and has no strange taste. Its taste is pleasant. The soup is fatty and tasty. This fish should be used for culinary preparations and smoked products.

Large finned jawfish—*Sciaena macropterus.* Commercially of little importance, widely distributed in the tropical parts of the Indian and Pacific Oceans. In catches off the coast of Pakistan in October, there is a predominance of fish 20 cm (18.5–24.0) long, weighing 150 g (95–200) and in catches from Algoa Bay in December there is an abundance of individuals 30 cm (25–31) long, weighing 500 g (380–720).

TABLE 203. WEIGHT PROPORTIONS OF INDIVIDUAL BODY PARTS, %

Head	Trunk	Meat with skin	Bones	Scales	Fins	Viscera
$\frac{24.2–31.4}{29.4}$	$\frac{50.7–60.0}{53.7}$	$\frac{42.2–50.8}{45.8}$	$\frac{7.5–9.8}{7.8}$	$\frac{1.3–3.5}{2.3}$	$\frac{1.6–2.6}{2.1}$	$\frac{9.0–13.0}{11.5}$

After cooking, the meat is white, soft and very tasty. The taste is typical of river carp. The soup is lemon yellow, rich and tasty. The fried meat is soft

TABLE 204. CHEMICAL COMPOSITION, %

Body parts	Moisture	Protein	Fat	Ash
Meat	78.2–79.5 / 78.9	18.3–19.1 / 18.7	0.2–1.1 / 0.7	1.2
Inedible parts (offal)	69.8–70.7 / 70.3	18.1–18.3 / 18.2	3.2–5.6 / 4.4	6.7–7.4 / 6.6

and very tasty. This fish should be used as table fish for first and second course dishes and hot smoked products.

Genus *Umbrina*

Corb—*Umbrina cirrhosa*. A commercial fish found in the Atlantic Ocean and in the Mediterranean and Black Seas. Attains a length of 1.5 m, weight 32 kg.

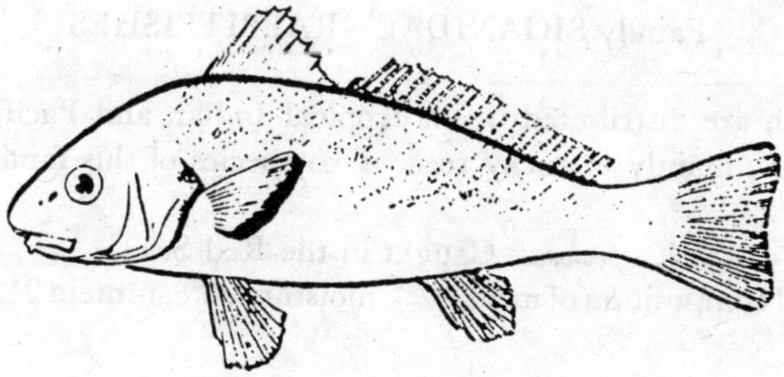

Fig. 179. Corb—*Umbrina cirrhosa.*

Size characteristics of corb caught in June on the Patagonian shelf: length 23 cm (21–26), maximum depth 8.2 (7.2–9.0), maximum thickness 3.3 (3.0–3.6), weight 300 g (260–460).

Chemical composition of meat (%): moisture 76.2, protein 17.9, fat 3.5, ash 1.2.

TABLE 205. WEIGHT PROPORTIONS OF INDIVIDUAL BODY PARTS, %

Head	Including meat of occiput and nape	Trunk	Including meat and skin	Scales	Fins	Viscera
29.5	6.3	51.4–59.6 / 54.5	40.4–49.1 / 45.1	6.3	1.9	7.5

This is an excellent table fish. After cooking, the meat is light colored, soft and tasty. It is very tasty when fried. This fish should be used for first and second course dishes, hot smoking products and canning in oil. The meat is somewhat dry, the taste is unpleasant and the sauce is dark when canning is done in tomato sauce.

Canary drum—*Umbrina canariensis.* Caught with other fish from the eastern Central Atlantic and along the southwestern coasts of Africa. Lives in the tropical zones of the Atlantic Ocean and along the western coast of Africa. Attains a length of 50 cm.

Size characteristics of fish caught in May: length 21–32 cm, maximum depth 8–12 cm, maximum thickness 3–5 cm, weight 0.2–1.0 kg.

Weight proportions of individual body parts (%): trunk 66–68, head 21–25, viscera 9–10. Output of fillets 43%.

Chemical composition of meat (%): moisture 74.8–75.2, protein 20.4–21.2, fat 2.5–3.5, ash 0.9–1.5.

This is a good table fish. The meat is soft and tasty. The meat is first frozen and subsequently used for culinary preparations.

Family SIGANIDAE—RABBITFISHES

These fish are distributed in the tropical Indian and Pacific Oceans, in coastal waters, mainly in rocky reefs. Two species of this family have been studied.

Siganus—*Siganus speciosus.* Caught in the Red Sea.

Chemical composition of meat (%): moisture 76.8, protein 21.0, fat 0.7, ash 1.5.

TABLE 206. SIZE CHARACTERISTICS OF SIGANUS SPECIOSUS CAUGHT IN NOVEMBER IN THE RED SEA

Length, cm	Weight, g	Maximum depth, cm	Maximum thickness, cm
24–27 / 26	320–470 / 400	9.8–10.2 / 10.0	2.9–3.5 / 3.2

The taste qualities of the meat are good. This fish should be used for table purposes and for hot smoked products.

Oramin rabbitfish—*Siganus oramin.* Found near Sri Lanka. Found rarely in catches.

Chemical composition of meat (%): moisture 76.4, protein 19.4, fat 1.2, ash 1.4.

TABLE 207. WEIGHT PROPORTIONS OF INDIVIDUAL BODY PARTS OF SIGANUS, %

Head	Trunk	Fins	Viscera
18.1–19.8 / 19.0	52.1–62.5 / 57.3	0.9–1.1 / 1.0	17.5–25.1 / 21.3

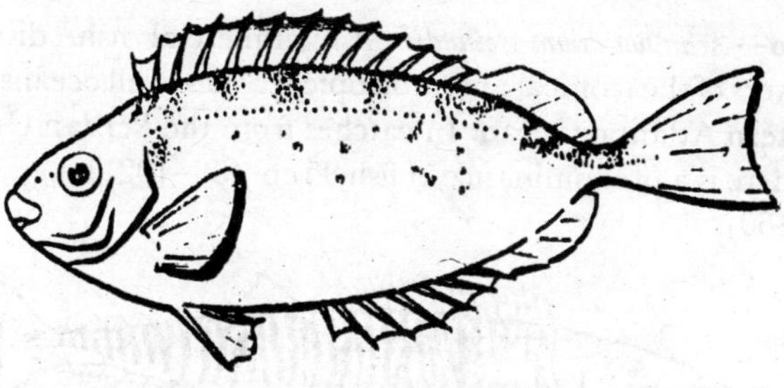

Fig. 180. Oramin rabbitfish—*Siganus oramin.*

TABLE 208. SIZE CHARACTERISTICS OF ORAMIN RABBITFISH CAUGHT IN JULY NEAR SRI LANKA

Length, cm	Weight, g	Maximum depth, cm	Maximum thickness, cm
21.0–26.3 / 23.0	235–475 / 300	7.8–9.0 / 8.3	2.6–5.0 / 3.5

After cooking, the meat is very white and tasty. The broth is rich and tasty. This fish should be used as a table fish for first and second course dishes. Since the spiny rays of the fins of this fish are poisonous, it should be dressed onboard the fishing vessels before freezing.

TABLE 209. WEIGHT PROPORTIONS OF INDIVIDUAL BODY PARTS, %

Head	Trunk	Meat with skin	Bones	Fins	Viscera
17.5–21.2 / 18.9	50.0–55.3 / 53.1	40.0–43.5 / 42.6	9.2–12.5 / 10.4	2.4	24.6

Family SCOMBEROMORIDAE—BONITOS

This family includes seven genera, of which six are fish of commercial significance. These fish are closely related to tuna and mackerel. They are widely distributed in the tropical and subtropical waters of all oceans. They have considerable commercial value.

Genus *Acanthocybium*

Wahoo—*Acanthocybium solandri.* A commercial fish, distributed in the coastal waters of the tropical and subtropical zones of all oceans. More common in the western Atlantic Ocean. In catches from the Persian Gulf and the Gulf of Aden, there is a predominance of fish 95 cm (89–102) long, weighing 6,600 g (4,850–8,460).

Fig. 181. Wahoo—*Acanthocybium solandri.*

The cooked meat is light colored and dense with a pleasant taste. After salt curing, it is pink and fatty with a pleasant taste. This fish should be used as a table fish for cured balyks, first and second course dishes and smoked products.

TABLE 210. WEIGHT PROPORTIONS OF INDIVIDUAL BODY PARTS, %

Head	Trunk	Including meat with skin	Fins	Viscera
$\frac{11.8–13.9}{13.1}$	$\frac{80.0–82.6}{80.9}$	$\frac{68.8–72.7}{70.0}$	0.9	$\frac{4.7–5.0}{4.9}$

TABLE 211. CHEMICAL COMPOSITION OF MEAT, %

Moisture	Protein	Fat	Ash
$\frac{74.1–75.7}{75.1}$	$\frac{21.1–21.7}{21.4}$	$\frac{0.9–2.5}{1.7}$	$\frac{1.3–1.5}{1.1}$

Genus *Orcynopsis*

Plain bonito—*Orcynopsis unicolor*. Distributed in the subtropical and tropical waters of the eastern Atlantic, along the coasts of Africa and southern Europe (Gulf of Cadiz) and in the Mediterranean Sea. Attains a length of 2 m and weight of 60 kg, usual weight 5–30 kg.

Weight proportions of individual body parts of plain bonito caught in June off the northwest coasts of Africa, 76 cm long, maximum depth 20 cm, maximum thickness 9 cm and weight 5.4 kg, (%): trunk with fins 70, head 20, viscera 10. Output of fillets 53%. Chemical composition of meat (%): moisture 74.2, protein 22.1, fat 2.4, ash 1.3.

This is a good table fish and can be used for culinary preparations.

Genus *Scomberomorus*

Indian Spanish mackerel—*Scomberomorus guttatus*. A commercial fish, distributed in the tropical and subtropical seas and in the coastal areas of the Indian and Pacific Oceans, near the Indo-Australian archipelago (excluding

Fig. 182. Indian Spanish mackerel—*Scomberomorus guttatus*.

TABLE 212. SIZE CHARACTERISTICS

Place of catch	Time of catch	Length, cm	Maximum depth, cm	Maximum thickness, cm	Weight, g
Persian Gulf	January–February	27–41 / 36	6.0–8.5 / 7.7	2.0–4.8 / 3.6	215–800 / 600
Western shell of India	January	78–103 / 90	—	—	7,000–16,000 / 12,000

TABLE 213. SIZE CHARACTERISTICS

Place of catch	Time of catch	Length, cm	Weight, g
Red Sea	November	90	6,000

TABLE 214. WEIGHT PROPORTIONS OF INDIVIDUAL BODY PARTS, %

Place of catch	Head	Trunk	Bones	Viscera	Fins	Output of fillets
Red Sea	10.6–14.6 / 13.6	80.0–83.5 / 81	—	3.2–5.5 / 3.6	0.9–1.7 / 1.1	—
Western shelf of India	13.0	66.7*	12.6	3.7	1.4	—
Indian Ocean	13.0	77.7	7.6	6.1	2.3	70.1
Persian gulf	10.1–15.6 / 12.9	74.5–83.5 / 80.9	5.6–8.9 / 6.8	3.6–4.4 / 3.7	—	—

* Output of meat with skin.

TABLE 215. CHEMICAL COMPOSITION, %

Place of catch	Body parts	Moisture	Protein	Fat	Ash
Persian Gulf	Meat	74.9–77.9 / 76.4	20.1–20.4 / 20.3	0.4–3.5 / 2.0	1.3
	Inedible parts (offal)	64.4	16.1	12.7	6.4
Western shelf of India	Meat	75.8	19.7	2.2	—
Red Sea, western shelf of India	—	68.7–73.9 / 71.3	19.8–20.5 / 20.2	2.4–9.7 / 6.1	1.4–1.5 / 1.5

Australian waters), along the coast of Asia north to Taiwan and west to Sri Lanka and India. Penetrates the Persian Gulf; sometimes enters rivers.

This is an excellent table fish. After cooking, the meat is white, soft and tasty. The soup is rich and tasty. Fried meat is dense and tasty. The products canned in oil have a good marketable appearance and taste. The products canned in tomato sauce are dark in color.

This fish should be used for table purposes for first and second course dishes and for cured balyks, smoked products and conserves in oil.

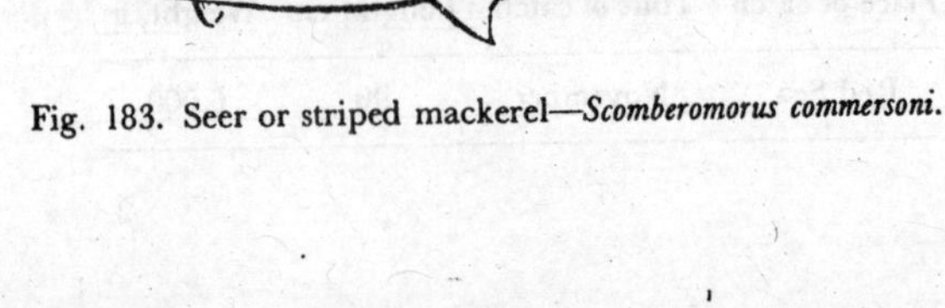

Fig. 183. Seer or striped mackerel—*Scomberomorus commersoni.*

Seer or striped Spanish mackerel—*Scomberomorus commersoni.* A valuable commercial fish. Attains a length of 180 cm and weight of 50 kg.

In catches from the Red Sea, Persian Gulf and Agulhas Cape, there is a predominance of individuals 70 cm (47–84) long, weighing 2,500 g (905–4,350).

TABLE 216. WEIGHT PROPORTIONS OF INDIVIDUAL BODY PARTS, %

Head	Meat with skin	Fins	Viscera
$\frac{11.0–15.9}{14.2}$	71.8	$\frac{0.9–1.5}{1.0}$	$\frac{3.2–6.3}{4.1}$

TABLE 217. CHEMICAL COMPOSITION OF MEAT, %

Place of catch	Moisture	Protein	Fat	Ash
Red Sea and Persian Gulf	$\frac{75.4–77.0}{76.2}$	$\frac{21.0–21.1}{21.1}$	$\frac{0.4–2.0}{1.2}$	1.5
Agulhas Cape	74.0	20.8	3.8	1.4

After cooking, the meat is light in color, dense with a meaty (non-fish) taste. The soup is lemon colored, very tasty with a meat-like taste.

This is an excellent table fish. It should be used for first and second course dishes, smoked products and cured balyks.

Spanish mackerel—*Scomberomorus maculatus.* Found in the western Atlantic Ocean from Brazil to north of Maine and in the east north of the equator. The fish are numerous in the Gulf of Mexico and north of the Gulf up to North Carolina. Usually this fish is found within the limits of the continental shelf. It has been reported in the eastern Pacific Ocean. Attains a length of 90 cm and weight of 45 kg.

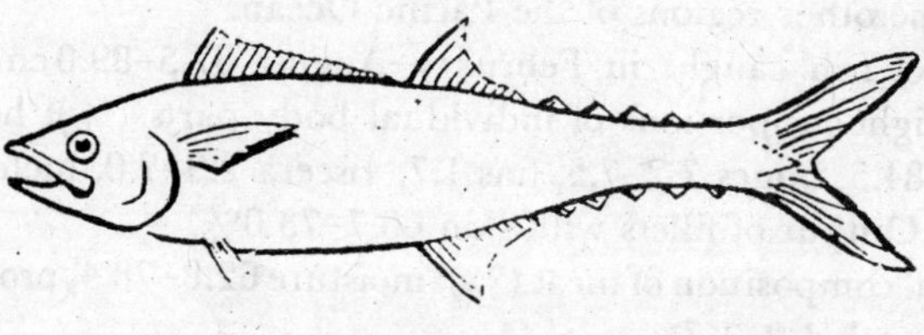

Fig. 184. Spanish mackerel—*Scomberomorus maculatus.*

TABLE 218. SIZE CHARACTERISTICS

Place of catch	Time of catch	Length, cm	Maximum depth, cm	Maximum thickness, cm	Weight, kg
Arabian Sea	September	60	15.5	5.5	2.0
Northwest African coast	—	50–62	10–11	5.0	1.5–2.5

In cooked form, the meat of the Spanish mackerel is characterized by soft consistency, pleasant taste and mild acidic flavor.

The Spanish mackerel is one of the best table fish. Cured balyks, smoked products and first and second course dishes can be prepared.

TABLE 219. WEIGHT PROPORTIONS OF INDIVIDUAL BODY PARTS, %

Place of catch	Head	Trunk	Meat	Skin	Bones	Fins	Viscera whole	Viscera including testes	Viscera including liver
Arabian Sea	13.5	—	71.2	2.4	7.6	1.0	3.3	0.6	0.5
Northwest African coast	13.0	78.0	62.0*	—	—	—	9.0	—	—

* Fillets

TABLE 220. CHEMICAL COMPOSITION, %

Place of catch	Moisture	Protein	Fat	Ash
Arabian Sea	73.8	21.6	3.8	1.2
Northwest African coast	72.8	21.1	4.7	1.4

Leopard mackerel—*Scomberomorus leopardus*. Found in the South China Sea and some other regions of the Pacific Ocean.

Length of fish caught in February–August 37.5–89.0 cm, weight 330–5,800 g. Weight proportions of individual body parts (%): head 10.3–17.8, trunk 75.0–84.5, bones 7.2–7.5, fins 1.7, viscera 2.9–7.0, including liver 1.2, gonads 0.9. Output of fillets with skin 66.7–73.0%.

Chemical composition of meat (%): moisture 62.8–78.4, protein 18.1–23.7, fat 0.7–16.6, ash 1.2–1.7.

The taste of the meat is mildly sour but good. It can be used for conserves to make snacks or frozen for later use in culinary preparations.

Japanese Spanish mackerel or Sawara—*Scomberomorus niphonius*. Lives in western part of Pacific Ocean and East China Sea. Length of fish caught in February was 45 cm and weight 687 g.

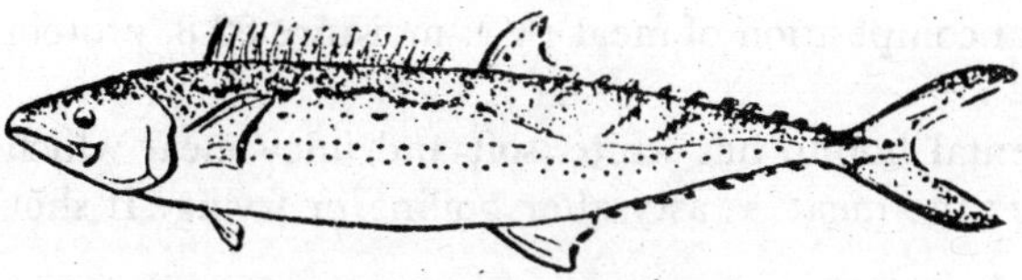

Fig. 185. Japanese Spanish mackerel or Sawara—*Scomberomorus niphonius*.

Weight proportions of individual body parts (%): head 16.7, trunk 75.0, viscera 5.3, bones and fins 12.3. Chemical composition of meat (%): moisture 76.3, protein 19.7, fat 2.5, ash 1.4. Culinary preparations can be made.

Genus *Sarda*

There are three species in this genus.

Atlantic bonito—*Sarda sarda*. A commercial fish, inhabiting the North Atlantic around the coast of England and in the North Sea on the east and New Scotland on the west. Also found in large numbers off the northern coast of Africa, in the Mediterranean Sea and along the American coast from Cape Cod to Brazil. Attains a length of 85 cm and weight of 4 kg.

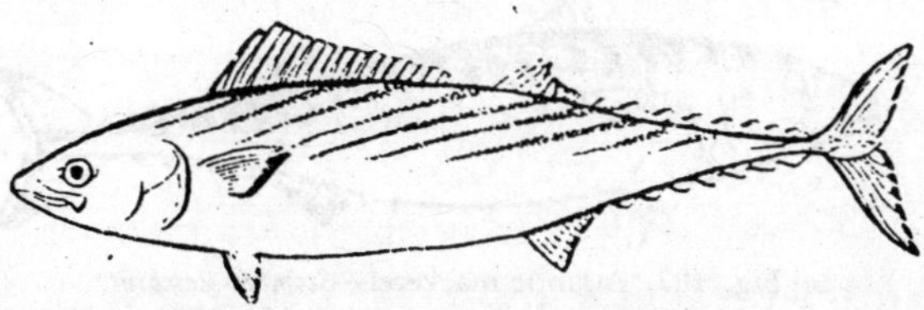

Fig. 186. Atlantic bonito—*Sarda sarda*.

Weight proportions of individual body parts of Atlantic bonito caught in May off the coast of northwest Africa 55 cm long, with maximum depth 14 cm, maximum thickness 9 cm and weight 3.1 kg (%): trunk 68, head 18, viscera 12, fins 2.

Chemical composition of meat (%): moisture 66.6–73.0, protein 22.8–23.1, fat 2.4–9.4, ash 1.2–1.5.

This is a good table fish which can be used for fillets.

Oriental bonito—*Sarda orientalis*. Found in the warm waters of the Pacific and Indian Oceans, south of Korea, near the middle of Honshu Island, along the eastern coast of Africa and in the waters of Australia, along the American coast from California to Peru.

Length of fish caught 80 cm and weight about 3 kg.

Weight proportions of individual body parts of fish caught in February in the Pacific Ocean 82 cm long, weighing 5,430 g (%): head 10.5, bones 13.1, fins 1.8, viscera 4.4, including liver 0.9. Output of fillets 70.2%.

Chemical composition of meat (%): moisture 62.8, protein 19.2, fat 16.5, ash 1.2.

The oriental bonito has white, soft and juicy meat which has a slightly acidic flavor; the meat is tasty after boiling or frying. It should be used for canning.

Family SCOMBERIDAE—MACKERELS AND TUNAS

Genus *Scomber*

The members of this family are commercially very important. They are widely distributed in all tropical and temperate zones of the oceans. This family consists of nine genera.

Atlantic mackerel—*Scomber scombrus*. Distributed in the northern Atlantic Ocean along the European coast from the White Sea (where it enters in warm years) to the Black Sea, along the American coast from Cape Hatteras in the south to the Labrador peninsula in the north.

Size characteristics of fish examined: length 16–39 cm, maximum depth

Fig. 187. Atlantic mackerel—*Scomber scombrus*.

TABLE 221. CHEMICAL COMPOSITION OF MEAT, %

Place of catch	Moisture	Protein	Fat	Ash
North American coast	66.8–74.1	19.0–22.8	2.4–11.4	1.4–2.1
Strait of La Manche and North Sea	63–67	16.5–19.5	11.9–18.3	1.2–1.6
Georges Bank, Sable	59.1–65.8	17.7–18.7	14.2–22.3	1.1–1.3
Gulf of Mexico	67.3	23.1	8.4	1.2
Patagonian shelf	73.3	24.2	0.9	1.6

TABLE 222. CHEMICAL COMPOSITION OF DIFFERENT PARTS, %

Body parts	Moisture	Protein	Fat	Ash
Head	61.3–66.6	12.3–15.3	14.1–19.5	4.6–5.0
Bones	56.7–61	15.3–18.4	14.8–20	5.4–6.4
Viscera	73.2–79.1	13.4–14	4.2–7.9	2–2.3
Liver	60.4–71.8	17.7–19.4	5.5–16.4	1.3–2
Roe	72.[illegible]–80.3	[illegible]	0.6–5.1	1.6
Testes	71.3–80.4	16.5	0.1–0.6	2.1

3–9 cm, maximum thickness 2.6 cm, weight 60–880 g. Weight proportions of individual body parts (%): trunk 60–75, head 16–30, viscera 8–20, bones 4.5–9, fins 0.6–1.7, scales 0.8, roe 0.7–5.0, testes 0.8–5.4, liver 0.7–1.7. Output of fillets 53–68%.

The meat of this mackerel is soft and tasty. Depending upon the fat content of the fish, taste properties vary from satisfactory to good. Canned and smoked products should be made from this mackerel.

TABLE 223. MACRO- AND MICROELEMENT CONTENTS OF MEAT, MG %

Place of catch	Potassium	Calcium	Magnesium	Phosphorus	Iron
African coast	240–330	29–45	68–73	230–330	4–8
North Sea	206	94	22	298	0.6

Chub or Pacific mackerel—*Scomber japonicus*. Commercially significant. Inhabits the western Pacific Ocean from Australia to the Sea of Japan and northern Sakhalin; possibly also found around the Hawaiian Islands. Also distributed in the Indian Ocean, Persian Gulf and Gulf of Aden and near Madagascar Island. Attains a length of 60 cm and weight of 1.5 kg. Fish 35–45 cm long, weighing 0.5–1.2 kg predominate in catches.

The fat content of this mackerel caught in the Pacific Ocean varies from

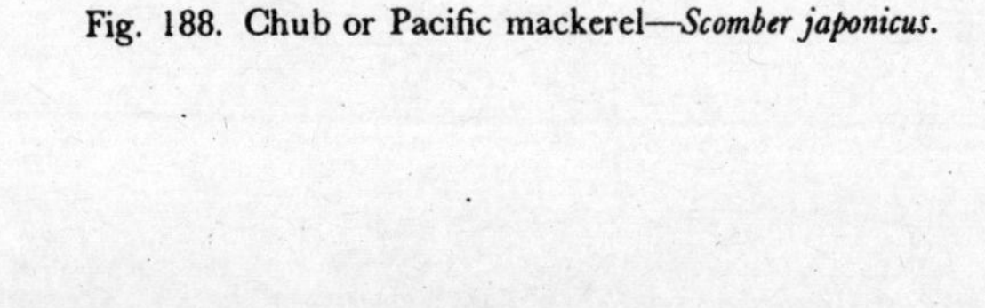

Fig. 188. Chub or Pacific mackerel—*Scomber japonicus*.

0.8–33.0%. It has been reported that the sexually mature and feeding mackerels are richest in fat content in the August–October period. Fish from different regions have variable taste qualities and therefore recommendations for their use will vary.

Fish from the Pacific Ocean have soft meat with a mildly sour flavor. They should be used for canned products, blanched in oil for natural, smoked and mildly salted products. Mildly salted mackerel deserves especially high grading.

TABLE 224. SIZE CHARACTERISTICS

Place of catch	Time of catch	Length, cm	Maximum depth, cm	Maximum thickness, cm	Weight, g
Gulf of Aden, St. Francis and near Madagascar Island	September–November	25.2–33.4 / 28.0	4.0–6.0 / 5.3	3.0–4.0 / 3.7	190–525 / 320
Persian Gulf	January	37.0–47.0 / 40.0	—	—	405–845 / 700
Sea of Japan and East China Sea	—	28–47	—	—	300–1,500
Bao Pash	May	26.8–28.0 / 27.5	4.8–5.5 / 5.0	3.5–4.2 / 3.9	255–310 / 278

In cooked form, the meat of these fish, from the Gulf of Aden in the Bao Pash region, Madagascar Island, is dark, varying in consistency from dense to hard, with a light acidic taste. There are many fine bones in the meat. The soup is sour. In products canned in oil the meat is pale pink, soft in

TABLE 225. WEIGHT PROPORTIONS OF INDIVIDUAL BODY PARTS, %

Place of cath	Head	Trunk	Meat with skin	Bones	Fins	Scales	Viscera
Gulf of Aden, St. Francis and near Madagascar Island	19.0–23.0 / 21.7	60.5–67.5 / 63.3	50.0–62.2 / 57.0	5.1–6.6 / 5.8	—	0.7–0.9 / 0.8	8.8–14.3 / 12.0
Persian Gulf	10.7*	85.1	81.8	—	0.6	—	3.3
Sea of Japan and East China Sea	13.6–18.3	—	52–65	5.3–10.0	—	—	9.6–20.1
Bao Pash	22.7	64.0	56.6	7.4	0.9	—	11.9

* Head separated by oblique cut.

TABLE 226. CHEMICAL COMPOSITION, %

Place of catch	Moisture	Protein	Fat	Ash
Gulf of Aden, St. Francis and near Madagascar Island				
Meat	69.3–72.7 / 71.8	21.4–22.4 / 21.8	3.3–7.3 / 4.8	1.3–1.5 / 1.4
Inedible parts (offal)	69.2–74.0 / 71.6	15.9–17.6 / 16.7	4.9–7.7 / 6.3	5.1–5.2 / 5.2
Persian Gulf				
Meat	73.6	19.8	5.0	1.3
Sea of Japan and East China Sea				
Meat	54.0–77.0	14.0–24.5	0.8–33.0	1.0–2.0
Bao Pash				
Meat	70.0	22.1	6.5	1.4

consistency with a pleasant taste. The products canned in tomato sauce have good external appearance and taste. The sauce is very light in color. In natural canned products, the meat is pink, soft, tasty and the bones become soft.

This fish should be used for canned products and natural canned products, for canning in oil and for canning in tomato sauce.

In salted forms the meat is light pink, with a soft consistency and the pleasant cured balyk-like taste.

Chub from these regions can be used for many kinds of products. As a table fish it can be used for smoked and sun cured products and canned preparations.

Spotted mackerel—*Scomber tapeinocephalus*. Found in the East China Sea. Length of spotted mackerel caught in March–April 27.5–30.0 cm, weight 245–330 g.

Weight proportions of individual body parts (%): head 18.2, trunk 64.3, including meat 57.4, fins 2.6, bones 6.9, viscera 13.3.

Chemical composition of meat (%): moisture 67.9, protein 22.8, fat 7.5, ash 1.8.

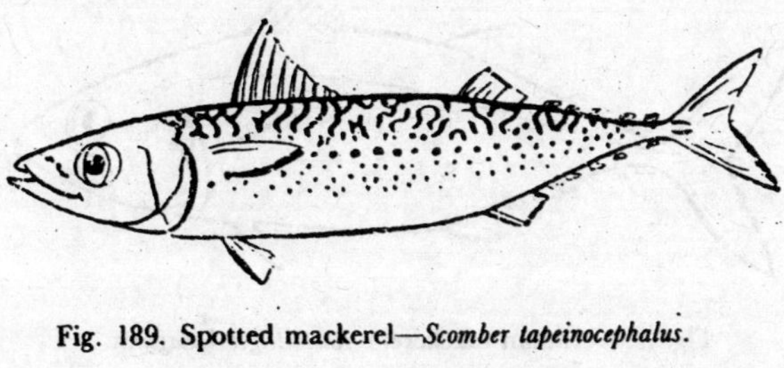

Fig. 189. Spotted mackerel—*Scomber tapeinocephalus*.

The meat is very tasty in fried and boiled forms. It can be used for natural canned products.

Eastern mackerel—*Scomber japonicus diego.* Distributed in the eastern Pacific Ocean.

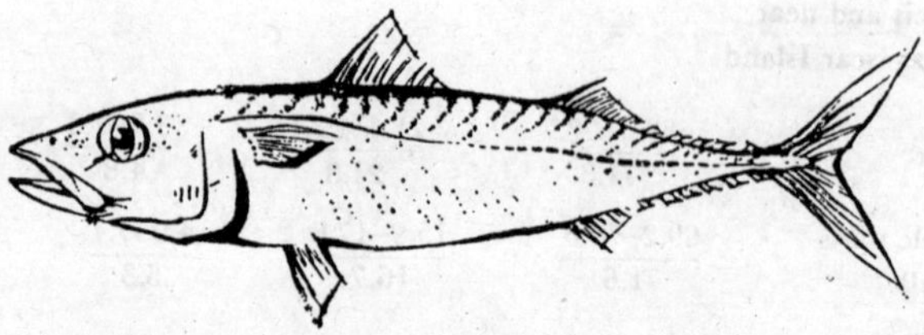

Fig. 190. Eastern mackerel—*Scomber japonicus diego.*

Weight proportions of individual body parts of mackerel caught in March 26–32 cm long, average weight 211 g, (%): head 26.2, trunk 61.8, bones 7.5, fins 1.4, skin 6.1, viscera 10.3, including liver 0.9. Output of fillets 46.5%.

Chemical composition of meat (%): moisture 77.9–79.5, protein 19.8–20.2, fat 0.1–0.3, ash 1.0–1.5.

The meat of eastern mackerel is dark with a somewhat dry consistency and light sour taste. This fish should be used for hot smoked products. Smoked fish should be used for canning in oil.

Genus *Rastrelliger*

Indian mackerel—*Rastrelliger kanagurta.* Commercially important in the territorial waters of India; in the open sea, caught with other fish. Distributed in the tropical and partly subtropical waters of the Indian Ocean and in the western Pacific Ocean. In the Indian Ocean lives along the coasts of Africa, India, Burma, near the Seychelles Islands and near Sri Lanka, in the coastal Indo-Pacific zones, along the western coast of Australia, in the Pacific Ocean from New Guinea to the coast of China and near the Hawaiian Islands.

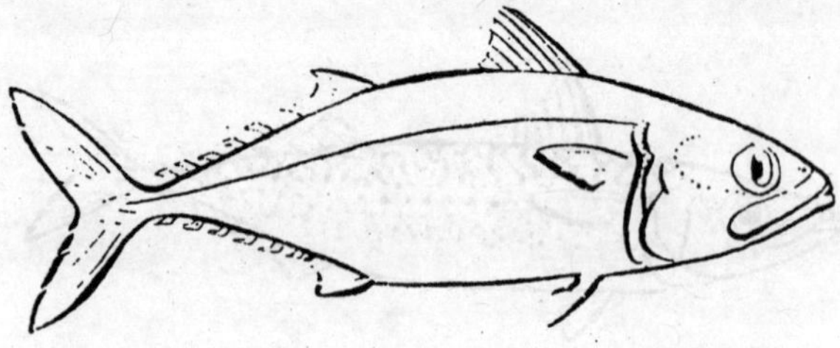

Fig. 191. Indian mackerel—*Rastrelliger kanagurta.*

TABLE 227. WEIGHT PROPORTIONS OF INDIVIDUAL BODY PARTS, %

Place of catch	Head	Trunk	Bones	Viscera	Fins	Output of fillets
Western shelf of India	17.5	75.8	—	3.5	1.6	—
East China Sea	18.2	64.3	6.9	13.9	2.6	57.4

In catches taken in February from the western shelf of India, there is a predominance of individual 18 cm (15.9–19.8) long and weighing 120 g (95–135). Length of mackerel caught in the East China Sea 19–22 cm, weighing 98–180 g.

TABLE 228. CHEMICAL COMPOSITION OF MEAT, %

Place of catch	Moisture	Protein	Fat	Ash
Western shelf of India	75.9	20.5	0.7	—
East China Sea	67.9–75.4	20.7–23.0	0.2–7.5	1.8–2.1

The meat is dark; after boiling, it acquires a dry and dense consistency. It has a pleasant fishy smell and light acidic flavour. This fish should be used for smoked products and conserves.

Genus *Thunnus*

Australian tuna—*Thunnus rara*. Inhabits the tropical and subtropical waters of the Indian and Pacific Oceans near the Indo-Australian archipelago, south to 37°S, north to the coast of Japan, west to Sri Lanka, the Maldive Islands, near the coasts of Kenya and Tanzania, in the Gulf of Aden, east to

TABLE 229. WEIGHT PROPORTIONS OF INDIVIDUAL BODY PARTS, %

Head	Meat*	Scales	Fins	Viscera
$\frac{19.9\text{–}25.0}{22.5}$	$\frac{47.6}{10.0}$	$\frac{0.4\text{–}1.2}{0.9}$	$\frac{1.4\text{–}9.2}{6.2}$	$\frac{2.2\text{–}8.2}{6.2}$

* The figures in the numerator pertain to light meat and in the denominator to dark meat.

TABLE 230. CHEMICAL COMPOSITION OF MEAT, %

Type of meat	Moisture	Protein	Fat	Ash
Light	$\frac{72.7\text{–}74.7}{73.7}$	$\frac{23.0\text{–}24.6}{23.8}$	$\frac{0.7\text{–}1.3}{1.0}$	$\frac{1.4\text{–}1.5}{1.4}$
Dark	$\frac{71.8\text{–}72.1}{71.9}$	$\frac{23.2\text{–}25.2}{24.2}$	$\frac{1.2\text{–}2.0}{1.6}$	1.4

New Guinea. In the Australian waters, found along the northern, western and eastern coasts. Forms large concentrations in the Moreton and Harvey Gulfs and along the eastern coast of Australia.

Attains a weight of 20 kg. Fish most often ranging in length from 70–80 cm and weight 6 kg.

The light colored meat of Australian tuna has a pinkish tinge and dense consistency. In cooked form, it has high taste qualities. Culinary products should be made from it and it can be used for natural canned products in oil.

Big eye tuna—*Thunnus obesus*. Distributed in the tropical and subtropical waters of all oceans. In the Atlantic Ocean found along the coasts of Morocco, up to Cape Huby, near Madeira, the Azores and Bermuda. Large concentrations of big eye tuna are found in the zones of the northern and southern equatorial currents and the Brazil current. In the Indian Ocean distributed in the zone of the southern equatorial current and north of it near the eastern coast of Africa and Madagascar Island and near the Indo-Australian archipelago. In the Pacific Ocean, found from the coast of Australia to 36° N. Reaches a length of 2 m and weight 100 kg and more.

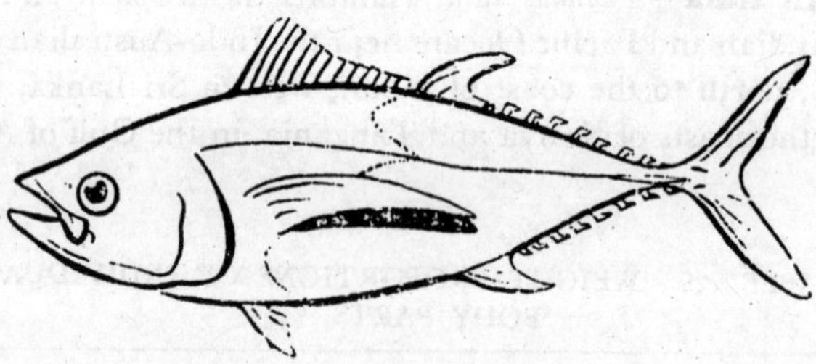

Fig. 192. Big eye tuna—*Thunnus obesus*.

This is a good table fish. After cooking, the light colored meat of big eye tuna becomes gray and dense with a pleasant taste resembling that of meat of terrestrial animals. The soup has very good taste. This fish should be used for natural conserves and those in oil and for making fillets and culinary preparations. The dark colored meat should be used for fish meal.

TABLE 231. WEIGHT PROPORTIONS OF INDIVIDUAL BODY PARTS, %

Place of catch	Length, cm	Weight, kg	Trunk	Meat	Head	Skin	Scales	Bones	Fins	Blood	Viscera
Mid-Atlantic	159.7	79.0	75.3	—	13.1	—	—	—	1.3	6.8	3.0
Indian Ocean, Mozambique	119	39.9	67.1	56.3*/5.6	26.9	1.7	1.9	3.5	2.3	—	1.0

*Numerator—light colored, denominator—dark colored.

TABLE 232. CHEMICAL COMPOSITION OF MEAT OF TUNA CAUGHT IN THE INDIAN OCEAN NEAR MADAGASCAR ISLAND, %

Meat	Moisture	Protein	Fat	Ash
Total	74.6–75.6	23.0–24.0	0.5–0.7	1.2–1.4
Light colored	73.7–74.6	22.2–23.5	0.3–1.1	1.4
Dark colored	73.3–74.6	21.4–22.4	1.0–1.2	1.3

Bluefin tuna—*Thunnus thynnus**. Distributed in the tropical, subtropical and in part in the temperate regions of all oceans. Attains a length of more than 3.5 m and weight of more than 700 kg.

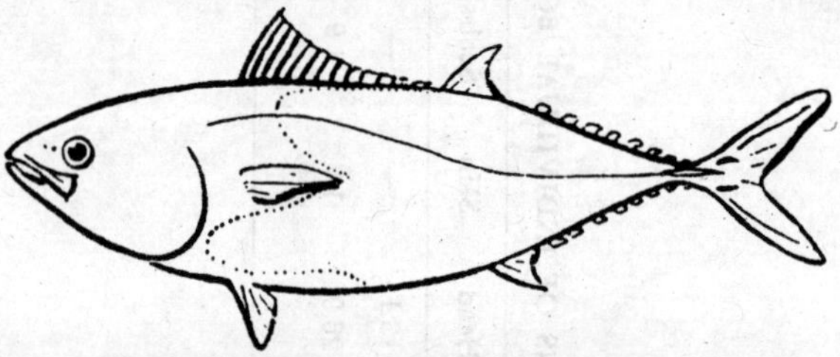

Fig. 193. Bluefin tuna—*Thunnus thynnus**

TABLE 233. WEIGHT PROPORTIONS OF INDIVIDUAL BODY PARTS, %

Meat	Including: light colored	Including: gray colored	Head with gills	Gills	Fins	Bones	Skin	Viscera: whole	Viscera: including liver
63.6	57.2	6.4	17.5	3.8	1.4	4.5	4.6	5.1	1.0

TABLE 234. CHEMICAL COMPOSITION OF MEAT, %

Moisture	Protein	Fat	Ash
68.4–71.2	17.6–20.0	7.4–10.5	1.1–1.4
69.8	18.8	8.9	1.3

The meat of bluefin tuna should be used for conserves.

Albacore or long finned tuna—*Thunnus alalunga*. Distributed in the tropical and subtropical zones of the Atlantic, Pacific and Indian Oceans. In the Atlantic Ocean, found in the east from the Gulf of Guinea to the Bay of Biscay, penetrating up to 53° N, in the west along the American coast, north of Florida to the state of Massachusetts, near the Bermuda and Bahama Islands and near Cuba. Found in large numbers around the Brazil current. In the Indian Ocean known near Madagascar Island and Reunion Island. In the Pacific Ocean found in the eastern part of the ocean, from Alaska to Chile, near the Hawaiian Islands, in the Sea of Japan, near Australia and New Zealand.

Fig. 194. Albacore or long finned tuna—*Thunnus alalunga*.

TABLE 235. CHEMICAL COMPOSITION OF LIGHT AND DARK COLORED MEAT AND THAT OF LIGHT COLORED MEAT FROM THE DORSAL AND ABDOMINAL PORTIONS, %

Type of meat	Moisture	Protein	Fat	Ash
Light colored	67.8	23.7	6.9	1.6
Dark colored	72.5	21.7	4.1	1.7
Abdominal	64.8	19.0	15.1	1.1
Dorsal	72.0	19.1	7.2	1.0

TABLE 236. WEIGHT PROPORTIONS OF INDIVIDUAL BODY PARTS, %

Meat	Head	Fins	Skin	Scales	Bones	Viscera
$\frac{61.8–77.5}{66.7}$	$\frac{11.4–22.4}{16.2}$	$\frac{1.2–3.3}{1.9}$	$\frac{1.3–4.5}{2.9}$	$\frac{1.8–2.2}{2.0}$	$\frac{3.5–4.6}{4.1}$	$\frac{4.6–7.3}{6.3}$

Attains a length of 1.5 m and weight of more than 45 kg. Usual weight of fish in catches 6–7 kg.

A characteristic property of the meat of albacore, as in the meat of other tuna species, is the natural acid which is not unpleasant. The boiled light colored meat of the albacore is white, soft and tasty (resembles the taste of chicken and the meat of the yellowfin tuna). Canned products in oil, natural canned products, smoked products and culinary preparations should be made from the light colored meat. The dark colored meat should be used for fish meal.

TABLE 237. CHEMICAL COMPOSITION, %

Body parts	Moisture	Protein	Fat	Ash
Meat	61.4–75.5 / 70.7	21.6–27.8 / 24.5	0.1–14.5 / 3.2	1.4–1.9 / 1.6
Head	48.3–62.6 / 55.4	20.0	11.1–20.7 / 15.9	6.6
Bones	62.4	20.4	3.0	14.2
Fins	47.3	—	4.2	8.3
Skin	59.5	29.2	9.0	2.3
Viscera	76.6	17.4	4.4	1.6

TABLE 238. CHEMICAL COMPOSITION OF LIGHT AND DARK COLORED MEAT AND LIGHT COLORED MEAT FROM DIFFERENT PARTS OF THE BODY OF ALBACORE, %

Type of meat	Moisture	Protein	Fat	Ash
Light	71.8	26.6	0.4	1.2
Dark	72.8	25.3	0.6	1.3
Light	60.4	13.5	24.9	1.2
Dark	65.6	10.2	23.5	0.7
Light	70.3	25.8	2.7	1.2
Dark	71.1	24.1	2.0	1.3
Light	72.5	25.1	1.3	1.2
Dark	71.2	24.0	2.0	1.3
Caudal	62.7	25.5	10.7	0.9
Dorsal	63.2	24.5	10.4	1.9
Abdominal	55.8	17.5	23.0	3.0

Yellowfin tuna—*Thunnus albacares.* A commercial fish, inhabiting the tropical and subtropical waters of all oceans. In the Atlantic Ocean, found in large numbers in the waters of Guinea, the Canary Islands and the North, Counter and South Equatorial Currents; in small numbers found in the waters of the Brazil and Benquela currents. In the Pacific Ocean, the yellowfin tuna is distributed in the waters of the equatorial currents, predominently in the west and south of the Hawaiian Islands as well as near the Galapagos Islands.

TABLE 239. WEIGHT PROPORTIONS OF INDIVIDUAL BODY PARTS, %

Head	Meat	Including dark meat	Skin	Scales	Bones	Fins	Viscera	Liver	Gonads
$\frac{15.6-23.3}{18.9}$	$\frac{60.4-68.8}{64.9}$	$\frac{3.5-6.4}{4.7}$	$\frac{1.3-3.3}{1.8}$	$\frac{1.2-2.4}{1.6}$	$\frac{3.6-5.5}{4.5}$	$\frac{0.5-2.8}{1.5}$	$\frac{2.0-10.4}{6.1}$	$\frac{0.4-1.4}{0.9}$	$\frac{0.6-2.1}{1.6}$

TABLE 240. CHEMICAL COMPOSITION, %

Body parts	Moisture	Protein	Fat	Ash
Meat	$\frac{71.7-75.3}{72.9}$	$\frac{22.0-26.2}{24.3}$	$\frac{0.1-3.7}{1.3}$	$\frac{1.2-1.7}{1.3}$
Head	$\frac{63.3-64.8}{64.1}$	$\frac{23.7-25.5}{24.8}$	$\frac{0.6-5.9}{2.8}$	$\frac{5.3-10.5}{7.9}$
Bones	49.1	24.2	10.0	16.3
Fins	58.3	27.5	3.2	11.0
Roe	67.4	24.4	6.4	1.8
Viscera	$\frac{76.3-77.3}{76.8}$	$\frac{18.4-19.6}{19.0}$	$\frac{1.6-2.5}{2.1}$	$\frac{1.4-1.8}{1.6}$
Liver	71.2	23.9	3.1	1.8
Heart	75.1	22.0	1.5	1.4
Testes	82.0	16.1	0.8	1.1
Swim-bladder	79.9	19.1	0.4	0.6
Kidneys	72.3	22.8	3.3	1.6
Skin	$\frac{58.5-60.6}{59.5}$	$\frac{33.9-35.4}{34.7}$	$\frac{2.8-4.5}{3.6}$	$\frac{1.6-2.7}{2.2}$

TABLE 241. CHEMICAL COMPOSITION OF MEAT, %

Type of meat	Moisture	Protein	Fat	Ash
Caudal	71.4	24.2	3.2	1.2
Dorsal	72.5	22.4	1.9	1.3
Abdominal	70.1	22.6	6.1	1.2

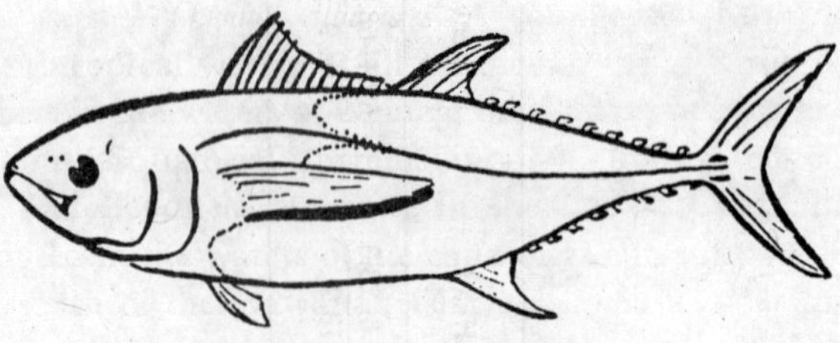

Fig. 195. Yellowfin tuna—*Thunnus albacares.*

Attains a length of 3 m and weight of 200 kg. Usually in catches from coastal regions there is a predominance of fish weighing 10–20 kg; in catches in the seas larger fish are found. Commercially exploitable length differs significantly, depending upon the area of catch.

Light colored meat of the yellowfin has a pinkish tinge and dense consistency. The dark meat unlike the light has a friable structure. After boiling, light meat becomes gray, with a dense, somewhat hard consistency. In taste, light meat resembles chicken and has a distinct acid flavor. Dark meat has low taste qualities.

TABLE 242. CHEMICAL COMPOSITION OF MEAT, %

Type of meat	Moisture	Protein	Fat	Ash
Light	67.8	24.7	5.7	1.6
Dark	70.2	24.0	4.2	1.6
Light	72.6	22.7	3.3	1.4
Dark	73.0	23.3	2.2	1.2
Light	73.3	24.8	0.3	1.4
Dark	72.2	24.6	0.6	1.3

The blood has an unfavorable effect on the color, taste and storage capacity of the meat. Therefore, carefully executed bleeding is an essential condition for processing yellowfin. The meat of this tuna spoils very rapidly with the formation of a toxic substance, histamine, which is produced from histidine under the action of bacteria. The yellowfin fat becomes oxidized quickly and very intense oxidation occurs in dark meat. The light meat of this tuna should be used for canning purposes (in oil, in tallow and naturally) and for culinary preparations, in particular sausages. The dark meat should be used for fish meal.

Genus *Auxis*

Frigate mackerel—*Auxis thazard*. A commercial fish, distributed in the warm waters of the Atlantic, Pacific and Indian Oceans. In the Atlantic Ocean, in the east found near the Canary Islands, the Madeira Islands, along the western coast of Africa and in the Mediterranean Sea. In the Pacific and Indian Oceans, found near Zond, Java, Celebes and New Guinea. Also occurs along the eastern, western and northern coasts of Australia. Attains a length of 30–40 cm and weight of 2.5–5.0 kg.

The frigate mackerel has soft and tasty meat. However, in its taste qualities it is inferior to the meat of big eye tuna. It can be used as a table fish. Fillets, smoked products and canned preparations can be made from it.

TABLE 243. WEIGHT PROPORTIONS OF INDIVIDUAL BODY PARTS, %

Place of catch	Length, cm	Weight, g	Head	Trunk	Fins	Meat	Skin	Scales	Bones	Viscera	Gonads	Liver
Mid-Atlantic coast of Africa	50	2410	13.3	68.5	1.0	56.0	—	—	—	16.5	—	—
Red Sea	37.5	1146	19.0	65.3	0.5	43.8*/13.5**	—	0.6	—	13.8	—	—
Arabian Sea	40.0	1600	19.0	—	0.7	59.1	—	0.8	5.4	12.1	5.5	3.0

* Light colored meat.

** Dark colored meat.

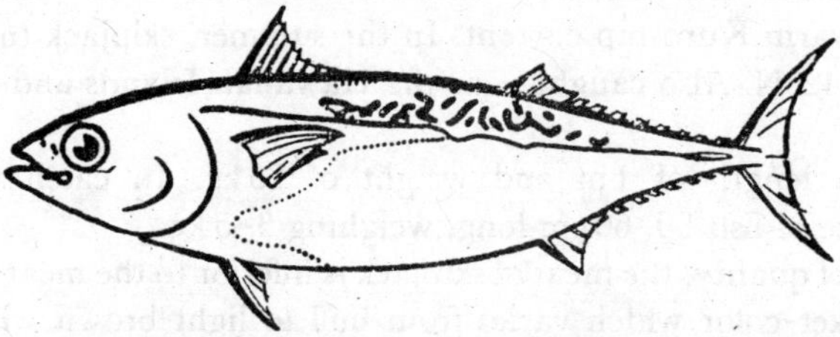

Fig. 196. Frigate mackerel—*Auxis thazard.*

TABLE 244. CHEMICAL COMPOSITION, %

Body	Place of catch	Moisture	Protein	Fat	Ash
Meat	Mid-Atlantic coast of Africa	61.3	22.0	15.8	0.8
Meat	Southern Pacific Ocean	72.6	18.5	5.1	1.2
Meat	Arabian Sea	72.7	23.3	2.4	1.5
Meat					
light	Indian Ocean	70.6	24.2	3.4	1.4
dark	Red Sea	69.2	26.0	3.1	0.9
Testes	Red Sea	78.5	18.1	1.6	1.8
Roe	Red Sea	71.2	23.0	4.6	1.2

Genus *Katsuwonus**

Skipjack tuna—*Katsuwonus pelamis.* Found in the Atlantic, Pacific and Indian Oceans.

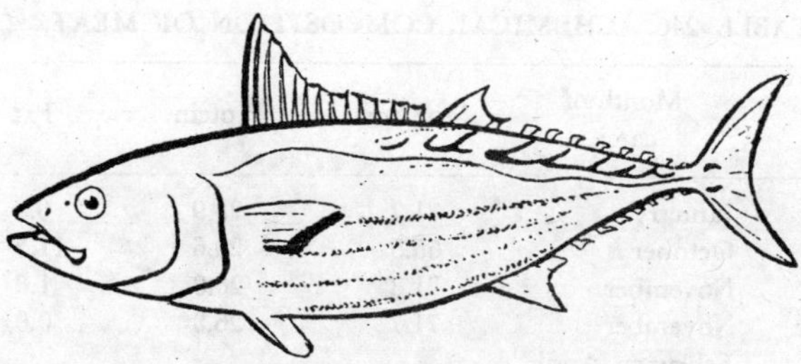

Fig. 197. Skipjack tuna—*Katsuwonus pelamis.*

In the Atlantic Ocean caught near the Azores, Madeira, Canary Islands, Cape Verde and in the Mediterranean Sea, along the western coast of Africa, south to Cape Town. In the Indian Ocean, found along the eastern coast of Africa, from Natal to Aden. Known to occur in the Red Sea, near India and Sri Lanka, in the Indo-Australian archipelago, near Sumatra, Borneo and

Celebes. In the Pacific Ocean, it is known to occur in the fishing areas in view of the warm Kuroship current. In the summer, skipjack tuna has been found up to 42° N. Also caught near the Hawaiian Islands and off the coast of Australia.

Attains a length of 1 m and weight of 25 kg. In catches there is a predominance of fish 50–60 cm long, weighing 3–5 kg.

In terms of quality, the meat of skipjack is inferior to the meat of yellowfin. It has a darker color which varies from buff to light brown with a pinkish tinge. After heat treatment it becomes slightly light. The consistency of the meat is harder and denser than the meat of yellowfin, although the taste and smell are quite pleasant. The meat of skipjack caught in the Indian Ocean near Madagascar does not have very high taste qualities; the meat was infested with parasites especially in the abdominal parts.

This fish should be used for natural conserves and conserves in oil and for semi-finished products and culinary preparations.

If there is heavy infestation of parasites in the meat, the skipjack should be used for fish meal.

TABLE 245. WEIGHT PROPORTIONS OF INDIVIDUAL BODY PARTS, %

Meat	Including dark meat	Head	Skin	Scales	Bones	Fins	Viscera	Including liver	Gonads
63.1–69.6 65.4	8.1	12.3–20.9 17.8	1.5–4.5 1.8	1.1	3.5–6.0 5.4	0.5–1.7 1.0	5.7–12.8 7.5	0.9–1.7 1.2	1.8

TABLE 246. CHEMICAL COMPOSITION OF MEAT, %

Place of catch	Month of catch	Moisture	Protein	Fat	Ash
Indian Ocean	January	71.7	25.9	0.6	1.4
Gulf of Mexico	October	65.5	20.6	11.5	1.6
South Atlantic	November	71.8	26.0	1.0	1.2
Central Atlantic	November	71.7	26.8	0.01	1.4
Pacific Ocean	February, April, June, December	70.8–74.5 71.4	23.0–28.0 25.7	0.3–4.1 1.5	0.9–1.6 1.4
	August, September	65.8–70.5 68.2	22.2–23.4 22.8	4.7–10.2 7.7	1.2–1.7 1.3

TABLE 247. CHEMICAL COMPOSITION OF MEAT, %

Meat	Moisture	Protein	Fat	Ash
Light	64.8	19.2	13.4	1.2
Dark	59.7	18.9	18.4	1.1
Light	66.8	22.9	9.0	1.3
Dark	72.4	21.5	4.8	1.3
Light	71.1	25.0	2.5	1.4
Dark	71.2	24.4	3.0	1.4
Light	72.2	25.6	0.8	1.4
Dark	69.7	26.8	1.9	2.0

TABLE 248. CHEMICAL COMPOSITION OF MEAT, %

Type of meat	Moisture	Protein	Fat	Ash
Dorsal	71.4	25.6	1.3	2.7
Caudal	70.3	26.4	1.5	1.8
Abdominal	69.4	25.7	2.9	2.0

TABLE 249. CHEMICAL COMPOSITION OF BODY PARTS, %

Body parts	Moisture	Protein	Fat	Ash
Head	60.3–66.2 63.3	18.4–21.5 19.9	1.2–13.1 7.2	6.1–9.0 7.6
Bones	56.4	17.5	12.5	11.4
Fins	44.2–49.3 47.3	22.5–30.3 26.0	1.1–14.4 7.6	13.3–24.9 18.0
Skin	51.7–64.1 55.2	26.0–30.6 27.5	0.4–19.7 13.2	3.1–6.2 4.1
Viscera	68.1	20.8	6.1	4.7

Genus *Euthynnus*

Little tuna—*Euthynnus alleteratus*. Found in the subtropical and tropical waters of the Atlantic Ocean, including the Mediterranean Sea, caught in small numbers in the east, from St. Helena to Senegal and west from the Cape Cod peninsula to the Gulf of Mexico.

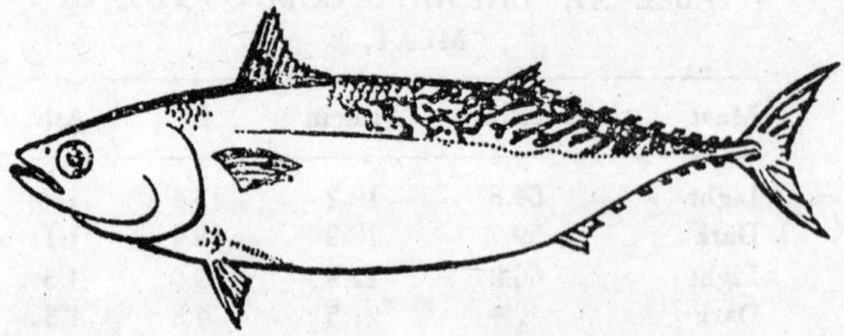

Fig. 198. Little tuna—*Euthynnus alleteratus*.

Attains a length of 120 cm, usual size in catches 40–80 cm.

In terms of quality, the meat of little tuna is inferior to the meat of albacore and yellowfin. The light meat ranges in color from buff to light brown. The dark meat is brown. After heat processing, the meat becomes somewhat light. After frying, it has a mild acid flavor, dense and hard consistency, but fairly pleasant taste and smell.

TABLE 250. WEIGHT PROPORTIONS OF INDIVIDUAL BODY PARTS, %

Head	Meat	Including dark meat	Scales	Fins	Viscera
17.1–24.0	52.7–56.7	12.5–13.8	11.5	0.5–2.5	8.0–14.2
20.7	54.7	13.1		1.4	10.6

TABLE 251. CHEMICAL COMPOSITION OF MEAT AND ROE, %

Body parts	Moisture	Protein	Fat	Ash
Meat	62.7–73.8	22.6–26.6	0.2–12.6	1.0–1.4
	70.9	24.5	2.9	1.2
Roe	71.7	23.5	3.3	0.9

TABLE 252. CHEMICAL COMPOSITION OF LIGHT AND DARK MEAT AND OF LIGHT MEAT FROM DIFFERENT PARTS OF LITTLE TUNA, %

Meat	Moisture	Protein	Fat	Ash
Light	71.0	25.9	1.7	1.4
Dark	73.2	23.3	0.3	1.4
Dorsal	74.4	24.0	0.4	1.3
Abdominal	71.6	25.8	1.3	1.3

This tuna should be used for canning and for culinary preparations.

Genus *Gymnosarda*

Plain bonito—*Gymnosarda unicolor.* A large and very valuable fish for use as food. Caught with other tuna. Average length of this tuna caught in the western open part of the Indian Ocean about 90 cm, weight about 10.5 kg, maximum depth 21 cm and maximum thickness 12 cm.

Weight proportions of individual body parts (%): head 14.3, including meat of occiput and nape 3.3, trunk 71.4, including meat 62.9, skin 3.8, bones 4.8, scales 1.3, fins 1.4, viscera 10.7, including liver 1.7.

Chemical composition of meat (%): moisture 61.3, protein 16.6, fat 20.3, ash 1.0.

Of all fish investigated from the Indian Ocean, the plain bonito has the highest fat content in the meat (20.3% fat).

This is an excellent table fish. After cooking, the meat is white and soft with a pleasant taste and mild acid flavor. The soup is colorless, with a strange flavor. The fried meat is very white, soft and very tasty. Meat from the abdominal portion is especially tasty. This meat should be used for second course dishes, canning in oil and for natural canned products (products canned in tomato sauce turn slightly dark and have a distinctly unpleasant taste).

Family SCOMBERESOCIDAE—SAURIES OR SKIPPERS

The members of this family live in the tropical and subtropical regions of the ocean. Only a few species enter the northern regions.

Saury pike—*Scomberesox forsteri.* Found in the Pacific Ocean along the South American coast. Outwardly resembles Pacific saury. Length of fish caught in February–June 21.7–36.0 cm, weight 30–175 g.

Canned products such as "blanched fish in oil" and "smoked fish in oil" can be prepared from the fresh and frozen fish. Glazed frozen fish has longer shelf-life than the Pacific saury because of less intense oxidation of the tissue fat.

TABLE 253. CHEMICAL COMPOSITION, %

Body parts	Moisture	Protein	Fat	Ash
Meat	70.7–72.9	23.0–24.9	1.7–3.2	1.4–1.3
Head	73.7	17.7	4.5	3.4
Viscera	73.3	16.0	6.9	1.8

Atlantic saury—*Scomberesox saurus.* Of great commercial value.

The fattiest fish in the northwest Atlantic were found in the beginning of

Fig. 199. Atlantic saury—*Scomberesox saurus.*

TABLE 254. WEIGHT PROPORTIONS OF INDIVIDUAL BODY PARTS, %

Parameters	Northwest Atlantic			Northeast Atlantic
	Sep. 29	Nov. 3	Nov. 25	November
Length, cm	—	—	—	26–30
Weight, g	79–90	100–170	70–130	65–90
Weight proportions of individual body parts, %				
trunk	78	79	79	77
head	10	9	11	12
viscera	10	11	8	9
fins	2	1	2	2

TABLE 255. CHEMICAL COMPOSITION OF MEAT, %

Moisture	Fat	Protein	Ash
72.7	3.0	23.0	1.3
65.2	11.0	22.5	1.3
70.9	5.1	22.7	1.3
72.2	1.5	24.6	1.7

November. Canned products like Pacific saury which were prepared from this fish received a high grade for quality. In November in the northeast Atlantic, the fish are lean with lower taste qualities than fish caught in the northwest Atlantic. The meat of this saury should be used for canning in oil.

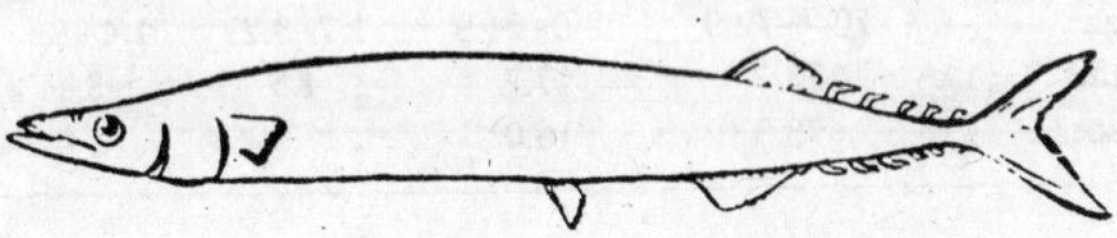

Fig. 200. Pacific saury—*Cololabis saira.*

Pacific saury—*Cololabis saira.* Commercially an important fish inhabiting the subtropical and temperate waters of the Pacific Ocean along the Asian and American coasts. From late July or August to November, the fish are found forming feeding concentrations near the southern Kuril Islands and in Central American waters. Length of Pacific saury caught 17–34 cm, weight 30–200 g. In dressing to prepare trunks there is about 25% wastage which includes heads 7–15%, viscera 6–12%, caudal fin 1%. Output of fillets 54–61%.

TABLE 256. FAT CONTENT IN MEAT OF PACIFIC SAURY IN DIFFERENT MONTHS, %

Place of catch and length of Pacific saury (cm)	June	July	August	September	October	November
Southern Kuril Islands						
28–31	5.9–15.0 / 10.4	10.5–22.1 / 15.4	18.6–25.0 / 21.0	17.2–27.7 / 23.4	25	—
23–27	8.9	10.0–13.1 / 11.5	6.7–12.1 / 10.8	5.9–18.7 / 14.2	16.2	—
19–21	4.2	4.7–6.2 / 5.3	3.8–7.7 / 5.3	4.2–9.6 / 6.6	11.0	—
North America						
large	—	—	—	9.1–9.3	7.0–10.9	4.0–4.7
medium	—	—	—	—	5.2–7.8	—

The quantity of proteinous matter in the meat of the Pacific saury is 19–23% and the mineral matter content is 1.0–1.3%.

The fat content in the viscera in the period of feeding is higher than in the meat and in some years in August–September large sized Pacific saury have a fat content of up to 40–50%. In the head, bones and fins, the fat content is 4–10%, protein content 17–24% and ash 4.5–7.0%.

The Pacific saury is mainly used for canned snacks such as "blanched Pacific saury in oil" and "smoked Pacific saury in oil". For canning purposes, frozen Pacific saury can be used if it has been stored not more than one month at temperatures of – 18° C, with glazing. Prolonged storage of the frozen Pacific saury is not recommended since its body fat is quickly oxidized and hydrolyzed.

Family SCORPAENIDAE—SCORPIONFISHES OR ROCKFISHES

These fish are found in the northern Atlantic Ocean and near the Arctic Ocean, in the Pacific Ocean and in the Mediterranean and Black Seas. This family includes many genera and species.

Genus *Helicolenus*

Uruguay rockfish—*Helicolenus lahilli*. Caught with other fish from the southwest Atlantic. Length of fish 17–22 cm, maximum depth 6–8 cm, maximum thickness 3–4 cm, weight 140–290 g.

Weight proportions of individual body parts (%): trunk 51–64, head 27–35, viscera 5–19.

TABLE 257. CHEMICAL COMPOSITION, %

Body parts	Moisture	Protein	Fat	Ash
Meat	76.8–78.0	18.0	3.1–3.8	1.5
Viscera	77.0–78.0	—	4.8–7.4	—
Liver	67	—	10.4	—
Bones, skin	70.8	15.5	6.9	6.9

The meat of fried fish is cream colored, juicy, soft and tasty. The Uruguay rockfish should be used for culinary products.

Blue mouth rockfish—*Helicolenus dactylopterus*. Caught with other fish from the northwest Atlantic. Distributed in the eastern Atlantic Ocean from the islands of Cape Verde and Madeira to the coast of Ireland. In small numbers the fish are found along the coast of Scotland and Denmark, along the Norwegian coast north to Tromso. Attains a length of 50 cm.

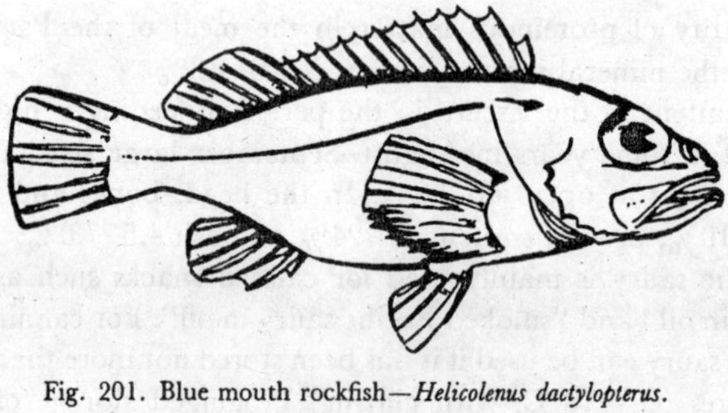

Fig. 201. Blue mouth rockfish—*Helicolenus dactylopterus*.

Weight proportions of individual body parts of blue mouth rockfish caught in November near 68° N 16–22 cm long, weighing 140–330 g (%): trunk with fins 52, head 36, viscera 12. Output of fillets 35%.

Chemical composition of meat (%): moisture 74, protein 17.3, fat 7.7, ash 1.1.

The meat of fried fish is white, juicy and very tasty. The fish should be used for culinary preparations and smoked products.

Genus *Sebastes*

Deepwater redfish—*Sebastes mentella*. Of considerable commercial value, distributed in the eastern and western North Atlantic at depths greater than 200 m. Attains a length of 72 cm. In catches taken in their feeding grounds there is a predominance of fish 24–48 cm long.

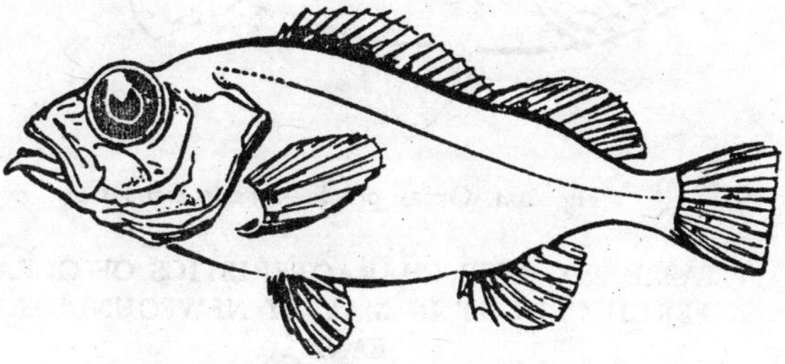

Fig. 202. Deepwater redfish—*Sebastes mentella*.

Weight proportions of individual body parts of a fish 37 cm long, weighing 1,028 g (%): trunk 62.3, head (separated along the cranial bones) 24.1, meat 47.3, skin 2.9, bones 10.6, fins 2.5, scales 1.0, viscera 8.9, including liver 1.7.

TABLE 258. CHEMICAL COMPOSITION, %

Body parts	Moisture	Protein	Fat	Ash
Meat	$\frac{73.6-78.6}{77.1}$	$\frac{16.4-18.6}{17.3}$	$\frac{2.4-6.4}{3.7}$	$\frac{1.0-1.4}{1.2}$
Liver	$\frac{53.2-73.7}{64.8}$	—	$\frac{6.7-33.7}{18.1}$	—
Head	$\frac{69.1-81.3}{70.9}$	$\frac{10.4-18.9}{13.2}$	$\frac{0.9-16.3}{8.1}$	$\frac{6.8-8.5}{7.2}$
Bones	$\frac{56.4-64.7}{61.6}$	$\frac{14.3-16.3}{15.6}$	$\frac{1.3-16.2}{10.2}$	$\frac{9.0-11.4}{10.1}$
Viscera	$\frac{54.7-78.4}{70.9}$	—	$\frac{6.6-33.7}{15.2}$	1.0

The liver oil of the deepwater redfish contains 128–6,500 iu/g of vitamin A and 79 γ/g vitamin E.

This is a good table fish and is especially valuable after smoke curing and sun curing.

Ocean perch—*Sebastes marinus*. Also commercially significant, inhabiting

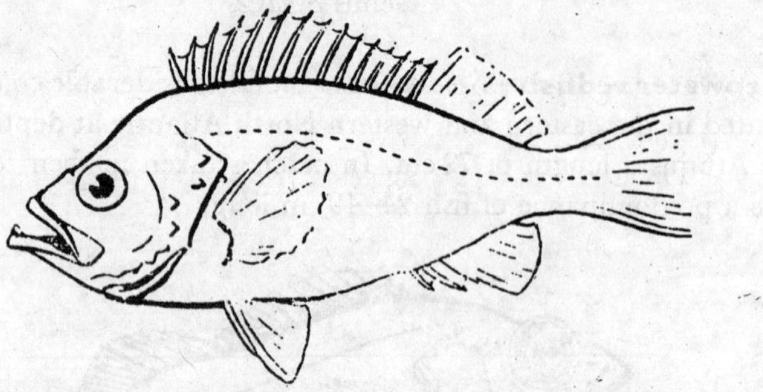

Fig. 203. Ocean perch—*Sebastes marinus.*

TABLE 259. SIZE CHARACTERISTICS OF OCEAN PERCH CAUGHT IN MAY AT NEWFOUNDLAND BANK

Length, cm	Maximum depth, cm	Maximum thickness, cm	Length of head, cm	Weight, kg
25	10	5	6	0.42
30	11	6	7	0.60
35	13	7	8	1.07
40	15	7.5	11	1.94
45	16	7.5	12	2.1
48	17	8	13	2.2
53	19	9	14	3.0

TABLE 260. CHEMICAL COMPOSITION, %

Body parts	Moisture	Protein	Fat	Ash
Meat	$\frac{69.6–77.5}{73.6}$	$\frac{17.7–18.5}{17.8}$	$\frac{2.2–10.3}{6.6}$	$\frac{1.4–1.5}{1.5}$
Head	$\frac{61.5–71.2}{65.6}$	$\frac{14.7–17.1}{15.8}$	$\frac{6.9–14.8}{11.0}$	$\frac{4.9–9.5}{7.2}$
Bones	$\frac{56.5–67.3}{61.8}$	—	$\frac{7.8–18.3}{13.6}$	—
Viscera	$\frac{30.1–67.3}{49.6}$	7.2–12.5	$\frac{5.2–62.7}{40.2}$	1.3
Liver	$\frac{45.7–78.3}{58.9}$	$\frac{9.8–14.1}{11.4}$	$\frac{10.0–41.1}{27.1}$	$\frac{1.0–1.9}{1.4}$

the North Atlantic along the European and American coasts. Attains a length of 90 cm.

Weight proportions of individual body parts of ocean perch 49 cm long, weighing 1,560 g, (%): head (separated along cranial bones) 21.5, meat 49.6, skin 2.8, bones 9.1, fins 2.9, scales 1.8, viscera 6.3, including liver 2.5.

The liver oil of this fish contains 170–8,600 iu/g of vitamin A. This is a good table fish. The meat is very tasty. It can be used for cold smoked products which are very popular.

Genus *Sebastodes*

This genus includes 14 species of which only two have commercial significance. The other species constitute only 1–10% of the perch catch of the genus *Sebastodes*.

The members of this genus occur in the Okhotsk and Bering Seas, along the coast of Alaska and the Pacific coast of America, south to California.

Blackthroat rockfish—*Sebastodes introniger*. Distributed in the Bering and Okhotsk Seas, in the Gulf of Alaska and along the Pacific coast of America to Santa Barbara.

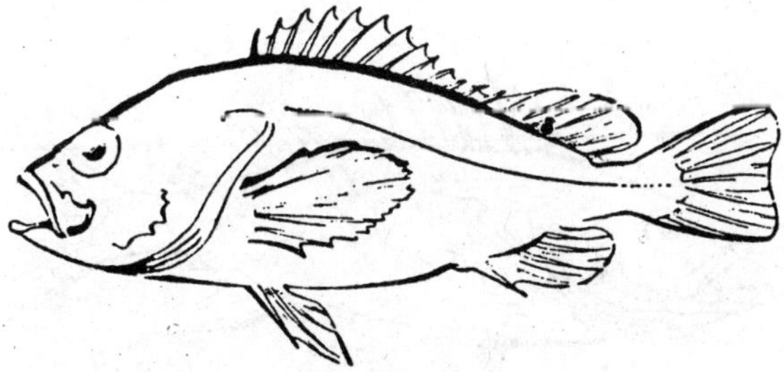

Fig. 204. Blackthroat rockfish—*Sebastodes introniger*.

Attains a length of 80–85 cm, weight 12–13 kg. Length of fish caught in September–May 12.0–38.2 cm, weight 1.8–2.0 kg.

Weight proportions of individual body parts (%): head 28.7–36.2, bones and fins 7.3–12.4, viscera 6.4– 17.1, including liver 1.6–3.3. Output of fillets with skin 39.5–46.7%.

Chemical composition of meat (%): moisture 76.7–80.1, protein 17.7– 19.6, fat 0.4–3.5, ash 1.4–1.6.

This is a table fish suitable for consumption in fried and boiled forms.

Blue rockfish—*Sebastodes mystinus*. Distributed in the Bering Sea, in the Gulf of Alaska and along the coast of North America.

Length of fish 27.5–35.0 cm and weight 450–930 g. In dressing blue rockfish caught in May, the output of the trunk was 56.6–61.8%. Output of fillets 48.8–51.2%.

Chemical composition of meat (%): moisture 78.2, protein 18.9, fat 0.8, ash 1.8.

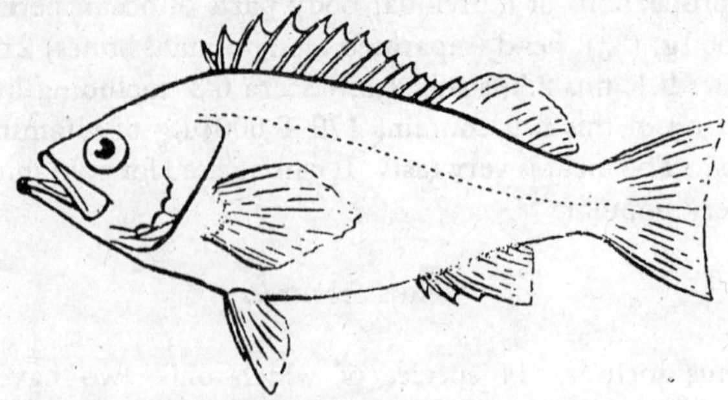

Fig. 205. Blue rockfish—*Sebastodes mystinus.*

This is a table fish. It has tasty white meat suitable for first and second course dishes.

Chilipepper—*Sebastodes goodei*. Species has little commercial importance. Distributed in the Bering Sea. Length of fish caught in May 29 cm, weight 510 g. Output of trunk after dressing 52.9% and output of fillets 44.5%.

Fig. 206. Chilipepper—*Sebastodes goodei.*

Chemical composition of meat (%): moisture 78.3, protein 19.0, fat 1.0, ash 1.7.

This is a table fish. It has tasty white meat.

Green striped rockfish—*Sebastodes elongatus*. Commercially of little importance, caught in the Bering Sea. Length of fish caught in May

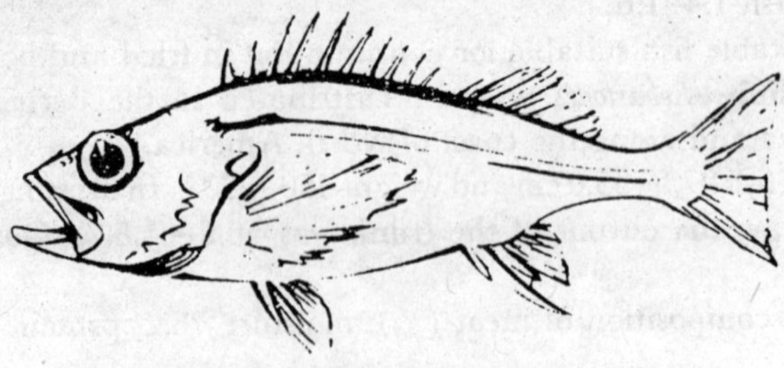

Fig. 207. Green striped rockfish—*Sebastodes elongatus.*

24.8–25.0 cm, weight 300–321 g. Output of trunks after dressing 48.3% and output of fillets 36.2%.

Chemical composition of meat (%): moisture 75.2, protein 21.5, fat 1.0, ash 1.5.

This is a table fish with tasty white meat.

Splitnose rockfish—*Sebastodes diploproa*. Commercially has little importance, caught in the northeastern Pacific Ocean. Length 22–30 cm, weight 270–705 g.

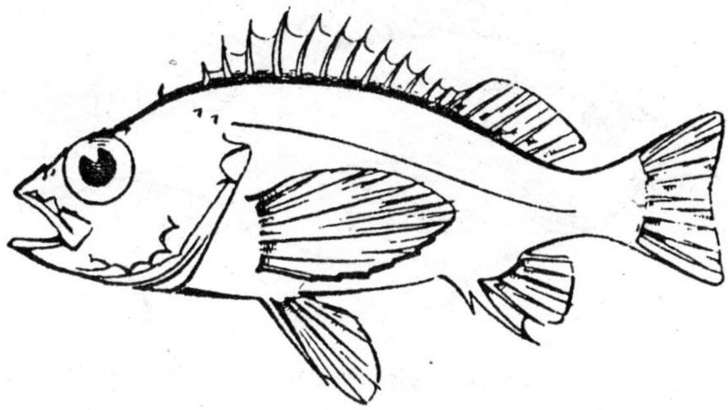

Fig. 208. Splitnose rockfish—*Sebastodes diploproa.*

Weight proportions of individual body parts of fish caught in May (%): head 35.4–39.2, trunk 44.7–48.6, fins 3.1–3.5, viscera 7.0–13.5, scales 3.1–3.5.

Chemical composition of meat (%): moisture 75.6, protein 16.4, fat 5.9, ash 2.1.

This is a table fish. The meat of this fish is used for first and second course dishes.

Flag rockfish—*Sebastodes rubrivinctus**. Commercially has little importance. Lives in the Bering Sea. Length of fish caught in May 28.5 cm, weight 590 g. After dressing, the output of trunks was 45.7% and output of fillets 36.4%.

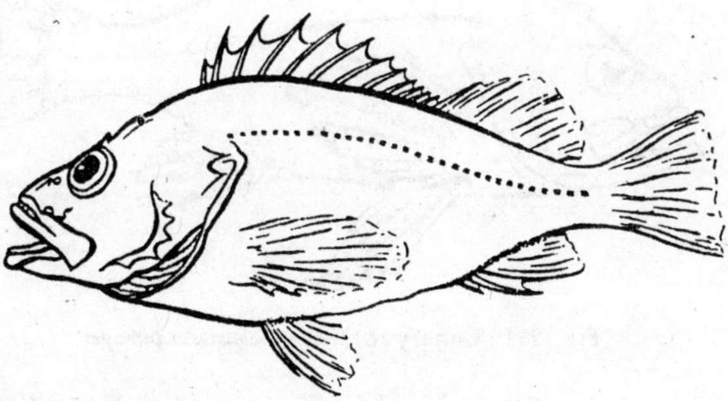

Fig. 209. Flag rockfish—*Sebastodes rubrivinctus.*

Chemical composition of meat (%): moisture 73.7, protein 19.6, fat 4.5, ash 2.2.

Northern (multispine) rockfish—*Sebastodes polyspinis.* Commercially a less important species, caught in the Bering Sea, near the Komander, Kuril and Pribilof Islands, along the eastern coast of Kamchatka, along the coast of Alaska and the Pacific coast of America south to California. Length of fish caught in February–May 30.6–37.1 cm, weight 630–1,050 g.

Fig. 210. Northern (multispine) rockfish—*Sebastodes polyspinis*

Weight proportions of individual body parts (%): head 27.4–27.8, bones and fins 10.0–12.6, viscera 13.4–16.3, including liver 2.1–2.2. Output of fillets 45.9–46.6%.

Chemical composition of meat (%): moisture 76.3–77.7, protein 16.5–20.8, fat 1.3–3.6, ash 1.4–1.8.

This is a table fish. It is used for first and second course dishes.

Canary rockfish—*Sebastodes pinniger.* This is commercially an unimportant species. It lives in the northeastern Pacific Ocean. Length of fish caught in January–May 46–48 cm, weight 1,980–2,110 g.

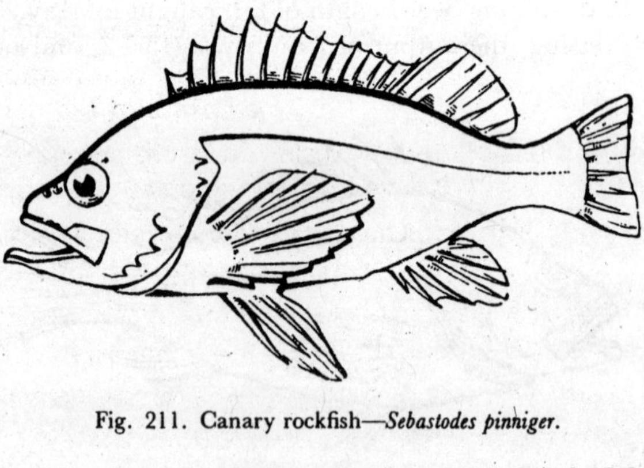

Fig. 211. Canary rockfish—*Sebastodes pinniger.*

Weight proportions of individual body parts (%): head 32.2–33.3, trunk 48.9–52.2, fins 4.5–7.4, scales 3.1, viscera 7.3–9.1.

Chemical composition of meat (%): moisture 77.4, protein 18.7, fat 1.8, ash 2.

This is a table fish. It is used for first and second course dishes.

Bocaccio—*Sebastodes paucispinis.* Commercially of little importance, caught in the northeastern Pacific Ocean. Length of fish caught in May 30.0–40.8 cm, weight 401–1,040 g.

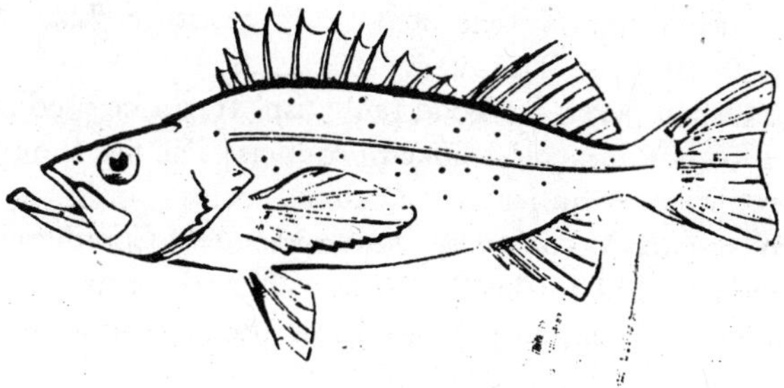

Fig. 212. Bocaccio—*Sebastodes paucispinis.*

Weight proportions of individual body parts (%): head 27.7–33.5, trunk 45.8–50.3, scales 4.0, fins 2.8–5.3, viscera 9.6–18.4.

Chemical composition of meat (%): moisture 77.9–78.2, protein 19.2–19.6, fat 0.2–10.5, ash 1.4–1.2.

This is a table fish. It has soft, white and juicy meat.

Gray rockfish—*Sebastodes clastodes.* Commercially of little importance, caught in the northeastern Pacific Ocean. Length of fish caught in January–May 460–540 cm, weight 1,955–2,550 g.

Weight proportions of individual body parts (%): head 31.7, trunk 48.3, fins 5.6, viscera 12.6.

Chemical composition of meat (%): moisture 78.9, protein 13.3, fat 5.9, ash 1.9.

This is a table fish. It is used for first and second course dishes.

Pacific Ocean perch—*Sebastodes alutus.* Commercially an important fish,

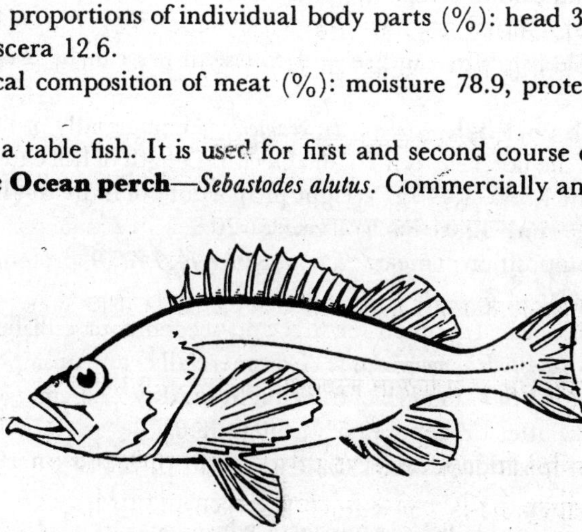

Fig. 213. Pacific Ocean perch—*Sebastodes alutus.*

caught in the Bering Sea, near the Komander and Kuril Islands, Pribilof Islands, along the eastern coast of Kamchatka, along the coast of Alaska and the Pacific coast of America south to California. Length up to 60 cm and weight 1,300 g.

Weight proportions of individual body parts of fish 33–43 cm long, weighing 500–700 g (%): head 28–37, bones and fins 10–14, viscera 3–12, including liver 1.3–3.8. Output of fillets 43–53%.

Chemical composition of the meat (%): moisture 72.2–80.9, protein 16.8–19.7, fat 1.5–10.6, ash 1.3–2.0.

The Pacific Ocean perch is a good table fish. It can be used in boiled and fried forms and for hot and cold smoked products. The fish should be dressed so that they are processed in frozen or canned forms.

Rosethorn rockfish—*Sebastodes helvomaculatus**. Commercially an unimportant species, caught in the Bering Sea. Length of fish caught in May 20.7 cm and weight 250 g. Output of trunks after dressing 56%, output of fillets 46%.

Chemical composition of meat (%): moisture 77.8, protein 19.1, fat 0.8, ash 2.3.

This is a good table fish. It has white and tasty meat.

Dusky rockfish—*Sebastodes ciliatus*. Commercially an unimportant fish. Caught in the Gulf of Alaska. Length of fish caught in February–April 29.8–51.5 cm, weight 560–2,180 g.

Weight proportions of individual body parts (%): head 23.8–28.0, bones and fins 10.5–11.6, viscera 9.5–9.8, including liver 1.4–1.5. Output of fillets with the skin 52.6–55.3%.

Chemical composition of meat (%): moisture 73.9–76.4, protein 19.5–20.5, fat 0.9–4.3, ash 1.4–1.5.

This is a table fish. First and second course dishes can be prepared from this fish.

Blackmouth rockfish—*Sebastodes crameri*. Commercially of little importance, caught in the northeastern Pacific Ocean. Length of fish caught in May 33–37 cm, weight 1,380–1,440 g. Weight proportions of individual body parts (%): head 31.0, trunk 42.0, fins 5.6, viscera 20.8.

Chemical composition of meat (%): moisture 74.9–76.9, protein 18.5–19.3, fat 2.8–4.3, ash 1.5–1.9.

This is a table fish. It is used for first and second course dishes.

Rockfish—*Sebastodes maculatus**. Commercially an unimportant fish, caught in the Bering Sea. Length of fish caught in May 22.5 cm, weight 195 g. Output of trunks after dressing 43.5%, fillets 33.8%.

Chemical composition of meat (%): moisture 78.0, protein 19.9, fat 0.2, ash 1.6.

This is a typical table fish. It has tasty white meat.

Genus *Sebastolobus*

Short spiny thorny head—*Sebastolobus alascanus.* A commercial fish, caught along the California coast, in the Bering Sea, near the Aleutian Islands. Length up to 61 cm, usual length 28.1–42.8 cm, weight 470–3,700 g.

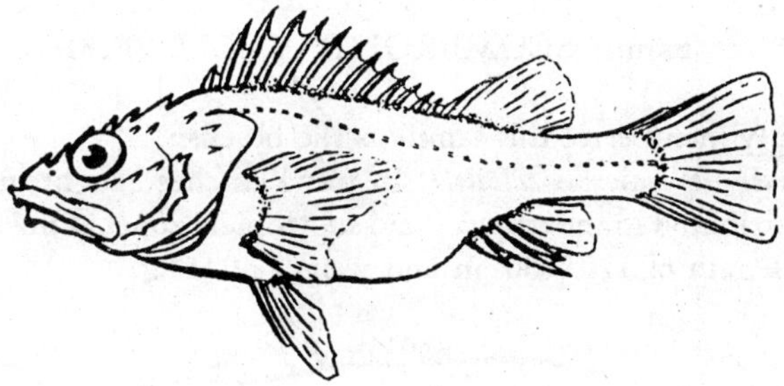

Fig. 214. Short spiny thorny head—*Sebastolobus alascanus.*

Weight proportions of individual body parts (%): head 34.0–41.6, bones and fins 8.9–10.7, viscera 7.1–12.8, including liver 1.9–5.3. Output of fillets 39.7–40.6%.

Chemical composition of meat (%): moisture 65.9–77.5, protein 13.2–16.0, fat 3.4–19.8, ash 0.9–1.2.

This is a table fish. First and second course dishes can be made from it.

Genus *Scorpaena*

Angola scorpionfish—*Scorpaena angolensis.* Weight proportions of individual body parts of fish caught in May along the coast of southwest Africa 29–30 cm long, with maximum depth 9–10 cm, thickness 6–7 cm and weight 690–920 g (%): trunk with fins 50, head 43, viscera 7. Output of fillets 34%.

Chemical composition of meat (%): moisture 78.8, protein 19.3, fat 0.4, ash 1.3.

The meat of fried fish is white and tasty. This fish should be used as a table fish prepared in beheaded and gutted forms.

Senegal scorpionfish—*Scorpaena senegalensis.* Weight proportions of in-

TABLE 261. CHEMICAL COMPOSITION, %

Body parts	Moisture	Protein	Fat	Ash
Meat	80.8	16.0	1.8	1.4
Bones, skin, fins	75.6	12.5	5.5	6.4
Viscera	80.0	10.2	7.9	1.9

dividual body parts of fish caught in June off the southwest African coast 20–23 cm long (length of head 8–10 cm), maximum depth 7–8 cm, weight 230–300 g (%): trunk 48, head 34, fins 9, viscera 7. Output of fillets 29%.

This is a table fish. The meat of fried fish is white and tasty resembling the meat of ocean perch.

Family SCOMBROPIDAE—BLUEFISHES

The only member of this family is the bluefish.

Bluefish—*Pomatomus saltatrix.* In trawl catches caught in an admixture. Widely distributed in the tropical and subtropical zones of the Atlantic Ocean. Attains a length of 120–130 cm and weight of 15 kg.

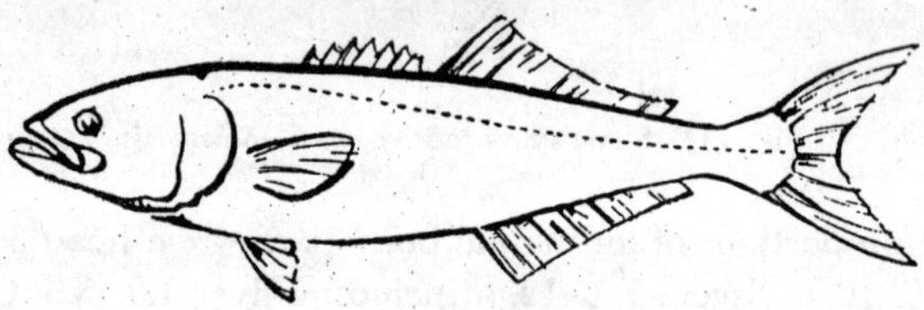

Fig. 215. Bluefish—*Pomatomus saltatrix.*

Size characteristics of bluefish: length 36–74 cm, maximum depth 10–18 cm, maximum thickness 5–9 cm, weight 0.8–4.0 kg.

Relationship between weight and length of bluefish

Length, cm	36	38	55	64	74
Weight, kg	0.8	1	2	3.8	4

Weight proportions of individual body parts (%): trunk 60–72, head

TABLE 262. CHEMICAL COMPOSITION, %

Body parts	Moisture	Protein	Fat	Ash
Meat	69–78	19.7	0.4–3.5	1.2–1.4
Head	59–60	18.7	13–17	5–7
Bones	51–57	17.8–18.7	18–21	5–10
Fins	44–50	19.3–20.5	12.6–15.3	14–21
Skin with scales	56	24.7	19	1.9
Liver	58.8	13.2	21.2	1.1
Roe	75.4	16.9	5.3	1.8
Testes	64.5	—	7.9	—
Viscera	73–75	15–17	6.3–6.8	1.5–1.9

18–33, meat 50–58, skin 2.4–3.4, bones 8–8.5, fins 0.9–2.7, gonads 0.8, viscera 8–13, liver 1.5–2.9.

Mineral composition of meat (mg, %): potassium 205, calcium 31, magnesium 36, phosphorus 167, iron 1.6.

The bluefish is a good table fish. Its meat is tasty with dense consistency. Bluefish are frozen in beheaded and dressed forms.

Family SCOPELIDAE—LANTERNFISHES

Shining anchovy—*Gymnoscopelus nicholsi*. Inhabits warm and temperate waters of all oceans.

Average length of fish caught in the Scotia Sea and in the Weddell Sea 18 cm, average weight 34 g. Weight proportions of individual body parts (%): head 17.5, meat 57.9, skin with fins 2.6, bones 5.0, viscera 14.1.

Chemical composition of meat (%): moisture 69.2, protein 14.7, fat 14.7, ash 1.2.

The fish seasons well in the process of salting; it is thus recommended to process these fish by salt curing.

Family SERRANIDAE—SEA BASSES

These fish are widely distributed in tropical and subtropical seas. They live in the Atlantic, Indian and Pacific Oceans. The family includes about 75 genera and more than 400 species of which some are fresh water fish.

Genus *Aciloperca*

Black bass—*Aciloperca rogad*. Weight proportions of individual body parts of fish caught in September in the Indian Ocean 53 cm long, maximum depth 22 cm, thickness 9 cm and weight 4.8 kg (%): head 24, fins 2.8, bones 8, scales 4, viscera 8, including gonads 3, liver 0.9. Output of fillets 52%.

Chemical composition of meat (%): moisture 71.3, protein 17.9, fat 9.7, ash 1.1. The mineral composition of meat (mg%): potassium 332, calcium 27, magnesium 34, phosphorus 164.

This is a good table fish. It is used for first and second course dishes.

Genus *Centropristes*

Black Sea bass—*Centropristes striatus*. Fish of commercial value. Weight proportions of individual body parts of the fish caught in April in the Atlantic Ocean on the shelf of the USA 20–33 cm long, weighing 250–1,050 g (%): trunk 56, head 31, viscera 8, fins 4.

Fried meat is white, dense, juicy and tasty. The soup is aromatic and

TABLE 263. CHEMICAL COMPOSITION, %

Body parts	Moisture	Protein	Fat	Ash
Meat	78.0	18.9	1.8	1.3
Viscera	65.4	16.0	17.2	1.3
Skin, bones, fins	64.3	19.6	5.3	10.8

tasty. This is a good table fish from which first and second course dishes can be made.

Genus *Epinephelus*

Grouper—*Epinephelus specilus.* Size characteristics of fish caught in February in the Persian Gulf: length 27–37 cm, maximum depth 8.5–12.4 cm, maximum thickness 5–7 cm, weight 0.4–1.3 kg.

Weight proportions of individual body parts (%): trunk 43–47, head 39–43, viscera 6–13. Output of fillets 34–39%.

The mineral composition of meat (mg %): potassium 266, calcium 41, magnesium 28 and phosphorus 288.

TABLE 264. CHEMICAL COMPOSITION, %

Body parts	Moisture	Protein	Fat	Ash
Meat	78	19.1	1.8	1.1
Total waste	70	16.4	5.1	7.6

The meat of this fish is white and soft with a pleasant, slightly sweet taste. The soup is rich with a meaty taste. This is a good table fish which should be used for culinary products and semi-finished preparations.

Grouper—*Epinephelus areolatus.* Weight proportions of individual body parts of fish caught in February and March in the Indian Ocean 60–71 cm long, weighing 3.1–5.5 kg (%): trunk 60, head 31, viscera 7, including liver 1.2, fins 1.5, bones 4. Output of fillets 49%.

Chemical composition of meat (%): moisture 77.5–79.7, protein 19.0–20.8, fat 0.2–0.7.

After boiling the meat of this fish is tasteless; it has a specific, poorly expressed fish smell. This is a table fish which can be frozen or used for culinary items.

Giant sea bass—*Epinephelus gigas.* Caught with other fish in the Atlantic Ocean along the western coast of Africa, especially in the Gulf of Guinea and in the northwest Atlantic near the Patagonian shelf.

Attains a length of about 1 m and a weight of about 50 kg. Usual length 29–69 cm, maximum depth 10–23 cm, maximum thickness 4–10 cm, weight 0.56–6.83 kg.

Weight proportions of individual body parts (%): head 23–38, meat 35–42, skin 2.5–3.5, bones 5–10, scales 1.2–1.9, viscera 9–18, including liver 1.0–5.8, gonads 0.5–0.6, fins 2–3.8.

Chemical composition of meat (%): moisture 77–80, protein 17.4–19.9, fat 2–9, ash 0.8–1.6.

TABLE 265. CHEMICAL COMPOSITION, %

Body parts	Moisture	Protein	Fat	Ash
Head	60–69	15–16.6	2.5–16	7–11
Bones	53–63	16–18	4–22	8–14
Skin	64–69	27–31	1–4.5	1–5
Fins	62–63	18.5–19.4	1–8	9–19
Gonads	70–82	14.5–20	1–4.8	1.7–1.8
Viscera	67–81	14–17	2–17	1–1.8
Liver	53–71	12–19	7–29	1.1–1.8

Mineral composition of meat (mg %): potassium 194, calcium 24, magnesium 49, phosphorus 129, iron 4.

Fried meat is white, juicy, soft and tasty. A good ukha soup is made from this fish. The soup is tasty and fatty. This is an excellent table fish from which culinary preparations are made.

Honeycomb grouper—*Epinephelus merra.* Length of fish caught in January and March in the Indian Ocean, near the shelf of Al Wajh and the Persian Gulf, 63–69 cm, maximum depth 22 cm, maximum thickness 12 cm, weight 5.1–7.0 kg.

Weight proportions of individual body parts (%): head 37–39, viscera 7–10, fins 2.1–2.4, bones 6.7, scales 1.9–3.9. Output of fillets 38–42%.

TABLE 266. CHEMICAL COMPOSITION OF MEAT, %

Body parts	Moisture	Protein	Fat	Ash
Meat	76–78	19.4–20.8	0.2–3.2	1.2–1.3
Entire offal	63	19.1	6.6	9.3

Cooked, the meat of this fish is light, very dense and tasty. The soup is tasty. The fried meat is very good. This is a good table fish which should be used for culinary items and semi-finished products.

Bronze sea bass—*Epinephelus aeneus.* Caught in the Atlantic Ocean, along the coasts of northwest Africa at depths of 250 m. Attains a length of more than 1 m.

Weight proportions of individual body parts of fish 32 cm long, with depth 11 cm, thickness 5.5 cm and weight 0.6–0.8 kg, (%): head 23–30, trunk 60–70, viscera 7–10. Output of fillets 47–51%.

Chemical composition of meat (%): moisture 74.5–77.8, protein 18.4–20.9, fat 0.1–5.8, ash 1.2.

The meat of this fish is light colored and tasty. This is a good table fish from which first and second course dishes can be made.

Trout reef—Cod—*Epinephelus fario*. Length of fish caught in November in the Indian Ocean off the coast of Oman 28–38 cm, maximum depth 8–13 cm, maximum thickness 3–6 cm, weight 0.46–1.0 kg.

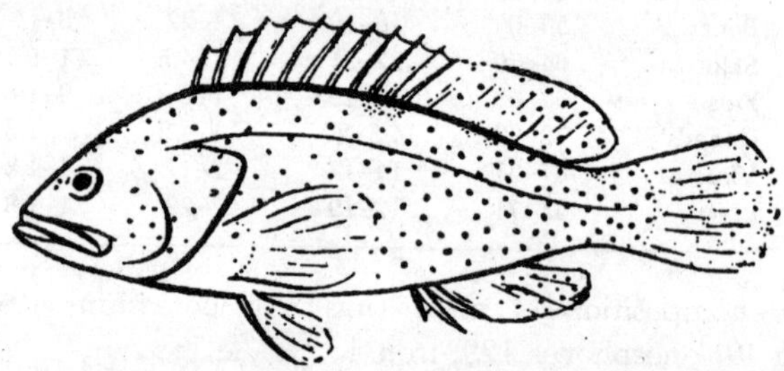

Fig. 216. Trout reef—Cod—*Epinephelus fario*.

Weight proportions of individual body parts (%): head 42–45, internal organs 3.5–5.2, fins 1.7–3.1, meat 34–42, bones 6.0–8.2, scales 2.0–2.9.

Chemical composition of meat (%): moisture 79.8, protein 18.3, fat 0.6, ash 1.2. Chemical composition of offal (%): moisture 68.3, protein 16.1, fat 3.2, ash 11.4.

The boiled meat is white, soft and tasty. The soup is tasty. This is a good table fish from which smoked products, culinary items and semi-finished products can be made.

Banded grouper—*Epinephelus fasciatus*. Commercially an unimportant

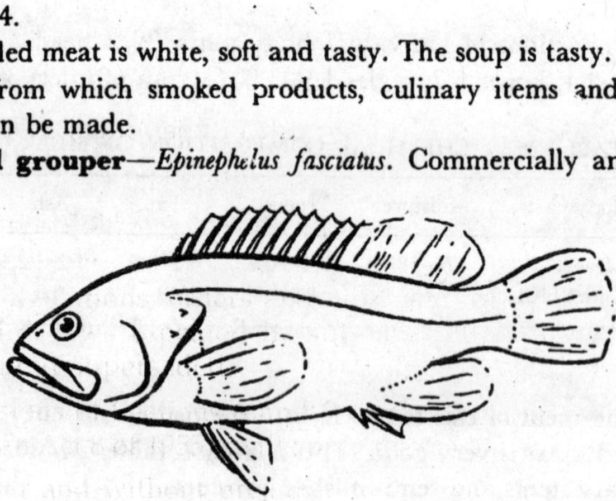

Fig. 217. Banded grouper—*Epinephelus fasciatus*.

fish, caught in the East China and South China Seas and near New Zealand. Length of fish caught in March–April 29–94 cm, weight 390–16,100 g. Weight proportions of individual body parts (%): head 16.2–38.2, viscera 3.5–11.5, including liver 0.5–2.0, fins and bones 8.0–15.1, meat with skin 56.0–66.2.

Chemical composition of meat (%): moisture 70.9–79.0, fat 0.2–6.7, protein 16.9–21.9, ash 1.3–1.8.

This fish has light colored meat which becomes dense after heat treatment; the meat is tasty in boiled and fried forms. In the winter months there is an accumulation of fat up to 5.7–6.7% in the meat. The soup prepared from the bony parts of the body is rich and gel-like.

This is a good table fish suitable for first and second course dishes.

Genus *Lateolabrax**

Japanese sea bass—*Lateolabrax japonicus*. Commercially an unexploited fish, inhabiting the Asian coast of the Pacific Ocean. Attains a length of 80 cm. Weight of the Japanese sea bass caught in October 150 g, length about 20 cm.

Fig. 218. Japanese sea bass—*Lateolabrax japonicus*.

Weight proportions of individual body parts (%): head 33.0, meat with skin 48.8, bones and fins 11.2, viscera 6.9.

Chemical composition of meat (%): moisture 79.3, protein 17.8, fat 1.8, ash 1.3.

This is a good table fish. It has dense white meat which is very tasty after boiling and frying.

Genus *Malacichthys**

Malacichthys—*Malacichthys griseus*. Commercially not very important, caught in the East China Sea. Length of fish caught in November–January 11–14 cm, weight 25–47 g.

Weight proportions of individual body parts (%): head 23.7, trunk 72.2, fins 0.3, viscera 3.0.

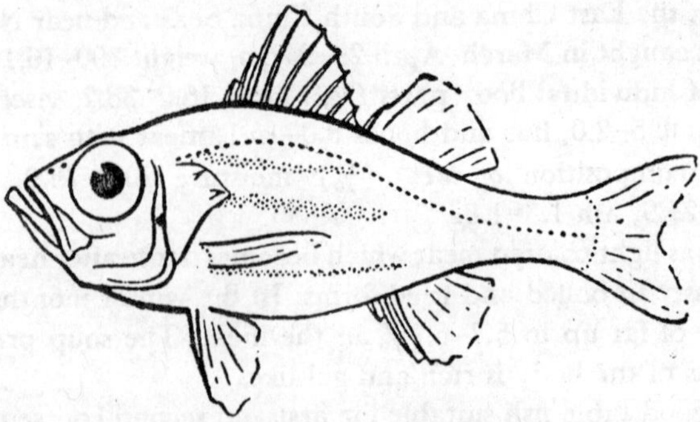

Fig. 219. Malacichthys—*Malacichthys griseus.*

Chemical composition of meat (%): moisture 79.1–79.6, protein 17.2–21.0, fat 1.0–1.5, ash 0.9–1.6.

The meat of this fish is bony and gray and has mediocre taste. It should be used for fish meal.

Genus *Mycteroperca*

Falcate grouper—*Mycteroperca falcata.* Weight proportions of individual body parts of fish caught in the Gulf of Mexico 71 cm long, with maximum depth 21 cm, maximum thickness 11 cm and weight 7 kg (%): trunk with fins 70, head 22, viscera 8. Output of fillets 55%.

Chemical composition of meat (%): moisture 75.3, protein 20.3, fat 3.2, ash 1.2.

Fried meat is juicy and tasty. This is a good table fish. It is used for fillets and culinary items.

Tusked grouper—*Mycteroperca rubra.* Found mixed in catches in the Central Atlantic.

Weight proportions of individual body parts of fish caught in September 60 cm long, weighing 10.2 kg (%): trunk with fins 69, head, 23, viscera 8. Output of fillets 42%.

Chemical composition of meat (%): moisture 77.1, protein 18.8, fat 2.9, ash 1.2.

After frying, the meat is white and tasty. This is a good table fish.

Genus *Niphon**

Spiny sander—*Niphon spinosus.* Commercially not very important, caught in the western Pacific Ocean and in the East China Sea. Length of fish caught in February 28.0–46.3, weight 240–1,420 g.

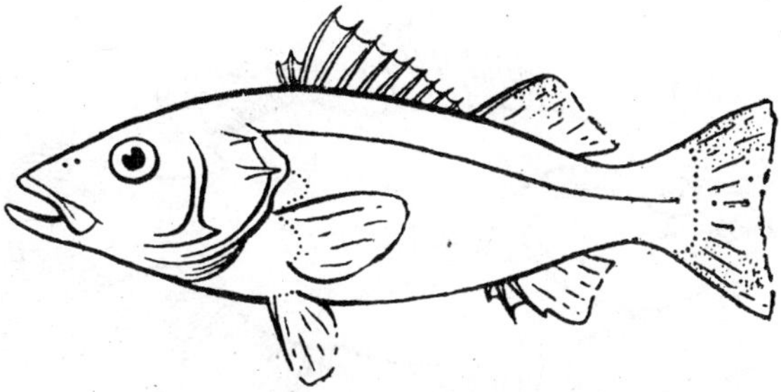

Fig. 220. Spiny sander—*Niphon spinosus.*

Weight proportions of individual body parts (%): head 28.6, meat with skin 52.7, viscera 6.3, including liver 1.3, bones and fins 9.8.

Chemical composition of meat (%): moisture 78.5, protein 19.1, fat 0.4, ash 1.8.

This is a fairly tasty fish; it is very good in boiled form. Fried, the meat is slightly hard. The bony parts of the body give a rich soup which forms a good gel. This is a good raw material for culinary preparations.

Genus *Polyprion*

Hapuka—*Polyprion oxygeneios.* Commercially not very important, caught in the shelf regions of the Indian Ocean. Length of fish caught in July–February 675–865 cm, weight 4,970–10,200 g.

Weight proportions of individual body parts (%): head 27.4–32.8, meat with skin 47.8–49.1, bones 9.5–11.4, fins 2.5–3.6, viscera 4.4–11.5, liver 0.7–2.0.

Chemical composition of meat (%): moisture 77.1–77.7, protein 19.5, fat 2.0–5.0, ash 1.2–1.3.

The meat has dense consistency in fried form, but it does not dry and it is pleasant in taste and white in color. Culinary products should be made from this fish.

Stone bass or Wreckfish—*Polyprion americanus.* Caught mixed with other fish from the eastern Central Atlantic.

Weight proportions of individual body parts of fish 71 cm long, with maximum depth 22 cm, maximum thickness 10 cm and weight 7.8 kg (%): trunk with fins 65, head 25, viscera 10%. Output of fillets 49%.

The meat of this fish in fried form is white, soft, fatty and very tasty. In taste this fish resembles the meat of fatty fresh water bream. This is one of the best table fish. Its smoked cured balyks have good taste qualities.

Fig. 221. Stone bass or Wreckfish—*Polyprion americanus.*

TABLE 267. CHEMICAL COMPOSITION, %

Body parts	Moisture	Protein	Fat	Ash
Meat	69.9	19.1	9.9	1.1
Viscera	64.0	—	6.7	—

Family SOLEIDAE—SOLES

These fish live mainly in equatorial and tropical waters, although some species occur deep in the northern region. They are typical shelf inhabitants and some may enter fresh water.

Lascar or French sole—*Solea lascaris.* Found along the European coast. Length up to 60 cm.

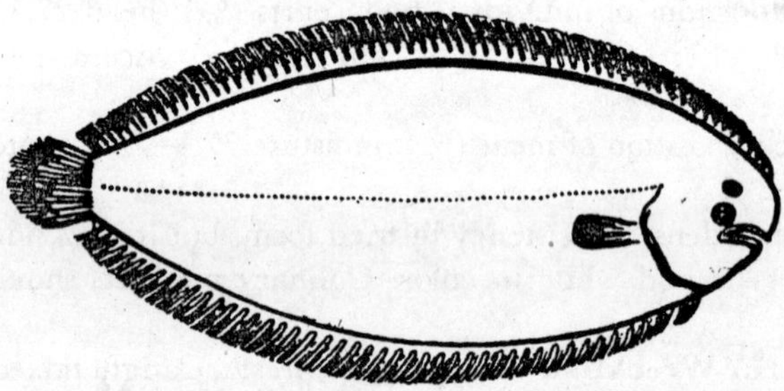

Fig. 222. Sole—*Solea lascaris.*

Weight proportions of individual body parts of fish 23.0–26.5 cm long, with maximum depth 8–10 cm, thickness 1.5–2.2 cm, head length 4 cm and weight 200–300 g (%): head 8–10, meat 60, viscera 4–7, skin 11, bones 11, fins 2.

Chemical composition of meat (%): moisture 78.6–80.9, protein 17.5–18.9, fat 0.1–0.4, ash 1.4–1.5. Mineral composition of meat (mg %): potassium 178, magnesium 41, iron 3.1.

The meat is soft and tasty. The fish should be used for culinary preparations and for canned products.

Sea sole—*Synaptura lusitanica*. Caught with other fish from the southeast Atlantic. Length 35–37 cm, maximum depth 12 cm, maximum thickness 2 cm, head length 8 cm, weight 350–390 g.

Weight proportions of individual body parts (%): trunk 78, head 14, viscera 7, fins 6. Output of fillets 48%.

Chemical composition of fish (%): moisture 80.8, protein 16.3, fat 1.4, ash 1.5.

The meat is white, soft and juicy. This is a good table fish.

Family SPARIDAE—PORGIES

This family includes more than 50 genera. They live mainly in the tropical and subtropical waters of all oceans and adjoining seas. The members of this family are called marine sparids.

Genus *Box*

Bogue—*Box boops*. Found in the Black Sea, Mediterranean Sea, along the southern coast of Italy and in the Adriatic Sea and in the Atlantic Ocean along the coasts of Africa from Casablanca to the Gulf of Guinea. Also found in the eastern Central Atlantic. Largest concentrations of this fish found from Senegal to Guinea and in the Gulf of Guinea. Usual length up to 20 cm.

Fig. 223. Bogue—*Box boops*.

The raw meat is dense and pinkish. The fried meat is white, somewhat sweet, but mildly fibrous with medium taste qualities. Because of the small size of this fish, it is used for fish meal. The meat of this fish could be used for mince and the offal could then be used for fish meal. It can also be used for canned products.

TABLE 268. SIZE PARAMETERS AND WEIGHT PROPORTIONS OF INDIVIDUAL BODY PARTS

Parameters	January	March	April
Length, cm	28	10–15	14–17
Weight, g	430	20–60	40–100
Weight proportions of individual body parts, %			
trunk	74	62	72–74
head	17	24	15–21
fins	1	2	—
viscera	8	12	7–11
output of fillets	—	—	55

TABLE 269. CHEMICAL COMPOSITION OF MEAT, %

Moisture	Protein	Fat	Ash
76.7	21.0	0.9	1.4
78.2	16.4	3.1	1.3
64–71	20.7–21.6	5.5–11.0	1.5

Genus *Boops**

Goldline or Salema—*Boops salpa.* Caught with other fish along the northwest African coast. Length of fish examined 33 cm and weight 0.76 kg.

Weight proportions of individual body parts (%): head 18, trunk 63, viscera 16, fins 3. Output of fillets 48%.

Chemical composition of meat (%): moisture 76.8, protein 21.5, fat 0.4, ash 1.3.

The meat of fried fish is white, juicy and soft. The taste qualities of this fish are good. It is a table fish.

Genus *Acanthopagrus*

Coarse scaled porgy—*Acanthopagrus latus.* Caught with other fish. In catches taken in September in the Arabian Sea, there was a predominance of fish about 35 cm long, with maximum depth 11.6 cm, maximum thickness 5.0 cm and weight around 950 g.

Weight proportions of individual body parts (%): head 32.0, meat 47.0, skin 2.1, scales 3.5, bones 8.8, fins 2.7, entire viscera 3.6, including testes 0.1 and liver 0.5.

Chemical composition of meat (%): moisture 77.7, protein 20.0, fat 1.0, ash 1.4.

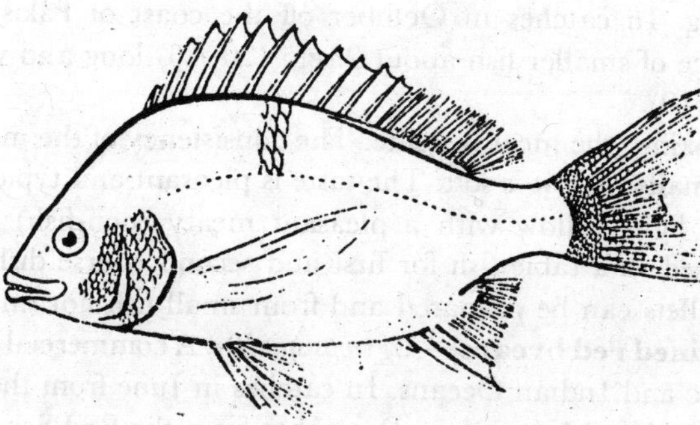

Fig. 224. Coarse scaled porgy—*Acanthopagrus latus.*

After cooking, the meat and broth are tasty with pleasant smell. This fish should be used for table purposes for first and second course dishes and hot smoked preparations.

Genus *Argyrops*

Filamented argyrops—*Argyrops filamentosus*. A commercial fish, widely distributed and attaining large size. In catches off the Travancore coasts (Indian Ocean) in May there is a predominance of large fish 45–47 cm long, with maximum depth 20 cm, maximum thickness 7 cm and weight around

TABLE 270. WEIGHT PROPORTIONS OF INDIVIDUAL BODY PARTS, %

Head	Inclusive of meat of occiput and nape	Trunk	Including meat with skin	Scales	Fins	Viscera
31.2–40.1 / 35.8	10	45.2–53.0 / 50.3	31.1–44.2 / 40.0	2.0–2.2 / 2.1	2.5–4.4 / 3.4	4.4–7.5 / 6.5

TABLE 271. CHEMICAL COMPOSITION, %

Body parts	Moisture	Protein	Fat	Ash
Meat	76.7–78.4 / 77.5	19.9–20.3 / 20.0	0.4–0.6 / 0.5	1.3–1.4 / 1.4
Inedible parts of body (entire offal)	61.5	19.5	3.9	13.5

2,700–3,500 g. In catches in October off the coast of Pakistan, there is a predominance of smaller fish about 23 cm (20–26) long and weighing about 400 g (270–610).

After cooking, the meat is white. The consistency of the meat in large fish is hard, in smaller fish it is soft. The taste is pleasant and typical of table fish. The soup is light yellow with a pleasant meaty (non-fish) taste. This fish should be used as a table fish for first and second course dishes. From large fish frozen fillets can be prepared and from small fish hot smoked products.

Long spined red bream—*Argyrops spinifer*. A commercial fish, distributed in the Pacific and Indian Oceans. In catches in June from the Gulf of Aden, in March–April and November–December from the Red Sea and in January from the western shelf of India, there was a predominance of individuals in two size groups: large fish averagely 47 cm (40–53) long and weighing around 2,400 g (1,300–3,375), maximum depth 17 cm, maximum thickness 8 cm and the small ones 26 cm (20–34) long and weighing about 600 g (300–1,000).

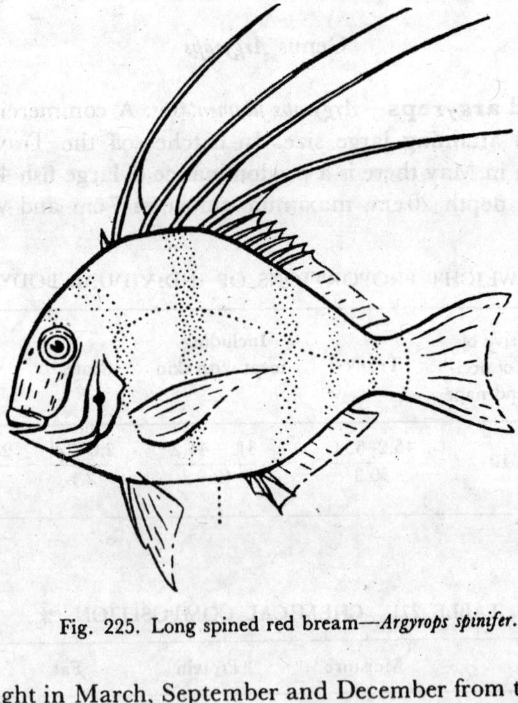

Fig. 225. Long spined red bream—*Argyrops spinifer*.

In fish caught in March, September and December from the Red Sea the meat was white after cooking; it disintegrated between the myosepta. In the meat and broth there was a trace of medicinal (iodine) flavor.

In fish caught in the Gulf of Aden and off the western shelf of India the medicinal flavor was not noticed and the taste qualities were better in the winter than in the summer. This fish should be used as a table fish for second

TABLE 272. WEIGHT PROPORTIONS OF INDIVIDUAL BODY PARTS, %

Head	Trunk	Including meat with skin	Scales	Fins	Viscera
$\frac{26.0–36.2}{28.8}$	$\frac{56.1–64.1}{59.5}$	48.2	$\frac{2.7–5.1}{3.3}$	$\frac{0.7–3.8}{2.1}$	$\frac{3.5–7.4}{5.6}$

TABLE 273. CHEMICAL COMPOSITION, %

Body parts	Moisture	Protein	Fat	Ash
Meat	$\frac{76.4–78.6}{77.6}$	$\frac{19.0–20.1}{19.6}$	$\frac{0.5–1.2}{0.8}$	$\frac{1.3–1.5}{1.4}$
Inedible parts of body (entire offal)	64.5	18.7	7.1	9.0

course dishes with the addition of condiments and sauces; hot smoked products can also be made from it.

Genus *Dentex*

Large eyed dogtooth—*Dentex macrophthalmus*. Encountered mostly in trawl catches. Distributed in the Atlantic Ocean along the coast of Africa and in the Mediterranean Sea. Also found in the eastern Central Atlantic, from the Straits of Gibraltar to Angola. Attains a length of 38 cm, usual length 16–24 cm.

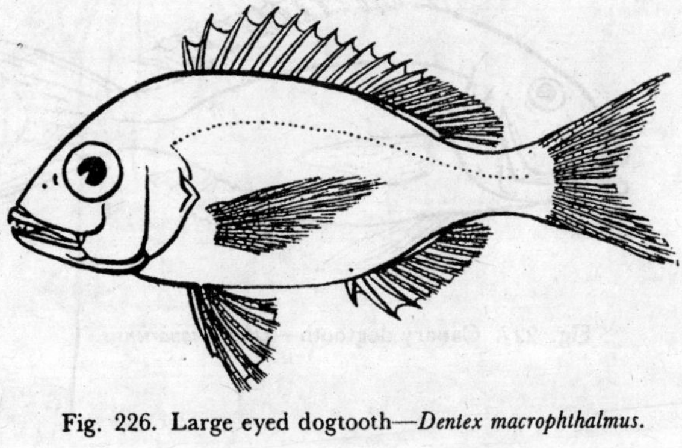

Fig. 226. Large eyed dogtooth—*Dentex macrophthalmus*.

TABLE 274. SIZE CHARACTERISTICS

Length, cm	Maximum depth, cm	Maximum thickness, cm	Weight, g
16	7	3	120–130
30–37	13–16	4.7–5.4	840–1,640

TABLE 275. WEIGHT PROPORTIONS OF INDIVIDUAL BODY PARTS, %

Trunk	Head	Viscera	Output of fillets
60	25	15	40
56	35	7	—

TABLE 276. CHEMICAL COMPOSITION, %

Body parts	Moisture	Protein	Fat	Ash
Meat	77.8–78.8	17.7–18.1	1.0–1.7	1.5–1.8
Head, bones, viscera	72.5	16.1	5.1	6.3

The meat is white and tasty. It is a good table fish.

Canary dogtooth—*Dentex canariensis*. Assumes commercial significance in catches in the Central Atlantic, along the northwest coast of Africa. Distributed in the Atlantic Ocean along the African coast and in the Mediterranean Sea. Length up to 60 cm, usual length 35–45 cm. Length of fish examined 24–33 cm, maximum depth 11–13 cm, maximum thickness

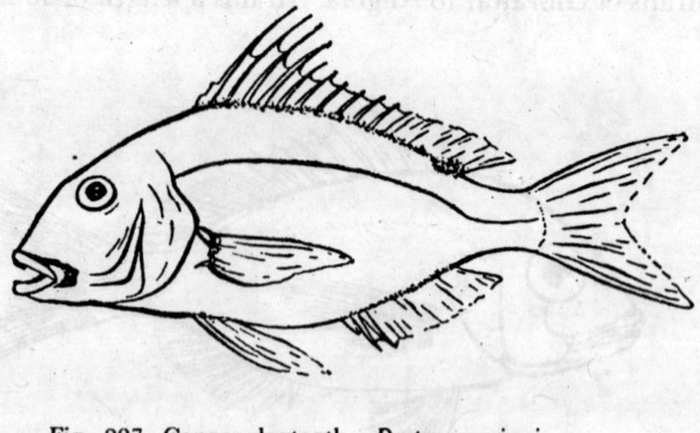

Fig. 227. Canary dogtooth—*Dentex canariensis*.

TABLE 277. CHEMICAL COMPOSITION, %

Body parts	Moisture	Protein	Fat	Ash
Meat	71.7–76.6	18.3–21.0	1.0–5.8	1.4–1.5
Head, bones viscera	72.5	16.1	5.1	6.3

4–5 cm, weight 880–1,240 g. Weight proportions of individual body parts (%): trunk with fins 60–67, head 25–27, viscera 8. Output of fillets 46%.

Mineral matter content in meat (mg%): potassium 174, calcium 25, magnesium 52, phosphorus 127, iron 1.9.

The meat is white and tasty. This is a good table fish.

Steep faced dogtooth—*Dentex filosus*. An important component in trawl catches in the eastern Central Atlantic. Distributed in the Atlantic Ocean along the African coast and in the Mediterranean Sea. One of the largest species among the marine sparids. Length of fish examined 56–58 cm, maximum depth 20–22 cm, maximum thickness 8 cm, weight 4,500 g.

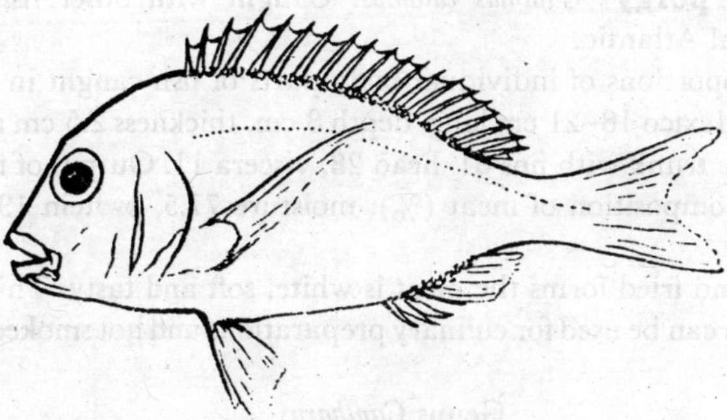

Fig. 228. Steep faced dogtooth—*Dentex filosus*.

Weight proportions of individual body parts (%): trunk 62–69, head 23, viscera 8–9. Output of fillets 47–52%.

Chemical composition of meat (%): moisture 71.4, protein 23.2, fat 3.9, ash 1.5. Mineral composition of meat (mg%): potassium 225, calcium 35, magnesium 37, phosphorus 146, iron 3.8.

The meat is white, tasty and slightly fibrous. This is a good table fish.

Moroccan dogtooth—*Dentex maroccanus*. Fish of great commercial significance, distributed in the Atlantic Ocean along the northwest coast of Africa. Length usually 16–20 cm, sometimes up to 25 cm.

Length of fish examined 15–17 cm, maximum width 7 cm, maximum height 3 cm, weight 100–150 g. Weight proportions of individual body parts (%): trunk with fins 67, head 22, viscera 11. Output of fillets 40%.

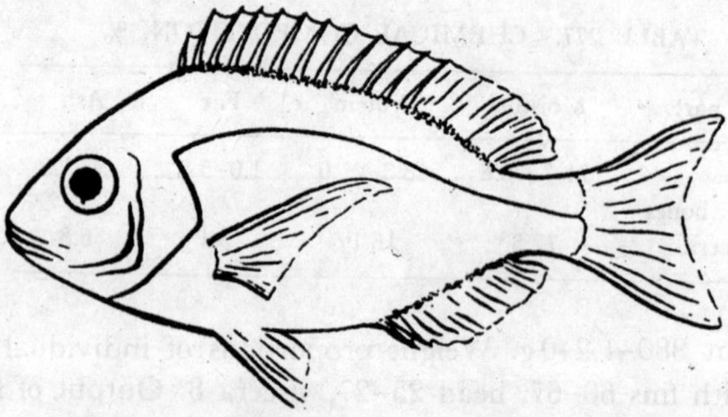

Fig. 229. Moroccan dogtooth—*Dentex maroccanus*.

Chemical composition of meat (%): moisture 73.7, protein 20.6, fat 4.3, ash 1.4.

The meat is white and tasty. It is a good table fish.

Genus *Calamus*

Saucereye porgy—*Calamus calamus*. Caught with other fish from the western Central Atlantic.

Weight proportions of individual body parts of fish caught in November in the Gulf of Mexico 18–21 cm long, depth 8 cm, thickness 2.5 cm and weight 180–200 g (%): trunk with fins 61, head 28, viscera 11. Output of fillets 39%.

Chemical composition of meat (%): moisture 77.5, protein 19.6, fat 1.3, ash 1.6.

In boiled and fried forms the meat is white, soft and tasty. This is a good table fish which can be used for culinary preparations and hot smoked products.

Genus *Cantharus*

Cantharus or Black Sea bream—*Cantharus cantharus*. Found in the eastern Central Atlantic. Usual length 20–40 cm, rarely more than 50 cm.

Weight proportions of individual body parts of fish caught along the northwest coast of Africa 34–37 cm long, depth 14–15 cm, thickness 5 cm and weighing 1.2–1.4 kg (%): trunk with fins 71, head 17, viscera 12. Output of fillets 50%.

Chemical composition of meat (%): moisture 73.5, protein 20.7, fat 4.6, ash 1.2.

The meat in fried and boiled forms is tasty and soft. There is both light and dark meat. The dark meat has a mild specific flavor. After the fish is boiled, the soup is fatty, tasty and rich. This is a good table fish which could be used for culinary products and hot smoked products.

Genus *Chrysoblephus*

Slinger—*Chrysoblephus puniceus*. A commercial fish distributed along the southern coast of Africa, from Cape Town to the northern part of the Mozambique Strait and near the southern and southeastern tip of Madagascar Island. In the Mozambique Strait it forms commercially exploitable concentrations. Length up to 82 cm. The September catches in the Arabian Sea had a predominance of individuals 40 cm long, with maximum depth 17.5 cm, maximum thickness 6.2 cm and weight around 1,500 g.

Weight proportions of individual body parts (%): trunk 60.5, head 29.0, meat 46.7, skin 2.5, scales 2.9, bones 11.3, fins 2.4, viscera 3.8, including liver 0.4.

Chemical composition of meat (%): moisture 76.7, protein 21.2, fat 1.2, ash 1.3.

After cooking, the meat and soup have pleasant taste and smell. The meat does not have small intramuscular bones. This fish should be used as a table fish for first and second course dishes and hot smoked products.

Genus *Gymnocranius*

Naked headed large eyed bream—*Gymnocranius griseus*. Caught with other fish. In the catches in August from the Persian Gulf there is a predominance of small fish around 20 cm long with maximum depth 10 cm, maximum thickness 3.7 cm and weight about 350 g.

After cooking, the meat is light gray and dense with the taste of a mediocre table fish. The soup is colorless, somewhat turbid and tasteless.

TABLE 278. WEIGHT PROPORTIONS OF INDIVIDUAL BODY PARTS, %

Head	Including meat of occiput and nape	Trunk	Including meat and skin	Bones	Scales	Fins	Viscera	Including roe
26.6–29.0 / 27.8	7.1	57.9–59.4 / 58.6	46.8–47.4 / 47.1	11.5	1.4	3.6	5.7	1.4

TABLE 279. CHEMICAL COMPOSITION, %

Body parts	Moisture	Protein	Fat	Ash
Meat	73.6	24.5	0.1	1.8
Inedible parts (waste after dressing)	64.3	20.7	4.7	8.9

This fish should be used as a table fish for second course dishes (with the addition of condiments and sauces) and for hot smoked products.

Genus *Leiognathus*

Golden striped ponyfish—*Leiognathus daurus*. Forms commercial concentrations. Widely distributed in the northwestern Indian Ocean, abundant in the shelf waters.

In catches in August in the Persian Gulf there is a predominance of individuals 6.4–7.4 cm long, with maximum depth 4 cm, maximum thickness 1 cm, weight 10–15 g.

Weight proportions of individual body parts (%): head 35.0, including meat of occiput and nape 5.0, meat with skin 42.0, bones 12.0, fins 2.5, viscera 7.0.

Chemical composition of meat (%): moisture 77.2, protein 19.6, fat 1.5, ash 1.5.

In cooked form the meat is white, tasty, sweet as in fresh water carp, but the fish are small with very fine bones. The soup is rich, sweetish-sour and tasty. This fish is best used for fish meal. In mechanized grading, larger fish weighing 15 g and more could be used for soup and sun cured products.

Genus *Chrysophrys*

Red Sea bream—*Chrysophrys major*. Commercially not very important, caught in the East China Sea, along the coast of Australia, in the East Korean Gulf and in the Sea of Japan. Length of fish caught in September–February 16–41 cm and weight 115–1600 g.

Weight proportions of individual body parts (%): head 24.3–34.2, bones 6.8–14.7, fins 1.5–4.2, scales 1.5–4.8, viscera 5.7–12.2, including liver 0.6–1.6. Output of fillets (with skin) 36.2–58.7%.

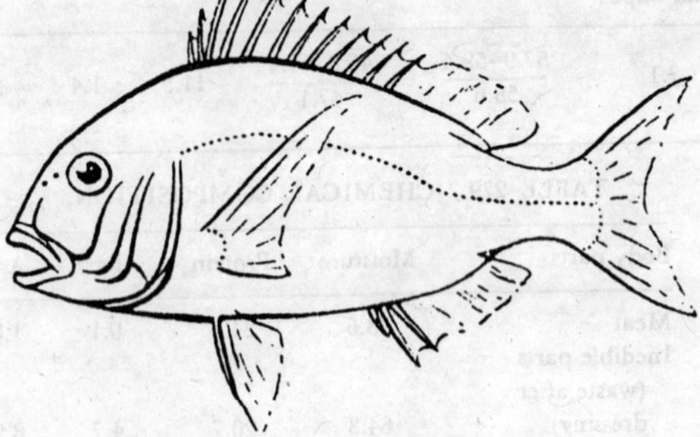

Fig. 230. Red Sea bream—*Chrysophrys major*.

Chemical composition of meat (%): moisture 72.7–81.8, protein 17.6–22.9, fat 0.2–3.1, ash 1.2–1.4.

The meat is white; it has excellent taste properties, especially in fried form. The ukha soup has good taste when prepared from the bones and head of the fish. This is a good table fish. It is best used for culinary preparations.

Genus *Petrus*

Red steenbras—*Petrus rupestris*. Caught with other fish along the coast of South Africa near estuaries. Attains a length of 180 cm and weight 70 kg. In the Gulf of Aden in June there is a predominance of fish 40 cm (31–49) long, weighing 1,600 g (715–2,600).

Weight proportions of individual body parts (%): head 31.9, meat 44.4, skin 3.6, bones 9.7, scales 2.9, fins 1.9, viscera 4.7, including liver 1.0.

TABLE 280. CHEMICAL COMPOSITION, %

Body parts	Moisture	Protein	Fat	Ash
Meat	77.6	20.8	2.1	1.4
Inedible parts (waste after dressing)	58.5	20.4	12.5	8.5

Boiled, the meat is white, somewhat hard and tasty. Fried, it is very tasty. The soup is rich and tasty. This is a good table fish. Products canned in oil have good taste qualities.

First and second course dishes can be made from this fish; it can be canned in oil.

Genus *Rhabdosargus*

White stumpnose—*Rhabdosargus globiceps*. A commercial fish, distributed in the southern Atlantic and Indian Oceans, along the South American coast; also found on the western coast of Australia and in the northeast along the Pacific coast.

The cooked meat is white, soft, tasty, slightly sweet (as in river crucian carp); however, it has fine bones. The soup is colorless, rich, turbid with a slightly sour taste. The fried meat is very tasty, light colored with soft consistency. This is a good table fish. Products canned in oil have good taste and appearance and soft consistency. In canned products prepared with tomato sauce, the meat and sauces are dark in color. This fish can be used for second course dishes, for fillets and conserves in oil.

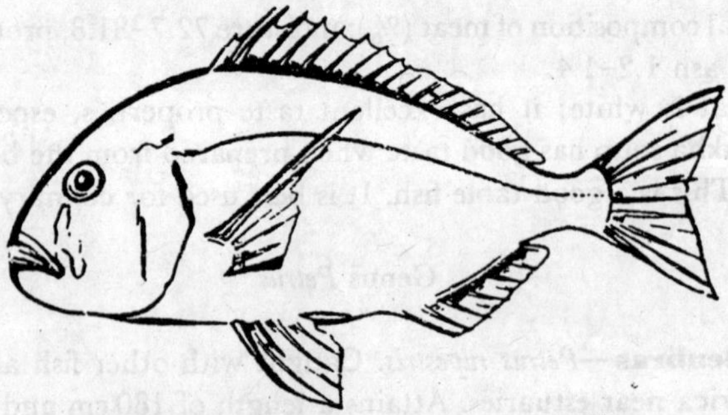

Fig. 231. White stumpnose—*Rhabdosargus globiceps.*

TABLE 281. SIZE CHARACTERISTICS OF WHITE STUMPNOSE CAUGHT IN DECEMBER IN THE ALGOA BAY (INDIAN OCEAN)

Length, cm	Maximum depth, cm	Maximum thickness, cm	Weight, g
32.0–33.5 / 34.0	12–14 / 13	3.2–4.0 / 3.7	920–1,200 / 1,000

TABLE 282. WEIGHT PROPORTIONS OF INDIVIDUAL BODY PARTS, %

Head	Meat with skin	Scales	Bones	Fins	Viscera	Including gonads
25.1–35.9 / 31.9	40.1–51.2 / 43.6	1.0–4.2 / 2.7	7.5–12.6 / 10.3	2.5–3.3 / 2.8	5.4–11.4 / 8.2	2.4

TABLE 283. CHEMICAL COMPOSITION OF MEAT, %

Body parts	Moisture	Protein	Fat	Ash
Meat	77.5	19.9	1.2	1.4
Inedible parts (waste after dressing)	67.1	16.7	7.5	8.7

Genus *Pagellus*

Porgy—*Pagellus natalensis*. Caught with other fish. Distributed in warm and temperate waters of all oceans.

In March and April in the Persian Gulf, on the shelf of Al Wajh and in the Saukara Bay (Indian Ocean), the length of the fish on an average is 18 cm (15–21), weight 180 g (100–310). At the same time, the length of fish in the Red Sea is about 32 cm and weight about 760 g.

Weight proportions of individual body parts of fish caught from different regions differ quite significantly (%): head on an average 29.9 (27.8–32.1), trunk 57.0 (42.3–61.7), meat with skin 47.4, scales 3.5, fins 2.0, viscera 5.4 (3.9–7.0).

TABLE 284. CHEMICAL COMPOSITION OF MEAT, %

Body parts	Moisture	Protein	Fat	Ash
Meat	76.4–78.1 / 77.2	19.2–21.1 / 20.8	0.5–0.9 / 0.7	1.4–1.5 / 1.5
Inedible parts (waste after dressing)	69.2–71.0 / 70.1	15.9–18.2 / 17.1	3.3–7.0 / 5.2	7.2

After cooking, the meat is gray. The consistency and taste properties differ in different regions: on the shelf of Al Wajh and in the Persian Gulf the meat is almost structureless and tasteless, the soup is sour and tasteless. The fish from these regions should be processed into fish meal. In sexually immature fish from the Saukara Bay and Red Sea the meat is dense and has a pleasant taste. The soup is pleasant and tasty. The fish from these regions should be used as table fish and for smoked products.

Canary bream—*Pagellus canariensis*. Caught with other fish. Inhabits tropical and subtropical waters of the East Atlantic.

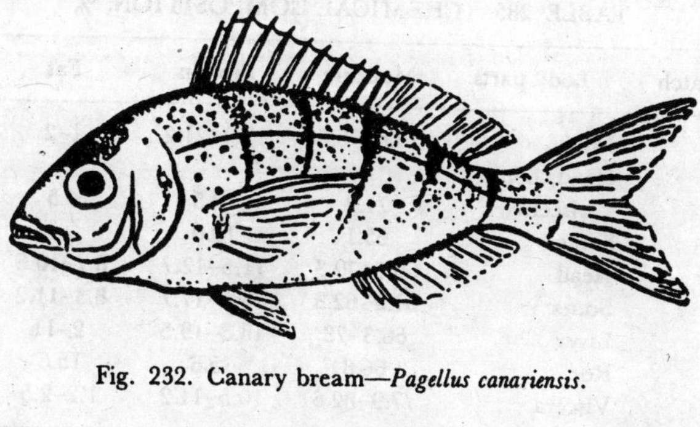

Fig. 232. Canary bream—*Pagellus canariensis*.

Length of fish caught in April along the coast of northwest Africa 15–23 cm, maximum depth 6–9 cm, maximum thickness 2.5–4.0 cm, weight 100–385 g. Length of fish caught in May 36 cm, maximum height 13 cm, maximum thickness 5 cm and weight 1,250 g.

Weight proportions of individual body parts of fish weighing 100–385 g (%): trunk 60–62, head 29–35, viscera 5–9. Output of fillets 40%.

Chemical composition of meat (%): moisture 71.5–73.9, protein 19.4–22.5, fat 3.7–4.7, ash 1.3–2.0.

The meat in fried form is white, juicy and tasty. This is a good table fish from which culinary preparations are made.

Gunner—*Pagellus centrodentus*. Commercial significance of this species not very great. Length 20–40 cm, sometimes up to 65 cm. Caught from La Manche to Morocco. Length of fish caught in May 25–36 cm, maximum depth 9–14 cm, maximum thickness 4–6 cm, weight 330–1,040 g.

Fig. 233. Gunner—*Pagellus centrodentus*.

Weight proportions of individual body parts of fish weighing 330–1,040 g (%): trunk with fins 66–69, head 20–27, viscera 7–10. Output of fillets 54%.

After further dressing of the fish, the proportions of different body parts are (%): meat 50.6, skin 2.9, bones 8.9, scales 3.5, roe 0.3–0.9, testes 0.2, liver 1.1–4.0.

TABLE 285. CHEMICAL COMPOSITION, %

Month of catch	Body parts	Moisture	Protein	Fat	Ash
May	Meat	75.6–78.6	17.0–21.2	1–2	1.2–1.5
June	Head, bones, viscera	71.2	17.5	4.5	6.8
May	Same	65.1	19.0	9.9	6.0
August	Head	69.6–70.3	11.5–12.7	8.2–10.6	6.6
August	Bones	61.2–62.8	16.1–17.7	8.3–11.2	9.9–10.2
August	Liver	66.3–73	18.8–19.5	2–11	1.8–1.9
August	Roe	66.8	15.6	15.6	2
August	Viscera	77.9–82.6	10.5–11.2	1.2–2.5	1.7–4.2

The meat of this fish in fried form is white, juicy and tasty. Fish weighing more than 0.5 kg are used for culinary products. After dressing the fat content of the waste—head and bones—is about 10%; this waste can be used to prepare ukha soup.

Axillary bream—*Pagellus acarne.* Commercially not an important fish. Found along the coast of northwest Africa. Length of fish caught there in

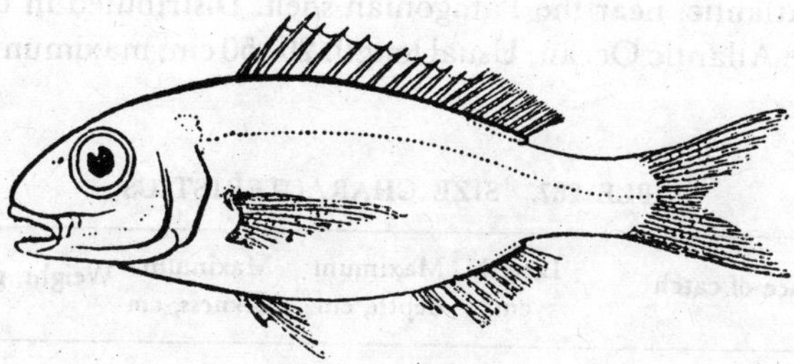

Fig. 234. Axillary bream—*Pagellus acarne.*

May 27–29 cm, maximum depth 10–11 cm, maximum thickness 4–5 cm, weight 500–880 g.

Weight proportions of individual body parts (%): trunk with fins 66, head 25, viscera 9. Output of fillets 40%.

The meat of this bream in fried form is white, juicy and tasty. Ukha soup can be prepared from the head and bones. Culinary preparations are made from the meat of the fish.

TABLE 286. CHEMICAL COMPOSITION OF MEAT, %

Body parts	Moisture	Protein	Fat	Ash
Meat	75.0	21.3	2.4	1.3
Head, bones, viscera	71.2	17.5	4.5	6.8

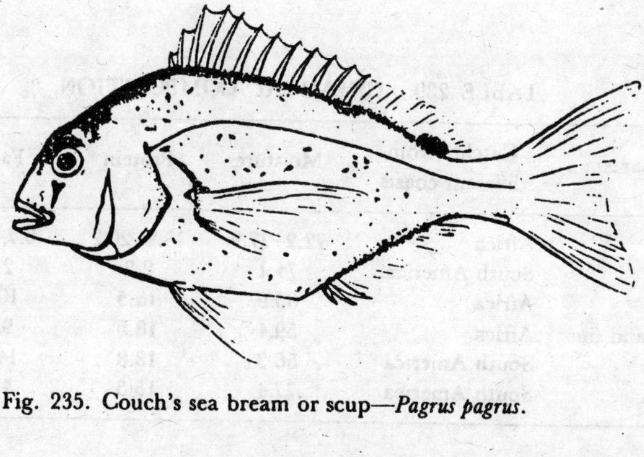

Fig. 235. Couch's sea bream or scup—*Pagrus pagrus.*

Genus *Pagrus*

Couch's sea bream or scup—*Pagrus pagrus.* Commercially important along the coasts of northwest Africa, caught in smaller quantities in the southwest Atlantic, near the Patagonian shelf. Distributed in the subtropical waters of the Atlantic Ocean. Usual length 20–50 cm, maximum length 65 cm.

TABLE 287. SIZE CHARACTERISTICS

Place of catch	Length, cm	Maximum depth, cm	Maximum thickness, cm	Weight, g
Northwest coast of Africa	17–31	7–12	2.5–5	120–705
Patagonian shelf	24–33	10–13	4–6	460–1,200

The meat of this fish is white, soft and juicy. This is a table fish. It is marketed in frozen form under the name "Atlantic scup".

TABLE 288. WEIGHT PROPORTIONS OF INDIVIDUAL BODY PARTS, %

Place of catch	Head	Bones	Fins	Scales	Viscera		Output of fillets
					whole	including liver	
Northwest coast of Africa	25–36	6–10	1.4–3.0	1.3–1.9	4–12	3–8	49–54
Patagonian shelf	25	—	—	—	16	—	—

TABLE 289. CHEMICAL COMPOSITION, %

Body parts	Catches from different coasts	Moisture	Protein	Fat	Ash
Meat	Africa	72.2–77.9	18.9–20.2	0.7–6.3	1.2–1.7
Meat	South America	74.1	2.2	2.5	1.4
Head	Africa	65.9	16.5	10.1	—
Bones and fins	Africa	59.4	18.5	9.7	9.7
-do-	South America	56.3	18.8	14.3	10.6
Viscera	South America	77.4	15.5	4.4	2.8

Genus *Sparus*

Humped porgy—*Sparus gibbiceps*. A commercial fish, caught in the Arabian Sea and the Gulf of Aden.

Chemical composition of meat (%): moisture 76.6, protein 20.5, fat 1.3, ash 1.4.

After cooking, the meat is light colored and dense in consistency. The meat and soup have a pleasant taste. The meat does not have small intramuscular bones.

TABLE 290. SIZE CHARACTERISTICS OF HUMPED PORGY CAUGHT IN AUGUST IN THE GULF OF ADEN AND IN SEPTEMBER IN THE ARABIAN SEA

Length, cm	Maximum depth, cm	Maximum thickness, cm	Weight, g
33.3–43.5 / 37.0	12.5–20.2 / 15.3	4.2–6.3 / 5.1	910–2,507 / 1,400

TABLE 291. WEIGHT PROPORTIONS OF INDIVIDUAL BODY PARTS OF HUMPED PORGY, %

Head	Trunk	Including meat	Scales	Fins	Viscera
30.5–37.2 / 34.0	51.6–60.1 / 55.0	41.9–48.5 / 44.8	2.0–3.2 / 2.7	2.1–3.6 / 2.8	3.2–9.4 / 4.5

This fish should be used as a table fish for first and second course dishes.

Porgy—*Sparus speciosus*. A commercial fish. Catches taken in January in the shelf zone of the Arabian Sea (Indian Ocean) have a predominance of fish about 28 cm (21–35) long, weighing about 600 g (260–1,060).

TABLE 292. WEIGHT PROPORTIONS OF INDIVIDUAL BODY PARTS OF SPARUS SPECIOSUS, %

Head	Trunk	Meat	Bones	Viscera	Scales	Fins
30.1–32.5 / 31.0	50.6–52.3 / 51.6	36.2–40.5–45.3* / 40.6	6.8–14.4 / 9.3	9.1–10.0–12.2** / 10.4	3.3–4.2 / 3.9	2.0–4.2 / 3.1

* 36.2 in large, 40.5 in medium and 45.3 in small fish.

** 9.1 in small, 10.0 in medium and 12.2 in large fish.

TABLE 293. CHEMICAL COMPOSITION, %

Body parts	Moisture	Protein	Fat	Ash
Meat	76.5–80.0 / 77.9	18.1–21.3 / 19.8	0.2–0.6 / 0.4	1.3–1.8 / 1.5
Viscera	76.1	not determined	0.1	1.3

The meat has many small intramuscular bones. After cooking, the meat resembles the meat of fresh water fish in taste. This fish should be used for table purposes.

Panga—*Sparus laniarius*. A widely distributed commercial fish. Length of fish caught in December in Algoa Bay and in February on the Agulhas shelf in the Indian Ocean on an average about 28 cm (21.6–31.2), maximum depth 11.4 cm, maximum thickness 3.2 cm, average weight about 600 g (300–830).

Fig. 236. Panga—*Sparus laniarius*.

TABLE 294. WEIGHT PROPORTIONS OF INDIVIDUAL BODY PARTS, %

Head	Meat and skin	Scales	Bones	Fins	Viscera	Including gonads
28.8–33.9 / 31.8	42.2	3.9–6.1 / 4.9	8.5–12.9 / 11.2	2.0–3.0 / 2.7	4.5–9.2 / 6.4	1.1

TABLE 295. CHEMICAL COMPOSITION, %

Body parts	Moisture	Protein	Fat	Ash
Meat	77.3–77.4 / 77.4	19.6–20.5 / 20.1	0.8–1.1 / 1.0	1.3–1.4 / 1.4
Inedible parts (offal after dressing)	65.3–67.1 / 66.2	17.8–17.9 / 17.8	4.8–5.2 / 5.0	9.9–11.0 / 10.8

After cooking, the meat is white and tasty, ranging from soft to dense in consistency. The soup is lemon colored with a pleasant taste; however, there is a mild taste of silt.

This fish should be used for table purposes for second course dishes and hot smoked products.

Long spined red bream—*Sparus spinifer.* A commercial fish. Widely distributed in the warm and temperate waters of all oceans. In catches in December in the Gulf of Aden there is a predominance of individuals about 20 cm (17–26) long, weighing about 360 g (140–588).

TABLE 296. WEIGHT PROPORTIONS OF INDIVIDUAL BODY PARTS, %

Head	Trunk	Scales	Fins	Viscera
$\frac{27.6–36.2}{33.5}$	$\frac{49.5–58.2}{54.7}$	$\frac{4.1–7.1}{5.1}$	$\frac{1.3–2.1}{1.7}$	$\frac{3.3–5.7}{4.9}$

Chemical composition of meat (%): moisture 74.6, protein 22.8, fat 0.3, ash 1.7.

Cooked meat is grayish, juicy, with the good taste typical of table fish. This fish should be used for second course dishes and for hot smoked products.

Silver bream—*Sparus sarba.* Found in the coastal regions of the Indian Ocean, in the Red Sea and in the western Pacific Ocean. A commercial fish. In catches from the Red Sea in December the predominant fish was 23 cm long and weighed 320 g and in the western Pacific Ocean the corresponding values were 30–41 cm and 875–1,700 g.

The meat is white and dense with a pleasant taste resembling that of chicken. The soup is light colored, transparent, with droplets of fat and slightly acid. This fish should be used for table purposes and for hot smoked products.

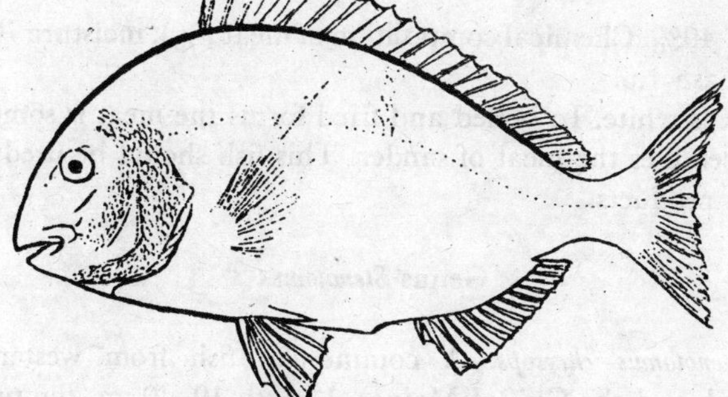

Fig. 237. Silver bream—*Sparus sarba.*

TABLE 297. WEIGHT PROPORTIONS OF INDIVIDUAL BODY PARTS, %

Place of catch	Head	Trunk	Scales	Bones	Fins	Viscera		Output of fillets (with skin)
						whole	including liver	
Red Sea	33.7	52.5	4.3	—	1.7	5.3	—	—
Western Pacific Ocean	24.3	—	—	11.2	11.2	4.3	1.4	58.7

TABLE 298. CHEMICAL COMPOSITION OF MEAT, %

Place of catch	Moisture	Protein	Fat	Ash
Red Sea	73.8	19.0	4.0	1.4
Western Pacific Ocean	73.8	22.5	2.8	1.1

Gilt head bream—*Sparus aurata*. Inhabits the eastern Central Atlantic. Output of trunks from fish 20–50 cm long, weighing 0.5–1.5 kg is 56–60%

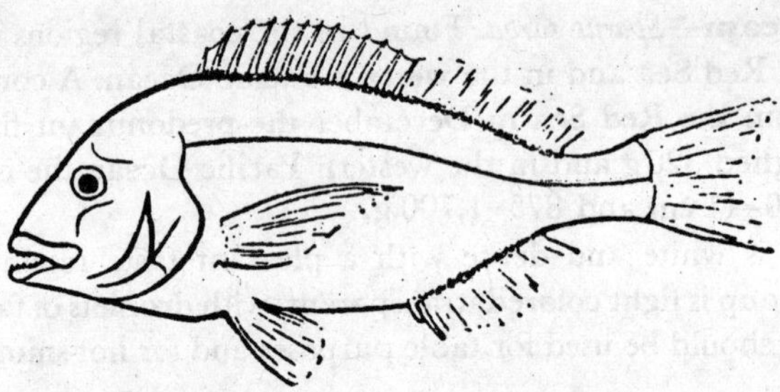

Fig. 238. Gilt head bream—*Sparus aurata*.

and fillets 42–49%. Chemical composition of meat (%): moisture 74.9, protein 19.7, fat 1.9, ash 1.6.

The meat is white. In boiled and fried forms the meat is somewhat dry; in taste it resembles the meat of sander. This fish should be used to prepare semi-finished products.

Genus *Stenotomus*

Scup—*Stenotomus chrysops*. A commercial fish from western Atlantic (Georges Bank and the Gulf of Maine). Length 10–40 cm, the predominant length 20–30 cm, weight 100–800 g.

TABLE 299. SIZE CHARACTERISTICS OF SCUP CAUGHT IN OCTOBER NEAR DELOVER

Length, cm	Weight, g	Number of fish in catch, %
15–20	100–140	7
21–24	240–420	33
25–28	490–760	37
29–32	830–1000	19
33–37	1200–1500	4

TABLE 300. CHEMICAL COMPOSITION, %

Body parts	Moisture	Protein	Fat	Ash
Meat	67.1–71.4	18–20.6	7.4–12.5	1.0–1.6
Head	53.9–63	13–14.5	13.4–21.8	6.2–8.3
Bones	49.4–58	13.7–18.3	15.0–26.5	6.3–10
Liver	46.2–59.9	13.8–16.6	21.6–37.2	1.1–1.4
Roe	80.8	15.1	1.4–1.5	2.0
Testes	76.1	18.2	3.1	2.3
Viscera	68.7–75.2	13.6–16.0	6.9–16.7	1.5–2.6

Weight proportions of individual body parts of scup 18–30 cm long, weighing 170–810 g (%): head 17–23, viscera 4–10, bones 11–12, fins 2.5–4.4, roe 1.7–2.2, testes 1.2, liver 2.6–3.1. Output of fillets 52–58%.

This is a good table fish. It can be used for canning, smoked products and culinary preparations.

Family SPHYRAENIDAE—BARRACUDAS

This family is represented by one genus *Sphyraena* and includes marine predatory fish widely distributed in equatorial and tropical waters. Some species attain fairly large size. The barracudas usually remain in small shoals.

Blunt jawed sea pike—Great Barracuda—*Sphyraena obtusata**. Commercially not very important, inhabiting the northwest Indian Ocean. Length of fish caught in May 54–89 cm, weight 1,000–5,600 g.

Weight proportions of individual body parts (%): head 16.0–19.0, trunk 68.0–72.0, viscera 5.5–11.0.

Chemical composition of meat (%): moisture 74.0–75.0, protein 21.2–24.2, fat 0.4–2.0, ash 1.4–1.6.

After cooking, the meat is light colored and somewhat hard with mediocre taste and strange medicinal taste. The broth is slightly bitter. According to reports in literature, the liver and gonads of this fish are poisonous. Its evisceration for fillets should therefore be done immediately after catching.

Great barracuda—*Sphyraena barracuda*. Commercially not very important, inhabiting the Central Atlantic. Length of fish caught in April, 37–40 cm and weight 330–495 g.

Fig. 239. Great barracuda—*Sphyraena barracuda*.

Weight proportions of individual body parts (%): head 13.2, trunk 67.5, fins 3.3, viscera 13.2. Output of fillets without skin 55.7%.

Chemical composition of meat (%): moisture 78.0, protein 19.5, fat 1.2, ash 1.3.

Culinary products can be prepared from this barracuda; they should be frozen immediately after catching.

Boreal barracuda or northern sennet—*Sphyraena borealis*. Also commercially not very important, inhabits the Atlantic Ocean.

Weight proportions of individual body parts of boreal barracuda 3.8–13.0 kg (%): head 11.5–19.5, trunk 72.6–80.0, viscera 7.7–8.1. Weight proportions of individual body parts of small boreal barracuda (%): head 13.5, trunk with fins 81.9, viscera 4.6.

Chemical composition of meat of small boreal barracuda (%): moisture 76.0, protein 21.5, fat 1.5, ash 1.0.

Culinary products (fried fish and cutlets) should be prepared from the boreal barracuda.

Guaguanche—*Sphyraena guachancho*. Commercially of little importance. Distributed in the Atlantic Ocean, South America. Length of fish caught in May–June 29–37 cm, weight 150–395 g.

Weight proportions of individual body parts (%): head 18.7, trunk 68.2, fins 4.5, viscera 5.7. Output of fillets (without skin) 58%.

Chemical composition of meat (%): moisture 76.9, protein 20.6, fat 1.0, ash 1.5.

Information about processing of this fish is not available.

Red barracuda—*Sphyraena pinguis*. Commercially of little importance, caught in the East China Sea and near Australia and New Zealand. Length of fish caught in February–May 33–34 cm and weight 220–230 g.

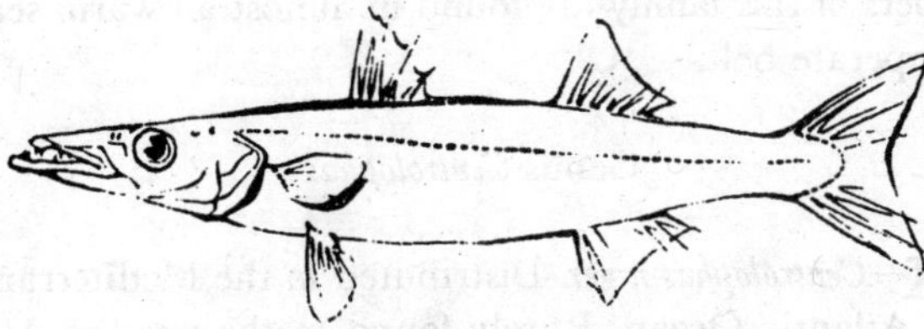

Fig. 240. Red barracuda—*Sphyraena pinguis.*

Weight proportions of individual body parts (%): head 14.4–22.1, meat with skin 60.6–67.0, viscera 6.2–6.4, bones and fins 7.8–12.4. Chemical composition of meat (%): moisture 64.0–78.2, protein 18.5–20.3, fat 0.4–16.1, ash 1.4–1.7.

The meat is juicy, soft, fatty and quite tasty. This is a good table fish which should be used for culinary preparations.

Short jawed sea pike—*Sphyraena langsar**. Commercially not very important, distributed in the Indian Ocean, near northern Australia. Length of fish caught in September 20–23 cm, weight 102–145 g.

Weight proportions of individual body parts (%): head 24.4, trunk 64.6, fins 3.5, viscera 3.9. The meat in boiled form is white and soft.

Data on processing of this fish are not available.

Banded or Indian barracuda—*Sphyraena jello.* Commercially not an important fish, distributed in the Indian Ocean. Length of fish caught in September–December 33.5–85.0 cm, weight 290–3,800 g.

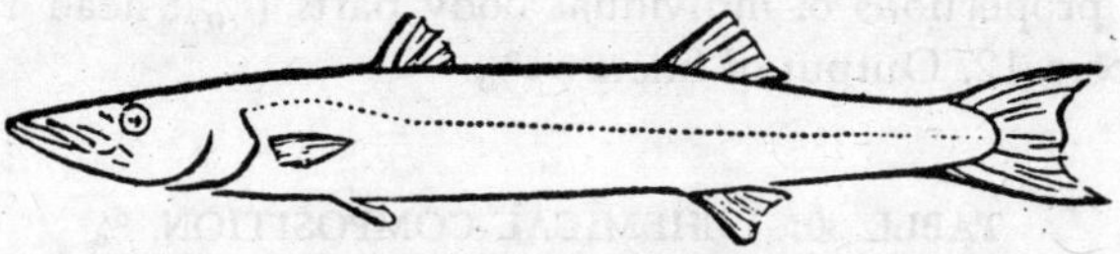

Fig. 241. Banded barracuda—*Sphyraena jello.*

Weight proportions of individual body parts (%): head 16.2–20.3, trunk 71.0–74.0, fins 1.3–2.5, viscera 6.2–8.6, liver 1.3, skin 4.0–6.8. Output of fillets 55.8–67.2%.

Chemical composition of meat (%): moisture 72.4–73.8, protein 22.0, fat 2.0–2.3, ash 1.4–1.6.

After boiling the meat is grayish and not very tasty. It is tastier after frying than in boiled form. It should be used for culinary products.

Family STROMATEIDAE—BUTTERFISHES

There are about 15 genera and many species in this family, of which some are widely distributed in all oceans, while others are localized forms.

The members of this family are found in almost all warm seas and in the seas of the temperate belt.

Genus *Centrolophus*

Black ruff—*Centrolophus niger*. Distributed in the Mediterranean Sea and in the eastern Atlantic Ocean. Rarely found in the western Atlantic, along the North American coast. Length of fish caught from the northern part of the area of distribution (in the Mediterranean Sea and along the Atlantic Coast of Europe) up to 50 cm and that from the southern part (along the Atlantic coast of South Africa) 120 cm. Usual length 90 cm.

Length of fish caught in December in the southeast Atlantic 46–70 cm, length of head 9–12 cm, maximum depth 11–17 cm, maximum thickness 5–8 cm, weight 1.5–3.4 kg.

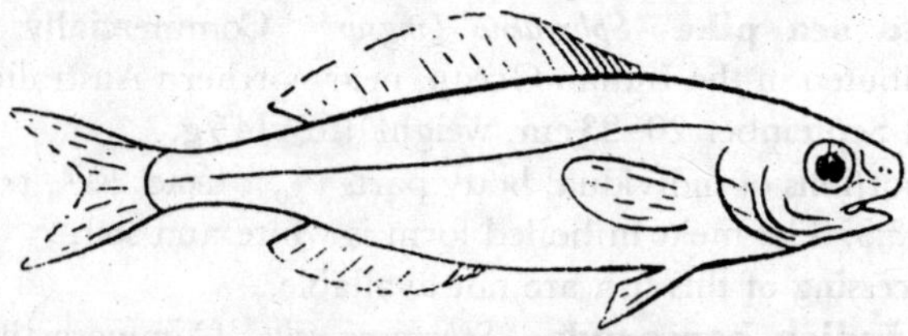

Fig. 242. Black ruff—*Centrolophus niger*.

Weight proportions of individual body parts (%): head 12, trunk with fins 75, viscera 12. Output of fillets 54%.

TABLE 301. CHEMICAL COMPOSITION, %

Body parts	Moisture	Protein	Fat	Ash
Meat	80.8	11.7	6.8	0.8
Viscera	82.4	9.7	7.3	0.6

The meat of this fish in boiled and fried forms is white, soft, slightly watery, sweet and tasty. The soup is tasty. This is a good table fish.

Genus *Cubiceps*

Natal butterfish—*Cubiceps natalensis*. Species of little commercial value, inhabiting the Indian Ocean. Length of fish caught in March 14.0–15.8 cm and weight 51–92 g.

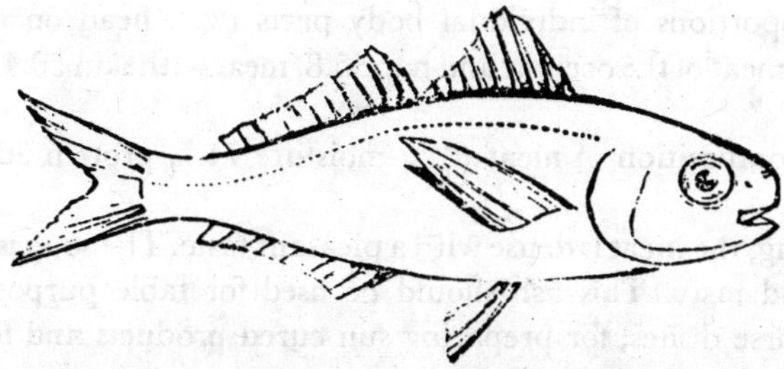

Fig. 243. Natal butterfish—*Cubiceps natalensis*.

Weight proportions of individual body parts (%): head 29.2, fins 1.7, scales 1.2, viscera 10.7, bones 7.0. Output of fillets with skin 47.7%.

Chemical composition of meat (%): moisture 70.2, protein 19.6, fat 8.8, ash 1.4.

The meat of this fish in cooked form is light colored, dense and tasty. The soup is yellow and tasty. This fish should be used for preparing sun cured products and culinary preparations. It can also be used for canned products such as "blanched fish in oil".

Indian butterfish—*Cubiceps indicus*. Caught with other fish. Widely distributed in the shelf waters of the Indian Ocean.

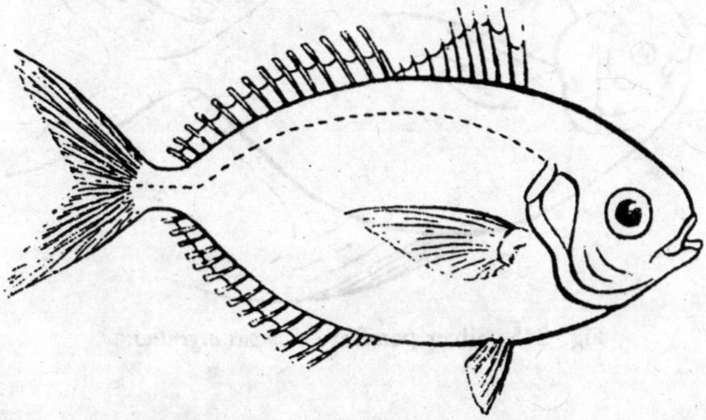

Fig. 244. Indian butterfish—*Cubiceps indicus*.

TABLE 302. SIZE CHARACTERISTICS OF INDIAN BUTTERFISH CAUGHT IN MAY AT THE AGULHAS SHELF (INDIAN OCEAN)

Length, cm	Maximum depth, cm	Maximum thickness, cm	Weight, g
12.8–13.7 / 13.0	5.8–6.7 / 6.2	1.7–2.2 / 1.9	71–100 / 85

Weight proportions of individual body parts (%): head on an average 27.8, including meat of the occiput and nape 5.8, meat with skin 50.4, bones 8.8, fins 9.9.

Chemical composition of meat (%): moisture 74.7, protein 20.9, fat 2.8, ash 1.5.

After cooking, the meat is dense with a pleasant taste. The soup is somewhat turbid, rich and tasty. This fish should be used for table purposes for first and second course dishes, for preparing sun cured products and for canning in oil.

Genus *Pampus*

Silver pomfret—*Pampus argenteus*. Lives in the East China and South China Seas and in the Indian Ocean. Length of individuals caught in March–November 15.5–32.0 cm, weight 165–1,240 g.

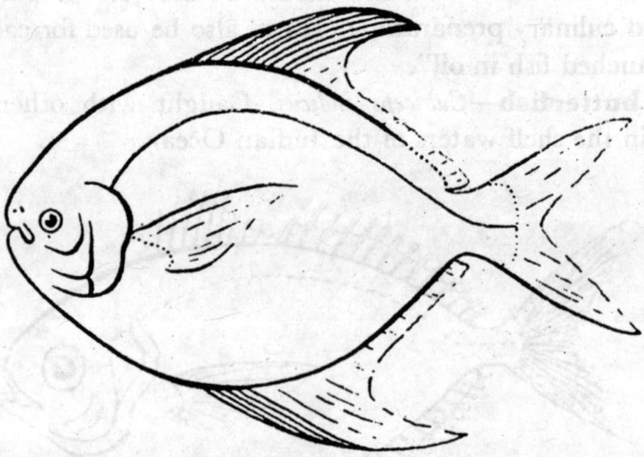

Fig. 245. Silver pomfret—*Pampus argenteus*.

Weight proportions of individual body parts (%): head 8.1–21.4, trunk 64.6–82.3, viscera 4.5–10.4, bones 4.5–10.9, fins 2.7–3.5. Output of fillets 56.6–71.4%.

Chemical composition of meat (%): moisture 69.5–79.6, protein 17.7–20.0, fat 0.3–10.5, ash 1.1–1.6.

The meat is soft, juicy and white with a pink tinge. Culinary preparations of this fish are high grade. The silver pomfret does not have a tendency for fragmentation of the meat and there is no visible separation along the myosepta in raw form or after heat processing. The soup is lemon colored, very tasty, resembling chicken broth in taste. The silver pomfret is an excellent table fish. Culinary products can be made from this fish.

Chinese or white pomfret—*Pampus chinensis*. Caught in the Red Sea.

Length 39.4–42.8 cm, weight 3,450–4,660 g. Weight proportions of individual body parts (%): head 27.3–32.8, trunk 51.5–57.6, scales 1.2, fins 2.5–3.4, viscera 9.3–11.2.

Chemical composition (%): moisture 76.0, protein 20.9, fat 1.6, ash 1.2.

After cooking, the meat is cream colored, soft, with juicy consistency, very pleasant taste resembling chicken meat. The soup has a chicken-like smell. This is an excellent table fish This fish should be used for culinary preparations.

Genus *Parastromateus*

Black pomfret—*Parastromateus nigre.*** Weight of black pomfret caught in February in the Indian Ocean 518–854 g.

Weight proportions of individual body parts (%): head 17.7, viscera 4.5, bones 13.1, fins 4.1, scales 2.8. Output of fillets 56.3%.

Chemical composition of meat (%): moisture 75.2–78.2, protein 18.9–20.1, fat 0.5–1.9.

The black pomfret should be used for canning and for culinary preparations.

Genus *Poronotus*

American butterfish—*Poronotus triacanthus.* Distributed along the Atlantic coast of North America, from Georges Bank to the Bahama Islands. Length 13–30 cm, average weight 200 g.

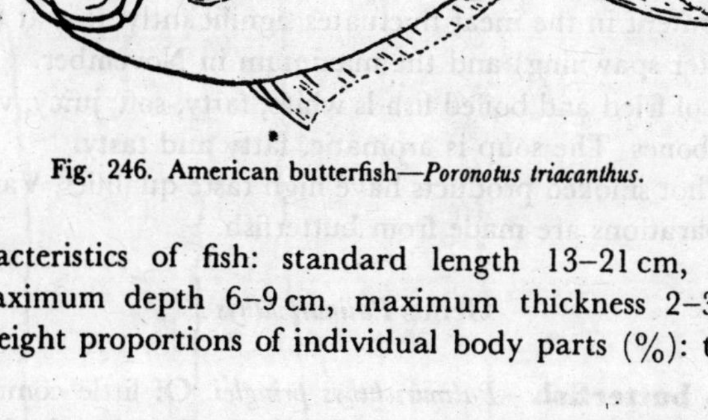

Fig. 246. American butterfish—*Poronotus triacanthus.*

Size characteristics of fish: standard length 13–21 cm, total length 16–25 cm, maximum depth 6–9 cm, maximum thickness 2–3 cm, weight 100–235 g. Weight proportions of individual body parts (%): trunk 64–74,

**Considered synonym of *Formio niger* which is described earlier—General Editor.

head 13–21, viscera 11–21, roe 0.8–1.0, liver 0.9–1.9, meat with skin 49–67, bones 6–14, fins 0.8–1.7.

Relation between length and weight of butterfish commonly caught

Length, cm	13	16	18	20	22	24	26	28
Weight, g	100	140	160	210	230	320	370	450

Fat content in meat of butterfish in different months of catch

Month	Jan.	May	June	July	Aug.	Sept.	Oct.	Nov.	Dec.
Fat, %	11	12	9	6–7	1.2–10	7–13	12–14	18	14

TABLE 303. CHEMICAL COMPOSITION, %

Body parts	Moisture	Protein	Fat	Ash
Meat	64–72	16–19	8–18	1.1–2.1
Head	68–72	11–13	10–17	3.5–4.4
Bones	58–66	13–16	12–25	5–7
Skin	57–63	22	14–20	1.4
Fins	41–48	11–13	35–38	7.7
Viscera excluding gonads	77–80	10–15	2–6	1.6–2.7
Roe	69–77	18–21	3–8	1.5–1.8
Testes	81–83	13–15	0.7–0.8	1.7–2.6
Liver	69–72	15–16	8–15	1.4–1.5
Fish as whole	69	16	10	2

The fat content in the meat fluctuates significantly: it is at the minimum in August (after spawning) and the maximum in November.

The meat of fried and boiled fish is white, fatty, soft, juicy, very tasty and contains few bones. The soup is aromatic, fatty and tasty.

Cold and hot smoked products have high taste qualities. Various kinds of culinary preparations are made from butterfish.

Genus *Palinurichthys*

Pringle's butterfish—*Palinurichthys pringlei*. Of little commercial significance, distributed in the southeast Atlantic. Also found in the Indian Ocean.

TABLE 304. SIZE CHARACTERISTICS, CM

Place of catch	Length	Maximum height	Maximum thickness
Southeast Atlantic	34–63	14–25	5–10
Indian Ocean	33–40	16–17	2.5–3.3

TABLE 305. WEIGHT PROPORTIONS OF INDIVIDUAL BODY PARTS, %

Place of catch	Weight, g	Head	Viscera	Liver	Bones	Skin	Fins	Output of fillets
Southeast Atlantic	1.1–6.4	17–22	13–18	—	4–5	—	—	51–54
Indian Ocean	1.5–1.7	25–27	11–13	1.1–1.2	—	2.2–2.6	1.9–2.2	—

This is a valuable commercial fish. The meat is white. The taste qualities of boiled, fried and smoked fish are high. Cured balyks should be prepared with cold smoking; the heads are used for preparing soup combinations.

TABLE 306. CHEMICAL COMPOSITION OF MEAT, %

Place of catch	Moisture	Protein	Fat	Ash
Southeast Atlantic	67.5–71.4	16.1–19.1	11.2–12.4	1.1–1.3
Indian Ocean	72.1	15.1	10.3	1.2

TABLE 307. MINERAL COMPOSITION OF MEAT, mg %

Place of catch	Potassium	Calcium	Magnesium	Phosphorus	Iron
Southeast Atlantic	191	31	34	424	3.3
Indian Ocean	189	51	13	122	1.4

Genus *Porona*

Palometa—*Porona signata*. Of commercial significance along the coasts of Brazil, Uruguay and Argentina. Length of fish caught in March near Patagonian shelf 31–34 cm, weight 0.5–0.8 kg. Weight proportions of individual body parts (%): trunk 72, head 18, viscera 7, fins 3. Output of fillets 56%.

Chemical composition of meat (%): moisture 73.9, protein 24.5, fat 0.5, ash 1.1.

The meat of this fish has medium taste qualities. It can be used for culinary preparations.

Genus *Psenopsis*

Japanese butterfish—*Psenopsis anomalus.* Commercially not very important, found in the East China Sea and in the western Pacific Ocean. Length of fish caught in January–November 17.5–20.0 cm, weight 170–330 g.

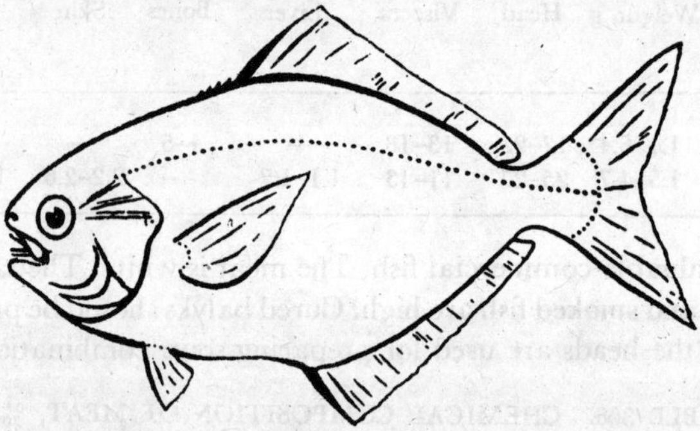

Fig. 247. Japanese butterfish—*Psenopsis anomalus.*

Weight proportions of individual body parts (%): head 13.0, viscera 6.8, bones and fins 9.4. Output of fillets with skin 70.6%.

Chemical composition of meat (%): moisture 67.2–70.1, protein 17.1–20.0, fat 11.5–13.3, ash 1.3–1.5.

The meat is very soft and juicy. The fat is uniformly distributed throughout the meat. The fish preserves well in frozen form. This is a good table fish which should be frozen and then used for culinary preparations.

Family SYNODIDAE—LIZARDFISHES

The members of this family are distributed in the tropical and subtropical waters of all oceans: in the Atlantic Ocean, along the eastern coast of America, from Cape Cod to the coast of Brazil and along the African coast; in the Indian Ocean it is found along the coasts of India and Indonesia; in the Pacific Ocean it occurs along the coast of Indonesia, in the Yellow Sea and along the coasts of Central and South America.

These fish do not form commercial concentrations, but they may constitute

TABLE 308. SIZE CHARACTERISTICS AND WEIGHT PROPORTIONS OF INDIVIDUAL BODY PARTS OF SLENDER LIZARDFISH, %

Place of catch	Month of catch	Length cm	Weight, g	Head	Trunk	Meat	Scales	Bones	Fins	Viscera	Including gonads	Including liver
East China Sea	January	23–36 / 32.6	197–465 / 362.4	10.5	—	72.5	—	10	—	4.0	1.0	2.0
Red Sea	February–April	35.2	446	16.7	67.6	—	8.5	—	1.8	5.4	—	—

an incidental catch in the trawls in quantities up to 200–300 kg, mixed with other fish.

Bombay duck (Bombil)—*Harpodon nehereus.* Distributed in the Pacific and Indian Oceans, from the coast of Zanzibar to the China coast, near the Indo-Malay archipelago. Lives in coastal shelf waters, frequently near estuaries. Attains a length of 40 cm. Length of fish caught in January in the Andaman Sea near Burma 18.1 cm, weight 55–135 g.

Weight proportions of individual body parts (%): trunk 73.4, head 13.2, fins 3.8, viscera 5.0.

Chemical composition of meat (%): moisture 90.7, protein 8.2, fat 0.1.

After cooking, the meal is jelly-like; it slips out easily and has a large number of small bones. It should be used for fish meal.

Slender lizardfish—*Saurida gracilis.* Distributed in the Indian Ocean, from the eastern coast of Africa to India including the Indo-Malay archipelago and the Sandwich Islands. Usually found in coastal waters, mainly near coral reefs.

In boiled form the meat is not tasty; it has a very faint iodine flavor and greasy consistency. It should be used for fish meal.

TABLE 309. CHEMICAL COMPOSITION, %

Place of catch	Month of catch	Body parts	Moisture	Protein	Fat	Ash
Gulf of Tonkin	July	Meat	76.7–78.4	21.5	0.1	23.3
East China Sea	January	Meat	75.5	—	2.5	—
Red Sea	March–April	Meat	79.2	19.4	0.1	1.4
East China Sea	January	Head	74.5	—	2.2	—
,,	January	Viscera	75.3	—	4.7	—
,,	January	Gonads	85.9	—	1.1	—

Indian lizardfish—*Saurida tumbil.* Found in the shelf waters from Madagascar and Zanzibar to the coasts of China and Japan and near the islands of the Indo-Malay archipelago, Australia, Sandwich Islands and the Philippines. Average length 31.9 cm (21.5–39.0), weight 454.7 g (96–780).

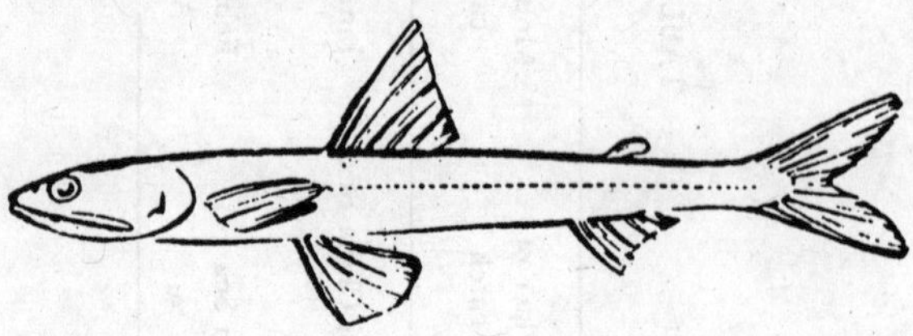

Fig. 248. Indian lizardfish—*Saurida tumbil.*

TABLE 310. WEIGHT PROPORTIONS OF INDIVIDUAL BODY PARTS OF INDIAN LIZARDFISH, %

Head	Meat	Skin	Scales	Bones	Fins	Viscera	Including	
							gonads	liver
11.9–26.2 18.5	43.3–62.7 53.7	4.6–6.5 5.6	0.9–4.2 2.3	5.7–9.7 8.2	1.3–2.1 1.8	9.8–10.0	3.4–5.0 4.2	1.7–3.0 1.8

TABLE 311. CHEMICAL COMPOSITION, %

Body parts	Moisture	Protein	Fat	Ash
Meat	72.9–79.5 76.9	17.8–23.0 20.2	0.3–3.5 1.2	1.2–3.5 1.6
Offal (head, viscera, bones, fins, scales, skin)	66.8–74.8 70.9	16.5–22.5 19.0	0.6–3.8 2.5	4.4–8.9 7.0

In cooked form, the meat is light gray, loose and watery; does not have very high taste qualities. Products canned "in their own sauce" and "blanched fish in oil" have unsatisfactory qualities. This fish should be used for fish meal and for canning.

Shortfin lizardfish—*Saurida elongata*. Length of fish caught in the East China Sea 32 cm and weight 330 g. Weight proportions of individual body parts (%): head 10.0, viscera 7.0, fins and bones 10.0. Output of fillets 73.0%. The meat contains 3.0% fat. The meat should be used for culinary preparations.

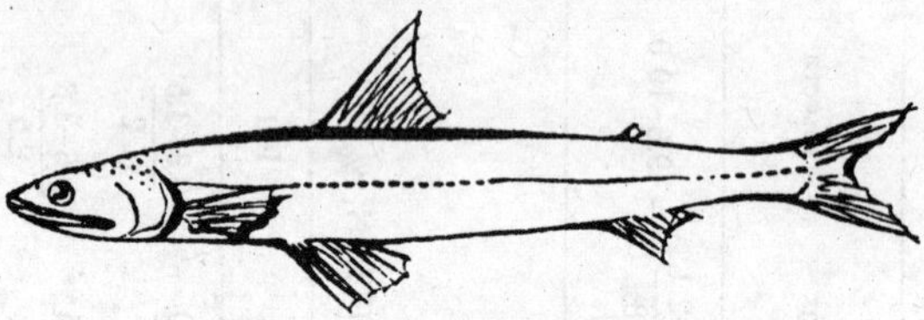

Fig. 249. Shortfin lizardfish—*Saurida elongata*.

Family THERAPONIDAE—TIGERPERCHES

Tigerfish (**Grunt**)—*Therapon theraps*. Caught in the South China Sea. Length of fish caught in April–October 10.5–17.5 cm, weight 30–92 g.

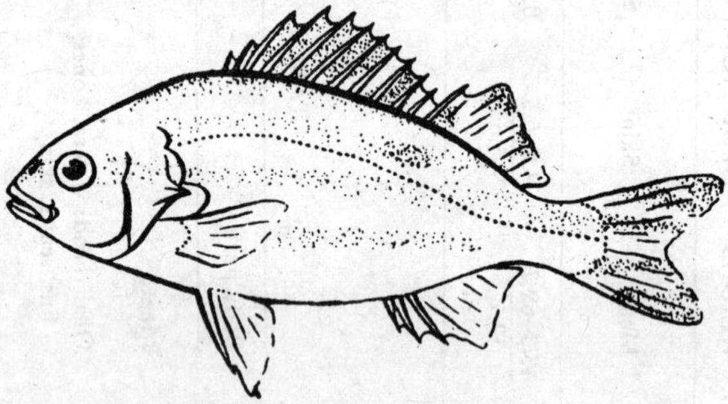

Fig. 250. Tigerfish—*Therapon theraps*.

Weight proportions of individual body parts (%): head 28.7–29.1, bones and fins 13.6–15.9, viscera 8.8–9.8, including liver 1.2. Output of fillets 43.0–46.2%.

Chemical composition of meat (%): moisture 77.4, protein 19.9, fat 0.8, ash 1.9.

The fish should be used for fish meal.

Family TRACHICHTHYIDAE—SLIMEFISHES OR SLIMEHEADS

Slimehead—*Gephyroberyx orbicularis**. Caught with other fish from the eastern Central Atlantic. Length of fish caught in June along the coasts of northwest Africa 34 cm, maximum depth 16 cm, maximum thickness 6 cm and weight 1.4 kg.

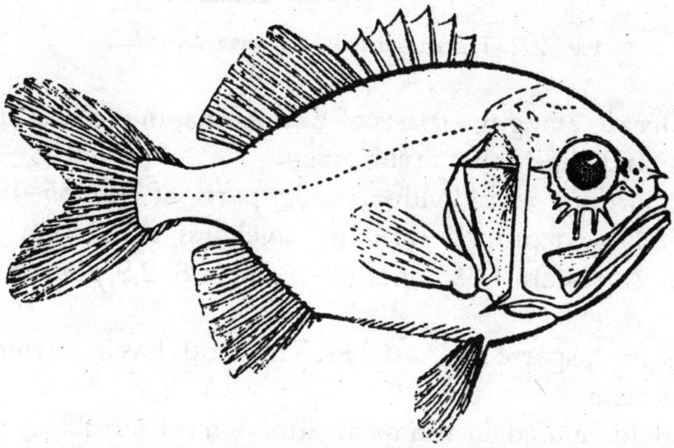

Fig. 251. Slimehead—*Gephyroberyx orbicularis*.

Weight proportions of individual body parts (%): trunk with fins 61, head 32, viscera 7. Output of fillets 42%.

Chemical composition of meat (%): moisture 79.0, protein 19.3, fat 0.6, ash 1.0.

The meat of this fish is white, juicy and tasty in fried form. The fish should be used for culinary items.

Gillchrist slimehead—*Hoplostethus gilchristi*. Caught with other fish from the southeast Atlantic. Average length of fish caught in May near Wolffish Bay 32 cm, maximum depth 13 cm, maximum thickness 5 cm and weight 0.92 kg.

Weight proportions of individual body parts (%): trunk with fins 49, head 39, viscera 12. Output of fillets 35%.

Chemical composition of meat (%): moisture 74, protein 15.5, fat 9.5, ash 0.8.

The fried meat is white and juicy.

Family TRICHIURIDAE—CUTLASSFISHES

This family includes two genera and a few species. The typical cutlassfish are distributed in the Atlantic, Indian and Pacific Oceans. They inhabit warm waters.

Indian hairtail or large headed ribbonfish—*Trichiurus haumela*. Found

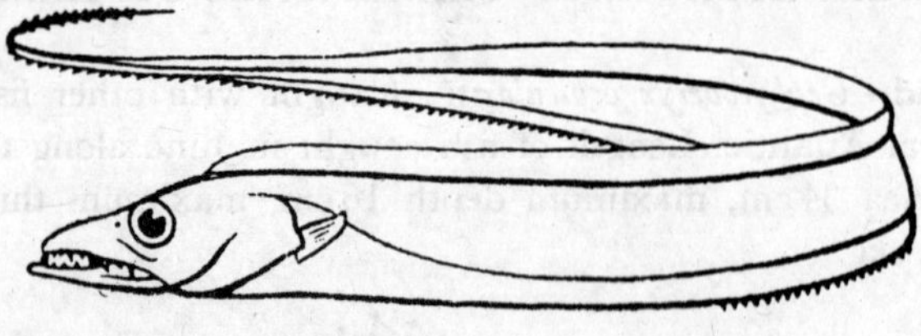

Fig. 252. Indian hairtail—*Trichiurus haumela.*

in the Indian Ocean, along the coasts of East Africa, in the seas and estuaries of India, China and the Malay archipelago.

Weight proportions of individual body parts of fish 55–102 cm long, weighing 225–885 g, depth 7.5–11.0 cm, thickness 2.0–3.5 cm (%): trunk 69–81, head 13–21, viscera 4–9, bones 12–16, fins 1.6–2.9, gonads 0.5. Output of fillets 58%.

The meat is gray, sparse and tasteless. The broth has a herring-like flavor and unpleasant taste.

This fish should be used for fish meal, although the possibility of preparing culinary items is not totally excluded.

TABLE 312. CHEMICAL COMPOSITION OF INDIAN HAIRTAIL, %

Body parts	Moisture	Protein	Fat	Ash
Meat	75.9–89.0	17.5–21.4	0.6–1.3	1.2–1.5
Offal	70.7–73.6	16.7–18.2	2.2–3.6	5.4–7.9
Viscera	76	18.8	3.7	1.4

Atlantic cutlassfish—*Trichiurus lepturus.* Found in the Atlantic Ocean, along the coast of America, near Florida and along the coasts of western and southern Africa. Found in small numbers in the Pacific Ocean, near California. A commercial fish in the Atlantic Ocean along the coasts of Africa and South America.

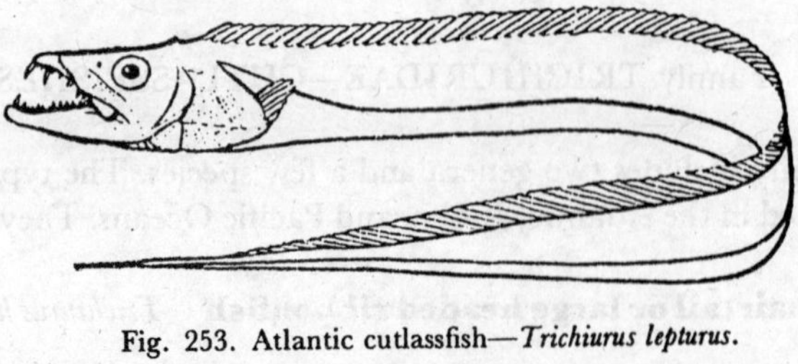

Fig. 253. Atlantic cutlassfish—*Trichiurus lepturus.*

Weight proportions of individual body parts of fish 80–115 cm long, with depth 7.5–11.0 cm, thickness 2.0–3.5 cm and weight 460–1,330 g (%): trunk 74–78, head 10–18, viscera 7–12, bones 15, fins 0.3, roe and testes 4. Output of fillets 58%.

Chemical composition of meat (%): moisture 69–76, protein 20.4–22.8, fat 1.2–9.4, ash 1.2–1.4. Mineral composition of meat (mg%): potassium 130, calcium 70, magnesium 55, phosphorus 140, iron 2.6.

In fried form, the meat is fatty, soft and tasty. It should be used for culinary preparations.

Frostfish—*Lepidopus caudatus**. Lives in the eastern Atlantic Ocean, including the Mediterranean Sea. Length 150–200 cm. The main commercial fishing areas of this species are near Cape Blanc and the estuaries of Senegal River.

Weight proportions of individual body parts of fish 148 cm long, with depth 13 cm, thickness 5 cm, weight 3.7 kg (%): trunk 69–76, head 18–22, viscera 5–9. Output of fillets 51–62%.

Chemical composition of meat (%): moisture 74.5–76.4, protein 19.5–21.1, fat 1.6–4.5, ash 1.3.

The taste properties of the meat of this fish are satisfactory. The frozen fish must be glazed well and stored at temperatures not higher than 18–20° C below zero. Culinary preparations should be made from this fish.

Family TRICHODONTIDAE—SANDFISHES

Pacific sandfish—*Trichodon trichodon*. **Japanese sandfish**—*Arctoscopus japonicus*. These species live in the East China Sea. Length of Pacific sandfish 10–20 cm, weight 16–100 g. Length of Japanese sandfish 17.0–26.5 cm, weight 50–180 g.

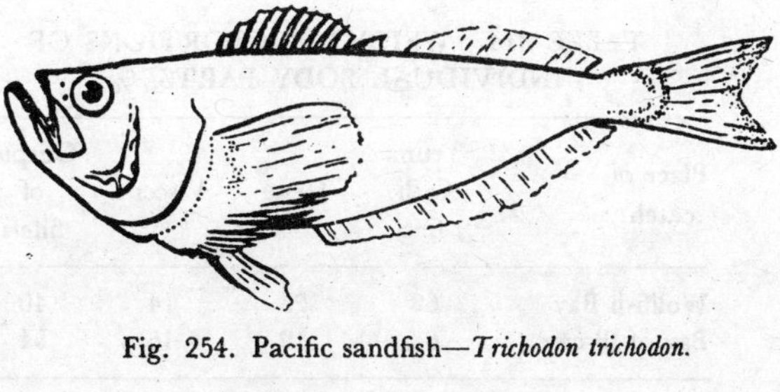

Fig. 254. Pacific sandfish—*Trichodon trichodon*.

Weight proportions of individual body parts (%): head 32.5–38.4, meat with skin 31.0–32.0.

Chemical composition of meat (%): moisture 80.1, protein 14.6, fat 3.6, ash 1.7.

The meat of the Pacific sandfish in boiled and fried forms is not tasty and therefore this fish should be used for fish meal.

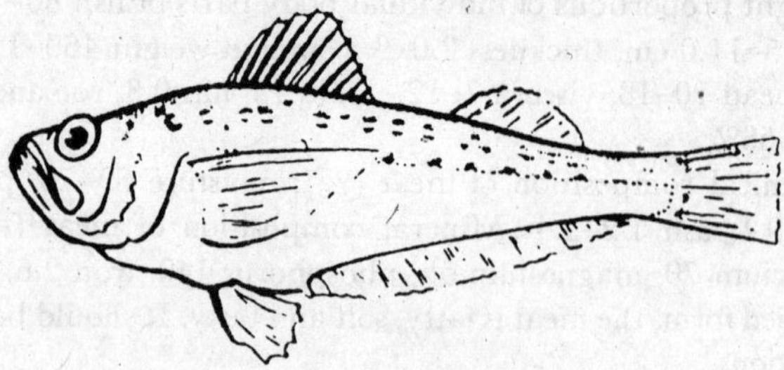

Fig. 255. Japanese sandfish—*Arctoscopus japonicus*.

Family TRIGLIDAE—GURNARDS

This family includes four genera. They are distributed in all temperate and subtropical waters. They inhabit coastal zones.

Red gurnard—*Trigla capensis*. Distributed in the temperate and tropical waters of the Atlantic Ocean.

TABLE 313. SIZE CHARACTERISTICS

Place of catch	Month	Length, cm	Depth, cm	Thickness, cm	Weight, g
Wolffish Bay	June	33	9	6.5	740
Bay of Biscay	April	36	8	8	970

TABLE 314. WEIGHT PROPORTIONS OF INDIVIDUAL BODY PARTS, %

Place of catch	Trunk with fins	Head	Viscera	Output of fillets
Wolffish Bay	63	23	14	40
Bay of Biscay	65	19	16	54

TABLE 315. CHEMICAL COMPOSITION, %

Place of catch	Moisture	Protein	Fat	Ash
Wolffish Bay	78.2	18.8	1.7	1.3
Bay of Biscay	66.1	21.5	11.1	1.3

The fried meat is tasty and sweet, but somewhat hard. It should be processed to make fish meal.

Gurnard—*Trigla hirundo*. Length of fish caught in January in the Atlantic Ocean near the South American shelf 24.0–24.6 cm, weight 290–320 g. Weight proportions of individual body parts (%): trunk 59–61, head 22–23, bones 12, fins 9.5, scales 0.8, viscera 12.5, including gonads 2.0, liver 2.5.

Chemical composition of meat (%): moisture 74.2, protein 19.9–20.3, fat 3.0–4.5, ash 1.4–1.6.

The meat in fried form is tasty and sweet, but slightly hard.

Shortfin gurnard—*Trigla microptera*. Commercially of little importance. Distributed in the Great Australian Bight. Length of fish caught in December–May 30–38 cm, weight 350–660 g.

Weight proportions of individual body parts (%): head 29.0, trunk 55.4, meat with skin 48.1, fins 5.4, bones 7.3, viscera 5.4, including liver 0.5, gonads 1.9.

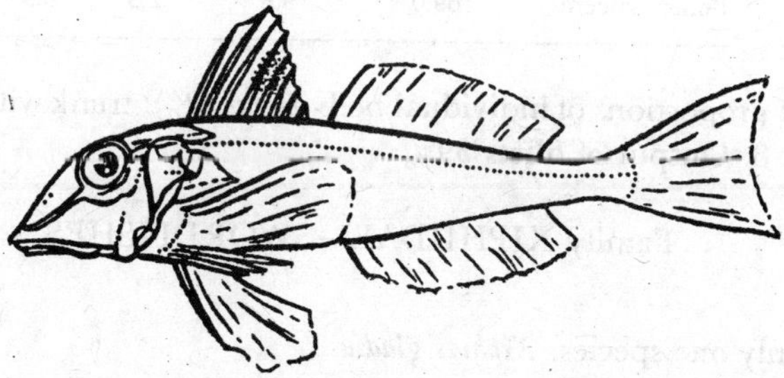

Fig. 256. Shortfin gurnard—*Trigla microptera*.

Chemical composition of meat with skin (%): moisture 78.2, protein 20.0, fat 0.1, ash 1.7.

The meat of this fish is somewhat dry. In fried and boiled forms it resembles in taste the meat of Bering Sea perch. Culinary items can be prepared from this fish.

Tub gurnard—*Trigla lucerna kumu*. Caught in the South China Sea.

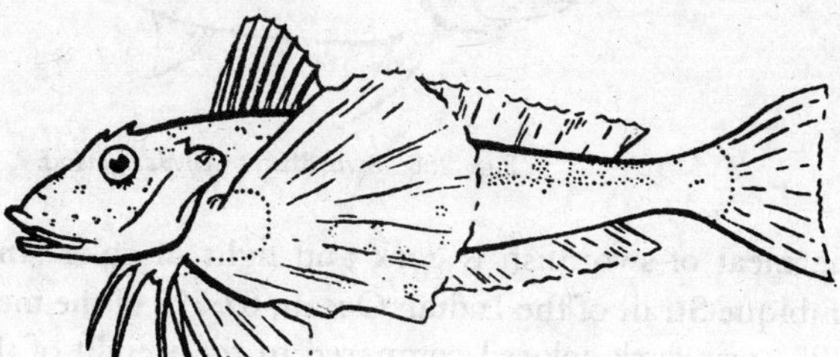

Fig. 257. Tub gurnard—*Trigla lucerna kumu*.

Length 9.5–16.0 cm, weight 15–49 g. After dressing the output of trunks 57.2%.

Chemical composition of meat (%): moisture 71.3–71.9, protein 20.6–21.3, fat 4.8–6.1, ash 1.1–1.9.

The meat should be processed into fish meal.

Family URANOSCOPIDAE—STARGAZERS

Stargazer—*Uranoscopus bufo*. Length of fish caught in May along the coast of southwest Africa 26 cm, maximum depth 5 cm, thickness 7 cm, weight 575 g.

TABLE 316. CHEMICAL COMPOSITION, %

Body parts	Moisture	Protein	Fat	Ash
Meat	75.6	21.9	1.5	1.0
Bones, viscera	69.0	19.8	2.3	8.9

Weight proportions of individual body parts (%): trunk with fins 55, head 37, viscera 8. Output of fillets 39%.

Family XIPHIIDAE—SWORDFISHES

There is only one species, *Xiphias gladius*.

Swordfish—*Xiphias gladius*. Distributed in temperate, subtropical and tropical waters, mostly in the high seas of the Atlantic, Indian and Pacific Oceans. Attains a length of 5–6 m and weight of 400–500 kg. Fish commonly caught weigh about 15 kg. Fish weighing more than 150 kg are rarely found.

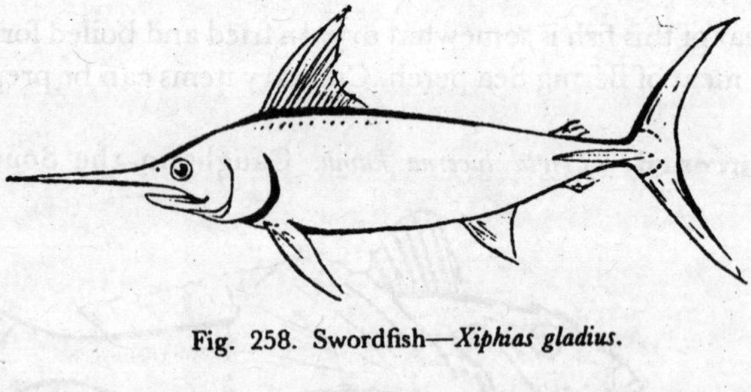

Fig. 258. Swordfish—*Xiphias gladius*.

The meat of swordfish is dark and light in color. In fish caught in the Mozambique Strait of the Indian Ocean, 62.7% of the meat was light colored and 5.2% was dark colored compared to the weight of the whole fish.

Chemical composition of light and dark colored meat of the same fish varies.

The light colored meat when cooked is white, soft and juicy with

TABLE 317. WEIGHT PROPORTIONS OF INDIVIDUAL BODY PARTS, %

Head	Meat	Skin	Bones	Fins	Viscera	Gonads	Liver	Blood
14.8–23.9 19.5	56.6–60.8 58.9	2.2–3.2 2.6	3.4–6.2 4.9	1.8–2.3 2.1	8.1–10.5 8.1	2.6	2.0–2.9 2.5	6.9–7.1 7.0

TABLE 318. CHEMICAL COMPOSITION, %

Body parts	Moisture	Protein	Fat	Ash
Meat	58.3–82.3 72.2	14.7–19.5 17.3	0.9–26.6 9.2	1.2–1.9 1.4
Head	32.7–57.2 45.0	16.6–33.6 25.1	19.3–30.2 24.8	3.5–6.3 4.9
Bones, fins	52.3	13.9	22.4	11.4
Viscera	75.3	16.7	2.5	1.9
Skin	59.7	32.9	0.5	6.8
Testes	76.5	15.8	4.0	1.8

TABLE 319. CHEMICAL COMPOSITION OF FISH CAUGHT IN DIFFERENT AREAS, %

Place of catch	Moisture	Protein	Fat	Ash
Indian Ocean, near Seychelles Islands	72.0	18.0	8.7	1.3
Central Pacific Ocean	81.5	15.7	0.9	1.9
Northern coast of Africa, the region of Bahama Islands	58.3	14.7	26.6	—
Indian Ocean, southeastern coast of Sumatra	76.0	19.5	4.5	—
Southern Pacific Ocean	70.3	17.4	9.6	1.8

TABLE 320 CHEMICAL COMPOSITION OF MEAT, %

Meat	Moisture	Protein	Fat	Ash
Dorsal	76.3	17.8	3.2	1.3
Abdominal	70.7	15.4	11.1	1.2

TABLE 321. CHEMICAL COMPOSITION OF MEAT, %

Type of meat	Moisture	Protein	Fat
Dorsal			
head	54.9	13.4	30.8
middle	55.7	15.6	28.7
tail	65.9	16.8	17.0
Abdominal	55.8	14.1	29.7

TABLE 322. CHEMICAL COMPOSITION OF LIGHT AND DARK MEAT OF SWORDFISH CAUGHT IN THE SOUTHERN REGION OF THE PACIFIC OCEAN, %

Meat	Moisture	Protein	Fat	Ash
Light	64.0–76.6 / 70.6	14.2–20.7 / 17.4	2.1–17.1 / 9.6	1.0–2.6 / 1.8
Dark	71.2–74.9 / 72.4	18.6–20.8 / 19.7	6.3–8.9 / 7.6	1.1–1.5 / 1.3

exceptionally good taste; the deep layers of the meat have a harder consistency than the surface layers. The abdominal muscles have an especially pleasant soft consistency. However, the meat has a specific flavor; this is especially so for dark colored meat. The smell of the meat is also somewhat sharp.

Swordfish should be used for cured balyks, conserves in oil, natural, frozen fillets and culinary preparations. The heads can be used for soup combinations.

Family ZEIDAE—DORIES

Only ten genera are known from this family. They live in the temperate latitudes of all oceans and in the adjoining seas. Dories are very popular with people in coastal regions.

John dory—*Zeus faber*. Found in the waters of the North Atlantic along the European coasts to Bergen in the north. Also found in large numbers in the Mediterranean Sea. In the Pacific Ocean, fish belonging to closely related species are found. Also found in the Indian Ocean. Caught with other fish. Length up to 75 cm.

Length of dories caught in eastern North, Central and South Atlantic 24–47 cm, maximum depth 13–25 cm, maximum thickness 3.6 cm and weight 0.4–3.0 kg.

In catches taken in February from the Indian Ocean at the Agulhas shelf

Fig. 259. John dory—*Zeus faber*.

TABLE 323. CHEMICAL COMPOSITION OF JOHN DORY, %

Place of catch	Body parts	Moisture	Protein	Fat	Ash
Indian Ocean	Meat	78.3–78.6 / 78.5	17.8–18.4 / 18.4	1.5–2.0 / 1.8	1.2–1.3 / 1.3
	Inedible parts of body (entire offal)	71.4–73.2 / 72.3	13.8–16.0 / 14.9	3.0–5.6 / 4.3	7.9–8.0 / 8.0
Atlantic Ocean	Meat	75–78	17.7–21.0	1.0–4.5	1.3–1.4
	Head	73.3	14.3	1.8	9.1
	Bones	55.9	14.1	13.5	14.5
	Skin	68.9	19.6	4.9	1.8
	Viscera	80.9	14.4	2.1	1.7
	Roe	71.5	22.8	2.1	1.1
	Liver	44–52	7.7	27–44	0.9

and those taken in November near San Francis there was a predominance of fish averaging in length about 30 cm (24–36), with maximum depth 15 cm (15–20), maximum thickness 3.7 cm (3.5–4.0) and weighing about 500 g (290–1,010).

TABLE 324. WEIGHT PROPORTIONS OF INDIVIDUAL BODY PARTS OF THE JOHN DORY, %

Place of catch	Head	Trunk	Meat with skin	Skin	Meat	Bones	Fins	Viscera	Including gonads	Including liver
Indian Ocean, Agulhas shelf and the San Francis region	26.3–34.5 / 31.1	43.1–61.7 / 51.2	33.8–40.4 / 39.2	2	—	10.4–14.0 / 11.2	1.5–1.9 / 1.8	14.7–17.0 / 15.7	—	2.0–3.2 / 2.7
Eastern North, Central and South Atlantic	21–28	55–58	—	3–6	35–40	9–13	0.8–1.2	11.2	1.2–6.0	2.2–6.0

After cooking, the meat of the dory is light colored, soft, pleasant in taste, typical of a table fish and resembling slightly the taste of shrimp, because of the peculiarities of its feeding habits. The soup is rich and tasty. The fried meat is very tasty. In salted form the meat matures well; in this process the meat acquires high taste qualities, typical of well cured fish. The dory is a good table fish for first and second course dishes and for making preserves in jars using pieces of mildly salted meat in mustard sauce and other sauces.

Deepwater dory—*Allocyttus verrucosus.* Found in the eastern Central Atlantic. Length of fish caught in May 20–23 cm, maximum depth 12 cm, thickness 4 cm, weight 0.26–0.39 kg.

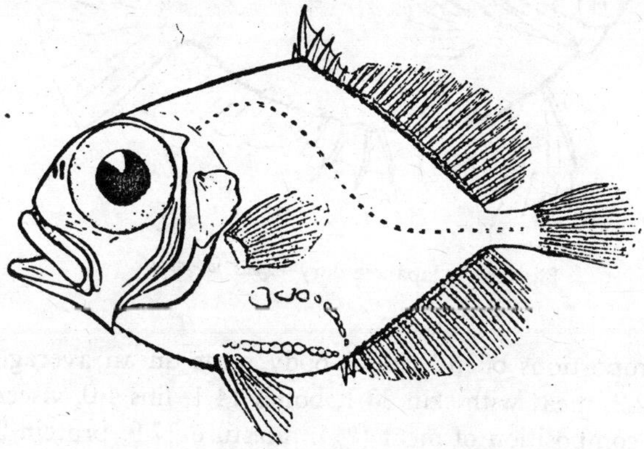

Fig. 260. Deepwater dory—*Allocyttus verrucosus.*

TABLE 325. CHEMICAL COMPOSITION OF DEEPWATER DORY (MAY CATCHES), %

Body parts	Moisture	Protein	Fat	Ash
Meat	74.9	16	8.0	1.1
Viscera	79.8	10.7	8.2	1.3

Weight proportions of individual body parts (%): trunk with fins 46, head 43, viscera 11. Output of fillets 31%.

The meat of this dory is fattier than the meat of the John dory. It should be used for canning.

Japanese dory—*Zeus japonicus.* A commercial fish. Length attained up to 45 cm. In catches taken in the Gulf of Australia there was a predominance of fish averaging in length about 34 cm (31–36), with maximum depth 17.5 cm (15–20), maximum thickness 3.7 cm (3.5–4.0), weighing about 900 g (570–1,160).

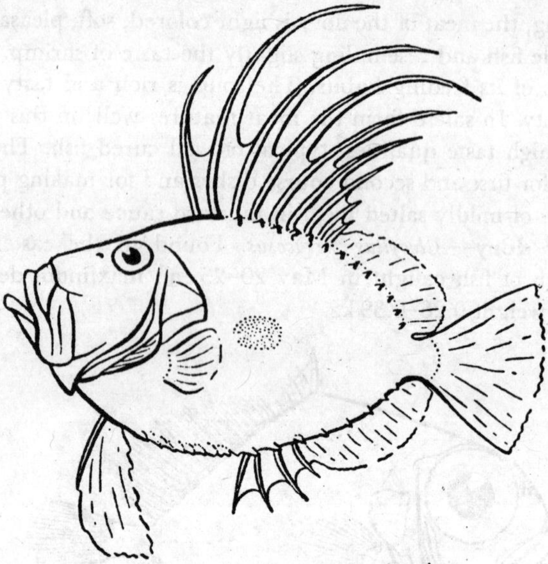

Fig. 261. Japanese dory—*Zeus japonicus*

Weight proportions of individual body parts on an average (%): head 28.3, trunk 52.2, meat with skin 38.1, bones 14.1, fins 4.0, viscera 15.3.

Chemical composition of meat (%): moisture 77.9, protein 20.0, fat 0.7, ash 1.4.

The meat in fried form is soft, juicy and pleasant in taste. The soup is rich, tasty and resembles chicken soup. The Japanese dory should be used as a table fish for first and second course dishes.

Family ZOARCIDAE—EELPOUTS

These fish live along the European coast in the White and Baltic Seas, in the northwest Atlantic (near the New Shetland shelf) and in the southwest Atlantic (near the Patagonian shelf).

Eelpout—*Zoarces anguillaris.* Usual length of fish 30–80 cm; length of fish caught near New Shetland shelf 67–81 cm, maximum depth 5.7–6.4 cm, maximum thickness 4.5–6.1 cm, weight 1.82–2.57 kg.

Weight proportions of individual body parts (%): trunk with fins 66, head 23, viscera 11. Output of fillets 48%.

Chemical composition of meat (%): moisture 76–77, protein 18–18.8, fat 2.6–3.6, ash 1.1.

The meat is white, soft, somewhat watery and tasty. This is a good table fish from which smoked products are prepared.

Ocean pout or American eelpout—*Macrozoarces americanus**. Caught in the Patagonian shelf. Length of fish 63 cm and weight 1.11 kg.

Weight proportions of individual body parts (%): trunk 67.5, head 20.8, viscera 9.5, fins 1.4, meat 53.8, skin 5.7, bones 8.

Chemical composition of meat (%): moisture 76.1, protein 18.8, fat 2.6.

The American eelpout should be used for culinary items.

Spotted eelpout—*Ilucoltes fimbriatus*. Three specimens of spotted eelpout caught in April at the Patagonian shelf were investigated. Length 30–37 cm, length of head 6–7 cm, maximum depth 4–6 cm, maximum thickness 3–4 cm, weight 200–260 g.

Weight proportions of individual body parts (%): trunk with fins 61, head 21, viscera 13, fins 5. Output of fillets 42%.

Chemical composition of meat (%): moisture 81.1, protein 15.8, fat 0.6, ash 1.3.

The meat in fried form is white with a bluish tinge, watery, with a specific flavor. The taste qualities of the fish are satisfactory. It should be used for smoked products.

INDEX OF COMMON FISH NAMES

INDEX OF LATIN NAMES OF FISH